RED DYNAMITE

A volume in the series
RELIGION AND AMERICAN PUBLIC LIFE
edited by R. Laurence Moore and Darryl Hart

RED DYNAMITE

Creationism, Culture Wars, and Anticommunism in America

CARL R. WEINBERG

CORNELL UNIVERSITY PRESS
ITHACA AND LONDON

Publication of this open monograph was the result of Indiana University's participation in TOME (Toward an Open Monograph Ecosystem), a collaboration of the Association of American Universities, the Association of University Presses, and the Association of Research Libraries. TOME aims to expand the reach of long-form humanities and social science scholarship including digital scholarship. Additionally, the program looks to ensure the sustainability of university press monograph publishing by supporting the highest quality scholarship and promoting a new ecology of scholarly publishing in which authors' institutions bear the publication costs.

Funding from Indiana University made it possible to open this publication to the world.

This work was (partially) funded by the Office of the Vice Provost of Research and the IU Libraries.

First published 2021 by Cornell University Press

Library of Congress Cataloging-in-Publication Data

Names: Weinberg, Carl R., 1962– author.
Title: Red dynamite : creationism, culture wars, and anticommunism in
    America / Carl R. Weinberg.
Description: Ithaca [New York] : Cornell University Press, 2021. | Series:
    Religion and American public life | Includes bibliographical references
    and index.
Identifiers: LCCN 2021020719 (print) | LCCN 2021020720 (ebook) |
    ISBN 9781501759291 (paperback) | ISBN 9781501759307 (pdf) |
    ISBN 9781501759314 (epub)
Subjects: LCSH: Anti-communist movements—United States—History—
    20th century. | Anti-communist movements—United States—History—
    21st century. | Evolution (Biology)—Study and teaching—Political
    aspects—United States. | Evolution (Biology)—Political aspects—United
    States.
Classification: LCC E743.5 .W34 2021 (print) | LCC E743.5 (ebook) |
    DDC 324.1/3—dc23
LC record available at https://lccn.loc.gov/2021020719
LC ebook record available at https://lccn.loc.gov/2021020720

For Beth

CONTENTS

RED DYNAMITE

Introduction

Belaboring Scopes

In the spring of 1925, in the sleepy, isolated, southern town of Dayton, Tennessee, a group of local notables gathered at Robinson's drugstore and hatched a plan to revive their town. Their group included George Rappleyea, manager of a local coal mining company that had fallen on hard times. The drugstore conspirators knew that the American Civil Liberties Union (ACLU) sought a teacher who could serve as defendant in a case testing the constitutionality of the Butler Act. Promoted by Christian fundamentalists led by William Jennings Bryan, the new law prohibited public school teachers from teaching "any theory that denies the Story of the Divine Creation of man as taught in the Bible, and to teach instead that man has descended from a lower order of animals." Rappleyea proposed the school's football coach and general science instructor, an agreeable fellow named John T. Scopes. Pulled from a game of tennis, Scopes joined the group and agreed to their plan. When the trial took place that July, Scopes, who may not have in fact taught evolution, never took the stand. He was utterly overshadowed by the legal titans contending in the

Rhea County Courthouse and outside on the lawn, where defense attorney Clarence Darrow made a fool of Bryan. Scopes was convicted, and the Butler Act was upheld. But Darrow's ruthless cross-examination of Bryan, broadcast on WGN radio, delivered the real victory to the evolutionary side. Science defeated religion. Ridiculed to great effect by acerbic journalist H. L. Mencken, the fundamentalists retreated and did not reemerge in American politics until the rise of the New Christian Right in the 1980s and '90s.

In a limited sense, this story is true.[1] Yet it is profoundly misleading. Based on the myth that Christian fundamentalists walked away from politics for decades after the Scopes trial, our thumbnail sketch neglects a continuous pattern of Christian conservative political activism from the 1920s through the 1970s. And it misses the true origin story of this activism by misconstruing the historical context of Dayton, Tennessee, itself and its best-known temporary resident, John Thomas Scopes. Dayton was neither sleepy nor isolated. Scopes was far from a passive, politically naïve victim of the trial. Both were tied to wider currents of radical labor and socialist activism, the explosive impact of industrial capitalism, and deep moral questions about the direction of American society at the turn of the twentieth century.[2] Seeing Scopes and Dayton in this light points to the central theme of this book—the deployment of anticommunist arguments by creationists from the Scopes trial to the present. At stake at the "Monkey Trial" were not only rival perspectives on natural history and the Bible, but conflicting politicized moral visions about how American society could and should evolve. The real story of Dayton and the Scopes family reveals a century-long explosive historical matrix that meshes with what I have called, following creationist George McCready Price, "Red Dynamite."

We begin thirty-four years before the "trial of the century." In the autumn of 1891, six hundred residents of Dayton, Tennessee, signed a petition to a special session of the state legislature. Speaking in the name of "miners, merchants, and citizens of all classes," the petition denounced Tennessee's convict lease system. After the end of the Civil War, the state of Tennessee had authorized coal mining companies to pay the state a fee in return for employing convicts in state prisons and paying them nothing. Labor activists viewed this system as a way for coal mine owners to lower wages, incite racial animosity between workers, and thereby undermine the power of labor unions. Just months earlier, three hundred armed coal miners in Briceville in nearby Anderson County had marched to the stockade that housed convict laborers brought in as strikebreakers at a local mine. They disarmed the mine guards and liberated the largely African American convicts. The Briceville action proved to be just one skirmish in an extended rebellion in 1891–92 by East and Central Tennessee coal miners against convict lease. Miners, including Blacks, engaged in peaceful lobbying, union-organized protest meetings, strikes, and disciplined armed actions to free convict laborers in 1891. The next year, miners fought in bloody battles against state militiamen, with a number killed on both sides. Warning of "dark and dire disaster" in Dayton if mine owners dared to import convict laborers there as well, the petition signatories begged state legislators to abolish convict lease so that there would be no "re-occurrence of trouble witnessed at our sister town of Briceville."

As they demanded relief from lawmakers, the Dayton petitioners were careful to stress that their protest had respectable goals. Convict lease threatened to undermine the ability of working people to be responsible property owners with a stake in the community. After all, they explained, "1,000 miners have provided for themselves little houses and have paid for them by honest toil."[3] If mine owners could employ convicts for practically nothing, then free wage earners would be in deep trouble, and who knows then what might happen. Daytonians knew that coal mining rebels would be demonized. After militia members were killed in action the following summer in Coal Creek, next door to Briceville, the editors of the *Knoxville Journal* denounced the rebels as "outlaws" and proclaimed that the "agitator, the anarchist" would be crushed. They favorably quoted a militia officer who described Coal Creek as "the place where the torch of the anarchist and communist lights the darkness of the mountain."[4]

Dayton mine owners chose not to take their chances with convict labor. Their reluctance may have sprung from the paternalistic inclinations of the outside investors who turned Dayton into an industrial boom town. Attracted to the Cumberland Plateau's vast iron ore, coal, and limestone deposits and newly built railroads, all in close proximity to the Tennessee River, English capitalists led by Titus Salt & Sons bought up twenty-five thousand acres of land, paid for mineral rights, and established the Dayton Coal and Iron Company (DCI) in 1884. A textile magnate who developed an English model community called Saltaire, Sir Titus Salt Sr. (1803–1876) fancied himself a true Christian humanitarian. Salt provided housing and plentiful cultural amenities for his textile workers, though he opposed labor unions and was alarmed at the "physical force" branch of the British Chartist movement fighting for working-class political rights. As Salt acquired land in Tennessee, he likely hoped that his American project would reflect those same philanthropic intentions and produce industrial tranquillity. Now led by his son Titus Salt Jr. (1843–1887), Salt's company built hundreds of homes for Dayton's burgeoning workforce and a two-story brick company store to supply their needs. By 1890, the tiny community of Smith's Crossroads (population 250) had become a bustling industrial

Figure 1. Dayton Coal and Iron facilities, c. 1915, Dayton, Tennessee. Sitting idle by the time of the 1925 Scopes "Monkey Trial," the company's bustling coal mines, coke ovens, and blast furnaces (shown here at the base of the smokestacks) powered an industrial boom that featured intense local labor battles and violent conflict in East Tennessee over the convict lease system. Courtesy of Brewer Collection, Rhea County Historical and Genealogical Society.

town of 5,000. In the hills around Dayton, miners dynamited and then shoveled coal to feed coke ovens dotting the landscape, fueling giant blast furnaces that smelted ore into pig iron, an essential ingredient for making steel. Hotels and factories followed, along with the city's first elementary school in 1895. It later became famous, and infamous, as Rhea County Central High School, where football coach and science teacher John Scopes allegedly taught evolution to his students.[5]

Dayton had arrived, but it was hardly a sleepy oasis of industrial peace. Quite the opposite. In the wee hours of Friday, October 16, 1896, a gigantic explosion threw slumbering Daytonians from their beds. According to police, a suspected incendiary had set a match to five fifty-pound cases of dynamite at the DCI warehouse, blowing the building "to atoms," sending "brick bats" everywhere, wrecking the Cincinnati Southern railroad station, and shattering windows all over town. Three years later, the facilities were once again destroyed, this time during a coal mining strike. Company officials accused striking miners of setting the power plant on fire, which reduced the company's entire aboveground facilities to a "mass of ruins."

Whether or not Rhea County coal miners were responsible, they were a strike-prone bunch. For all of Titus Salt's fine intentions, Dayton Coal and Iron was forced to compete with other producers and lower costs wherever possible. Salt's managers routinely cut tonnage rates (miners were paid by the weight of coal mined) or docked miners' pay for "excessive" amounts of slate in their coal when the market for iron was weak. The company also irked miners by paying them in "scrip," forcing them to shop at the company store, until protests ended the practice. All in all, the result was a seemingly endless string of walkouts; miners struck in 1897, 1898, 1899, 1903, and 1904. In that latter year miners organized Local 1117 of the United Mine Workers of America (UMWA) in Dayton. In an attempt to cripple the 1904 strike, Dayton Coal and Iron sought an injunction from the district federal court in Chattanooga against local UMWA leaders. In the eyes of the company owners, since they sold pig iron across state lines to steel companies, the walkout interfered with interstate commerce. Dayton Coal and Iron accused the union men of violent attacks against nonunion miners: they allegedly had dynamited a nonunion miner's house, fired shots into a trainload of nonunion miners, and threatened the local sheriff with his life.[6]

The intensity of these conflicts may have reflected the deadly risks that Rhea County shared with miners everywhere. The decade leading to the 1904 strike was spectacularly bad for Dayton. In 1895, five days before Christmas, miners arrived for work at the DCI's Nelson mine. Located two miles west of the city up on Walden Ridge, the "slope" mine tunneled on a gradual downward incline, snaked two miles into the mountain, and employed over one hundred men.[7] Known as a "gassy" mine, Nelson held so much methane (called "firedamp" by the miners) that it bubbled up through nearby Richland Creek, which periodically caught fire. That day, the mine received its regular early morning inspection, but as several dozen miners walked down one of the main alleyways ("entries") to their working "rooms," a gigantic explosion occurred. Flame shot throughout the one-mile entry, fueled by firedamp and coal dust, and then spread to other entries. The shock of the explosion brought down tons of slate and killed twenty-nine men. Dayton was "convulsed in horror and grief," wrote a reporter for the *Chattanooga Times*. The "shrieks of bereaved wives and mothers" rang throughout the town. The disaster, the reporter concluded, "stands without a parallel in all its horrible details in the mining history of Tennessee."[8] Two parallels were yet to come. In 1901, a gas explosion at DCI's Richland mine killed twenty-one miners. The next year, the Nelson mine once again became a death trap. A gas explosion killed twenty-two.[9] Unlike the 1895 disaster, in which responsibility fell to "nature" for the deaths of twenty-nine men, the 1901 and 1902 blasts led to a spate of lawsuits against DCI. Taken together, according to one press report, these cases represented "one of the most stubbornly fought legal battles in the court annals of Rhea County."[10]

Among the Dayton dead in 1895 and 1902 were free African American miners. The racialized politics of post–Civil War Tennessee had powerfully shaped the convict lease system. African Americans were disproportionately arrested, convicted, and then leased out to coal mining companies. It could seem easy for white Tennessee miners to conflate black skin with a threat to their livelihood even if their Black coworkers were free men. In the early days of DCI, when Dayton miners came to the company office to pick up their pay, the waiting room leading to the company store had both "white" and "colored" sections.[11] But a growing number of free African American miners in East Tennessee sunk roots in their communities, belonged to the fledgling mine workers' unions, which preached racial equality in the name of working-class solidarity, and took part in battles to end convict lease. Thus, in the early 1890s, William Riley, elected and then reelected secretary-treasurer of District 19 of the UMWA (covering East Tennessee and southern Kentucky), was an African American miner and preacher.[12] For all of its contradictions and limitations, the East Tennessee coal mining region surrounding Dayton in the late nineteenth century could be a surprisingly progressive place.

In 1885, the year that Dayton appeared on the Tennessee map, a young progressive-minded English-born apprentice machinist named Thomas Scopes, father of future accused evolutionist John Scopes, arrived in the United States. Born into a working-class London family in 1860, Scopes was literate and a freethinker. When he walked off the ship in New York, he carried with him four books: the Bible, a hymnal, a volume on the French Revolution, and *On the Origin of Species* by Charles Darwin.[13] The young Scopes worked his way across the US in waterfront towns from Philadelphia to Galveston, Texas, and then to Union City, Tennessee. In Union City he met his future wife, Mary Alva Brown. Her Kentucky family roots included a tobacco farmer who helped Confederate general Nathan Bedford Forrest, and a Cumberland Presbyterian preacher. The new couple moved to Mound City, Illinois, where Scopes applied his skills to the booming railroads.[14] In 1894, during the nationwide Pullman strike led by Eugene Debs and his newly created American Railway Union (ARU), Scopes was working as a roundhouse foreman. Like Sir Titus Salt, Illinois railroad car manufacturer George Pullman had created a company town that aimed to keep his workers—whom he called his "children"—content but under his thumb. When the Panic of 1893 sent the American economy into a tailspin, Pullman cut wages but not rents. His workers rebelled, and railroaders far and wide stopped work in sympathy. Their action set the stage for a major confrontation, as Pullman obtained a federal injunction, granted by the Democratic Grover Cleveland administration against Debs and the ARU. Sitting in a federal prison, disillusioned with what seemed like a sham of American democracy, Debs advanced down a path toward socialist politics.[15]

Ordered by his boss to take the trains out and serve as a Pullman strikebreaker, Thomas Scopes refused and was fired. Shortly thereafter, he joined the International Association of Machinists (IAM), and following the lead of Eugene Debs, who became his lifelong hero, Scopes also became a socialist. Moving next to the Ohio River town of Paducah, Kentucky, where John Scopes was born in 1900, Thomas Scopes worked for Illinois Central Railroad as a roundhouse foreman.[16] He ran the local lodge of the IAM and campaigned for a socialist "cooperative commonwealth." He ran for city judge in 1901 on the Socialist Party ticket and served as a presidential elector for the Socialist Labor Party in 1904. When Eugene Debs came to Paducah in 1910 and spoke to an audience of three thousand, Scopes introduced him from the stage.[17]

Reflecting decades later on the forces that led him to challenge Tennessee's Butler Act, son John Thomas Scopes, known by his family as JT, could not help thinking of the example set by his mother and father. His parents, Scopes wrote, created "a tolerant environment that taught me early in life to revere truth and love and courage." But the influence of Thomas Scopes loomed especially large on the young JT's thinking. "Dad taught me always to stand up for what I thought was right," Scopes wrote. Beyond this affirmation of free thought, Thomas Scopes conveyed more specific lessons. When America entered World War I, young Scopes was nearly draft age, and the family had relocated across the Ohio River to Salem, Illinois. "I want you to understand this much, J.T.," said his antiwar Socialist father. "*This war is none of your business.* It is strictly a fight . . . for control of the world commercial markets. . . . This is nothing but an all-out economic struggle into which they're trying to drag every last workingman of Europe and America."[18]

Thomas Scopes also shaped John's critical perspective on Christianity. Once an elder in the Cumberland Presbyterian Church of Paducah, Thomas Scopes left because he disagreed with the church's "holier-than-thou" approach to prostitutes working the river town's red-light district. Opposing efforts to ban them or to kick them out of town, Thomas Scopes thought they needed good-paying jobs, not lectures. "I believed, as Dad did, that economics was the most important factor in a person's well-being," John Scopes recalled. "For a society to be righteous, it must cherish and work for economic justice first of all." Thus the teenage Scopes learned from his father to accept the moral teachings of Jesus, minus "the myths and miracles of Christian dogma."

Dad also got his son a summer job. After JT finished Salem High School in the spring of 1919, with none other than native son William

Jennings Bryan as commencement speaker, he joined Thomas Scopes in the Chicago and Eastern Illinois Railroad roundhouse, shoveling coal into boilers for the summer.[19] Given that 1919 saw four million workers go on strike across the US, in the politically explosive aftermath of the Bolshevik Revolution in Russia, one imagines that the Socialist railroad machinist and his son had plenty to talk about.

By 1913, most Dayton miners had stopped shoveling coal. The big steel companies found a cheaper supply of pig iron. DCI's plant and technology were growing obsolete. Sir Titus Salt's successors in London and Glasgow were growing desperate. When it became clear that their investment, which looked so promising in the 1880s, was now going bust, London-based company president Peter Donaldson took his own life by driving his motorcar into the Thames River. Over the next twelve years, a procession of investors tried and failed to make DCI's mines, coke ovens, blast furnaces, foundries, and quarries turn a profit.[20] By 1925, the mines now owned by Cumberland Coal and Iron were deserted, full of water, and unguarded aside from some barbed-wire fences.[21]

Shortly after newly arrived Daytonian John Thomas Scopes agreed to serve as a test case of the antievolution Butler Act, he traveled back home in early summer to visit family and talk strategy with his father. Thomas Scopes was proud his son was standing up for his beliefs—"freedom of religion, freedom to teach, freedom to think and to believe." The elder Scopes identified so much with his son's case, in fact, that "he considered the trial to just as much be his as mine." JT clearly valued his father's advice. And so, now as a retired railroad machinist with some time on his hands, Thomas Scopes followed JT back to Dayton and stayed in a local hotel for the entire course of the trial. On at least one of those hot July days in 1925, a news photographer captured father and son in deep conversation during a break from the trial. Many years later his obituary in the *New York Times* described John Scopes as a "shy, clean-cut, young man who never uttered a word at the trial" and was "clearly overwhelmed in the carnival-like circumstances under which it was held."[22] The photo suggests a man fully engaged in the moment and in control. After all, when JT graduated from the University of Kentucky in 1924, with a minor in geology, his bachelor's degree was in law.

John Scopes's legal battle was also an occasion for Thomas Scopes to see old friends. As JT recalled, one of these was E. Haldeman-Julius of Girard,

Figure 2. Thomas Scopes, *left*, and John Scopes, July 1925, Dayton, Tennessee. Taking a break from his trial for teaching evolution at Rhea County High School, science teacher John Scopes confers with his father, a longtime Socialist and labor union activist. Photo by Louis Van Oeyen / Western Reserve Historical Society. Courtesy of Getty Images.

Kansas, a "good friend" of his father whom the younger Scopes had once met in Paducah. A quirky socialist writer and editor hailing from an immigrant working-class Jewish family in Brooklyn, Haldeman-Julius served during these years as editor of *Appeal to Reason*, the best-known socialist newsweekly. He was notorious for promoting advanced but scandalous ideas about sex and marriage—Julius hyphenated his last name when he married Marcet Haldeman. And he pioneered a series of five-by-three-and-a-half-inch "Little Blue Books" whose millions of sales at twenty-five cents a copy paved the way for the rise of the modern paperback.

In the year before John Scopes went on trial, Haldeman-Julius published *Darwin and the Theory of Evolution* (also known as *Little Blue Book No. 568*) by Carroll Lane Fenton (1900–1969). A budding geologist and paleontologist with a journalistic bent, Fenton would receive his PhD in geology from the University of Chicago in 1926, just months before John Scopes arrived there post-trial, fully funded by his supporters, to pursue his own graduate geological studies.[23] Haldeman-Julius enlisted Fenton because he not only ably summarized Darwin but also did battle with the antievolutionists. Decrying their "willful ignorance and distortion of facts," Fenton noted that they attributed various social and political evils to Darwin, including Bolshevism and "weakening morals."[24]

Though Fenton ridiculed this claim, Haldeman-Julius provided fuel for the creationist fire in his own *Studies in Rationalism*, published in 1925. In this blistering series of atheist essays, Haldeman-Julius raised the hackles of Christian fundamentalists when he asked, "Is Religion the Necessary Basis of Morality?" Haldeman-Julius argued instead that morality had a "material basis" that enabled human society to evolve upward. One illustration of the positive "evolution of society," in the author's eyes, was "feminism triumphant." Haldeman-Julius celebrated the erosion of Christian social taboos related to sex, including for women. Sex was no longer shameful; it could now assume its "wholesome, natural, joyous place." And women would be "companions of man in moral freedom and equality."[25] Evolution was not just about fossils and finches. Morality could evolve, too.

The economic and moral battle waged by working people against the leasing of convicts to coal mining companies in Tennessee was almost surely common knowledge to both Scopes and Haldeman-Julius. Not only was it a big news story around the nation and a subject of ongoing attention in the labor movement, but Eugene Debs addressed what he called the "Tennessee tragedy" in the pages of *Appeal to Reason*. In a widely reprinted 1899 speech, Debs explained that convict lease was an injustice to both the convicts and all those who toiled in the coal mines: "the convicts, themselves brutally treated, were used as a means of dragging the whole mine-working class down to their crime-cursed condition." Debs also paid tribute to the willingness of East Tennessee miners to rebel. Once they took up arms, he noted, politicians changed their laws "in a twinkling," and hundreds of convict miners were set free. As Debs also observed, however, Tennessee ended convict lease without ending convict labor. It built a *state*-run coal mine employing convicts that would operate for decades.[26]

As the trial of John Scopes approached in early July 1925, it was almost inevitable that someone out there would connect the dots: John Scopes, Thomas Scopes, Socialist, freethinker, friendly with militant atheists, sexual immorality. Clean-cut? Hardly. Indeed, July 10, the opening day of the trial, newspaper readers in Chattanooga got the dirt on the defendant's family. Thomas Scopes had left the Presbyterian Church and become a devoted Socialist. The elder Scopes often spoke against "the religious and political systems of America" and with his wife raised young John in "an unchristian and socialistic environment." There was more. George Rappleyea, the mine manager who worked with John Scopes to bring about a test case of the Butler Act, was a member of the American Civil Liberties Union. The ACLU conspired with the communists and aimed to bring about a violent general strike that would overthrow the American government. That secret and unsavory agenda explained why ACLU attorneys had partnered with Clarence Darrow, known for his defense of left-wing radicals, and socialist Dudley Field Malone, whose high-profile divorce had scandalized the Catholic Church. In the eyes of Rev. Timothy W. Callaway, who provided this revelatory material to the *Chattanooga Times*, the challenge to the Butler Act posed by John Scopes was not just a secular, legal slap at Christian theology.[27] It was part and parcel of a broad-based, immoral, and communistic attack on American institutions.

Ten days later, as if to validate Callaway's conspiratorial suspicions, John Scopes wrote his own news story published in the *Daily Worker*, the newspaper of the fledgling American Communist Party. Scopes wrote that he was not surprised the jury found him guilty, but looked forward to winning the case on appeal.[28] Scopes was no Communist, but the early American communists, like their socialist predecessors, joined Scopes in a fervent defense of evolutionary science. They ran numerous articles on the trial and took great pains to explain to revolutionary workers what was at stake in the "trial of the century."

The violent, conflicted, rich international industrial history of Dayton, Tennessee. The very real and hidden labor and socialist history of the Scopes family. The red-baiting of John Scopes. The appearance of an article by Scopes in a communist publication. All these point to an important untold story: how antievolutionists throughout the twentieth century mobilized their followers by linking evolution, communism, and immorality. Labeled "Red Dynamite" by the pioneering creationist geologist George McCready Price just months after the Scopes trial, this potent political mix helps to answer a question still nagging at us today: Why has creationism persisted into the twenty-first century in the most scientifically advanced country in the world? The commonsense answer revolves around the strength of American religious belief. Creationists contend that evolution undermines the authority of the Bible, a dynamic reflected in our understanding of the Scopes trial as a battle between science and religion. This book advances a different explanation for why and how Christian conservatives have succeeded in demonizing Darwin: they convinced their followers that evolutionary thought promotes immoral social, sexual, and political behavior, undermining existing God-given standards and hierarchies of power.

While "scientific" creationists have trumpeted the intellectual inadequacy of biological evolutionary theory and the superiority of a Bible-based model, their real concern is not exclusively scientific or religious. What alarms them is the concept of social evolution.[29] If moral standards can change over time, as E. Haldeman-Julius freely acknowledged, then "man" and not God becomes the ultimate authority, and "anything goes," a phrase with both violent and sexual overtones. Since the Marxist founders, Russian Bolsheviks, and their American socialist and Communist successors joined their promotion of an ever-changing class-based standard of morality to their embrace of Darwin's discovery, they drew regular fire from creationists. From this standpoint, the stakes in the battle against evolutionary science could not be bigger. The fundamental issues are not biological but social. The ultimate question is not, narrowly speaking, religious, but political: Whose morality will prevail during our time on this earth? As maverick Christian evangelist Francis Schaeffer titled his bestselling book, *How Should We Then Live?* (1976).

For George McCready Price and his creationist successors, evolutionary science not only raised questions about the central theological, otherworldly question of salvation—whether Christian believers have access to eternal life and death—but also generated deep concern about its this-worldly social and political repercussions, or what Jesus called the "evil fruits" of a "corrupt tree" in his Sermon on the Mount: "Beware of false prophets, which come to you in sheep's clothing, but inwardly they are ravening wolves. You shall know them by their fruits. Do men gather grapes of thorns, or figs of thistles? Even so every good tree brings forth good fruit; but a corrupt tree brings forth evil fruit."[30] Unlike the other two standard creationist mantras—evolution is "bad science," and it contradicts the book of Genesis—the "fruits" argument has a unique ability to speak to the mundane struggles of ordinary Christian believers. Instead of paying attention to complex critiques of evolutionary science or detailed analysis of biblical texts, Christians need only know that if we teach people they are descended from animals, they will act like animals. Recent "evil fruits" that creationists attribute to the "corrupt tree" of evolution include school shootings, gay marriage, and abortion. This moral consequentialist political logic can be neatly summed up by the title of an influential conservative manifesto authored by Richard Weaver some eighty years ago and still invoked by creationists today: *Ideas Have Consequences* (1948).[31] *Red Dynamite* highlights a key part of that creationist mobilizing strategy: the argument that socialism and communism, along with their alleged allied immoralities—centered on sex and death—are among the "evil fruits" produced by evolutionary thinking.[32]

These arguments had a populist flavor. They told ordinary Christian believers that they did not need to know any arcane details of biology to judge the validity of evolutionary ideas. Simply apply the "fruit test."[33] If evolution and its "culture of death" had spurred Stalin to kill millions, then regardless of any evidence about the merits of evolutionary science, it must be invalid. As one creationist skeptically asked, "What appreciation of the truth has emerged in the mind of the common man from all this profound probing into rocks and fossils, into the anatomy of the ape?"[34]

The degree to which such arguments avoided any need to judge an idea based on scientific evidence was striking. What mattered was the practical effect of evolutionary concepts. Given creationists' long-standing hostility to John Dewey and "progressive education"—which some also drew into the anticommunist net—it is ironic that they lent support to an essentially pragmatic idea.[35]

Even while Red Dynamite creationist thinkers focused on the here and now, they did not lose sight of the Christian theological stakes: evolution and its atheistic and communist associations were often linked to Satan, the ultimate false prophet—the great tempter and deceiver.[36] The focus on deception amplified the impact of conspiratorial claims in which people were not who they seemed to be. The false prophet argument dovetailed with the end-times theology of premillennial dispensationalism, in which an attractive, convincing, compelling leader turns out to be the Antichrist.[37] If evolution could be linked to Satan—a claim made repeatedly by Henry Morris, the founder of post–World War II "scientific" creationism— then Christians must stop it at any cost.

Over the last century, historians have enriched our understanding of the antievolutionary impulse in a variety of ways but have tended to ignore or neglect its anticommunist dimension.[38] Yet creationist anticommunists correctly saw that evolution and communism were allies. To take one surprising example, *Evolution: A Journal of Nature*, the first popular monthly magazine in the United States to promote the cause of evolutionary science, was founded and edited during the 1920s by a central leader of the US Communist Party.[39] The historian Richard Hofstadter taught generations that the primary "social Darwinists" were conservative, individualistic robber barons, obsessed with "survival of the fittest." Yet left-wing "social Darwinism," with an emphasis on the collective good, was just as real.[40] And despite the disastrous experiments of the Soviet agronomist Trofim Lysenko, who championed Lamarck's theory of inheritance of acquired characteristics over Darwin's emphasis on natural selection, "Marxist-Darwinism" was the framework in which Soviet science developed in the 1920s, at the same time that evolution disappeared from American biology classrooms and textbooks.[41]

The difficulty that academics have in taking creationists seriously stems from yet another one of Richard Hofstadter's influential intellectual creations—the "paranoid style in American politics." Prompted by the 1964 Republican Party presidential race, Hofstadter was reacting mainly to "extreme right-wingers" such as members of the conspiracy-minded John Birch Society, who were stumping for Barry Goldwater. Their "paranoid style" featured "heated exaggeration, suspiciousness, and conspiratorial fantasy." Such thinkers claimed not only that there were isolated conspiracies, but that they formed the "motive force" of history. Conspirators embodied "demonic forces of almost transcendent power."[42] While Hofstadter has been rightly criticized for dismissing conspiracy theorizing as an ill-defined mental illness, his description of conspirators as nearly omnipotent is helpful in underlining the essential continuity of conspiracy theorizing with religion.[43] Conspirators, whether communists or capitalists, seem to have supernatural command over events. Speaking of communists, this point helps explain why Marxists are not conspiracy theorists: while they recognize the immense power wielded by ruling wealthy classes (who sometimes meet and plot in secret), they also argue that working people have tremendous power, at least in potential form.[44]

In a number of respects, creationist anticommunism does seem to perfectly embody the conspiratorial "paranoid style." A number of antievolutionist crusaders pinned Darwinism on the Illuminati. They made plentiful use of the fabricated *Protocols of the Elders of Zion*, which blamed a Jewish plot for evolutionary science. And creationists regularly adopted anticommunist language that Hofstadter described as "apocalyptic and absolutistic." Such rhetoric may well have led evolutionary biologist Richard Dawkins to write these unfortunate words some thirty years ago: "It is absolutely safe to say that if you meet somebody who claims not to believe in evolution, that person is ignorant, stupid or insane (or wicked, but I'd rather not consider that)."[45]

But Dawkins and like-minded secular Americans dismiss creationist conspiracism far too easily. If creationist anticommunists tended to exaggerate the degree to which communists were lurking behind every corner and secretly directing events, there was a kernel of truth in the idea. To paraphrase Joseph Heller's *Catch-22*, just because you are a "paranoid" creationist does not mean that they are not organizing against you. Not only did socialists and communists promote evolutionary science (both natural and social), but they were also in the forefront, at critical moments, of campaigns for women's political and reproductive rights, the African American freedom struggle, the battle to establish industrial unionism through the Congress of Industrial Organizations (CIO) in the 1930s and '40s, innovations in sex education, and the gay rights movement. They gravitated toward these movements not because they were Satan's minions, but because they had a vision of forging a united working class that surmounted divisions based on gender, race, nationality, skill, and sexual orientation. In this regard, the "culture war" issues that are often considered to be separate from class conflicts are inextricably bound up with them.[46]

Without propounding a conspiratorial viewpoint, historians have increasingly recognized the key roles played by a variety of left-wing radicals in twentieth-century nonrevolutionary social movements. They include, for example, Betty Friedan, the central founder of the National Organization for Women (NOW), whose labor activism in Communist Party circles in the 1940s was crucial to her postwar feminist vision. Earlier generations of socialist feminists included birth control pioneer Margaret Sanger, who was inspired in part by Darwin's writings to challenge traditional women's roles. This does not mean that abortion is a communist evolutionary "plot." But the history of birth control cannot be fully understood without the role played by "reform Darwinist" socialist feminists.[47] Exaggeration, elevated suspiciousness, fantasy—all of these did characterize those preaching and writing about evolutionist and communist conspirators. Nevertheless, those in the creationist conspiratorial camp were expressing in a distorted form the real social, political, and economic conflicts that swirled around them.[48] This dynamic is no less true of recent rounds of conspiracy mongering, from anti-Obama birtherism to pro-Trump QAnon.[49]

As a metaphor explaining the creationist anticommunist tradition, dynamite is remarkably apt. For this we can thank Swedish inventor, chemist, and engineer Alfred Nobel, who lodged his new explosive squarely in evolutionary history. The key ingredient Nobel added to nitroglycerine to make the compound stable was diatomaceous earth, a sedimentary deposit made up of fossilized diatoms, a single-celled aquatic algae that evolutionary geologists date as far back as the Jurassic period.[50] Nobel also had a way with words. Patenting the new concoction in 1867, Nobel coined his world-famous neologism based on the Greek δύναμις (*dunamis*), usually translated as "power," a word that can mean a force for good or for evil.[51] This internally contradictory word perfectly conveys the struggle over morality and power that creationism expressed. Dynamite means mortal danger. It arouses an unreasoning, primeval fear. It poses a lethal threat that justifies any means of escape or resistance, as the Nelson and Richland miners knew all too well. The repeated denunciations of evolution and its baleful effects did tend to have an all-or-nothing quality, a sense that at stake was nothing less than the existence of human civilization. Paradoxically, dynamite also can mean exactly the opposite—"terrific," "wonderful," and "impressive." Not for nothing did a string of conservative Christian preachers who denounced evolution in the strongest terms also boast that their own sermons were "dynamite." While George McCready Price never explicitly connected the explosive power of dynamite with the human orgasm, the metaphor also inevitably resonates with sex. In so many ways, the anxieties raised by the teaching of evolution—whether they revolved around "free love," fluid gender roles, abortion, divorce, homosexuality, racial mixing, dancing, or petting parties—came down to fear of the power of sexual arousal to disrupt the established social order.[52] Fundamentalist preacher and Moral Majority founder Jerry Falwell captured this duality when he warned about evolution-induced "sexual anarchy" in the nation's schools.[53]

The "red" half of the equation is similarly multivalent. In Price's era and for decades thereafter, "red" signified communists and socialists, workers in revolt against the capitalist order. Depending on one's political sympathies, red could be a badge of revolutionary honor or a mark of shame, and in the eyes of the dominant thread of Christian premillennialism, a "mark of the beast." As the original basis of the choice of color for the flag of revolt, red also meant blood sacrificed in a righteous cause and—as any fan of the musical version of *Les Misérables* knows—a measure of the dedication of its determined defenders.[54] In the eyes of antievolutionists, it could connote the massive volume of blood spilled by evolution-induced mass murder in the twentieth century. For those who stood opposed to "Darwinism," sex and death fairly well sum up the supposed effects of teaching evolution to the nation's youth.

While I have uncovered plentiful evidence of Red Dynamite creationist politics, both scientific creationists and conservative evangelicals have consistently downplayed their political aims. Each group had distinctive reasons for doing so. Henry Morris and his counterparts at the Institute for Creation Research sought to reinforce their standing as "real" scientists, who presumably abjured the rough-and-tumble of the political world. Fundamentalist ministers of the Christian gospel attempted to uphold Jesus's injunction not to mix religion and politics: "Render to Caesar the things that are Caesar's, and to God the things that are God's."[55] That distinction can be notoriously difficult to define. As Jerry Falwell wrote in 1979, "homosexuality, abortion, pornography are not political issues, they are moral issues that have become political."[56] Contrary to Falwell's claim, my book proceeds on the contention that religion and politics—in the broadest sense—have always been inextricably intertwined.[57] In that respect, it builds on a scholarly foundation laid by historians Darren Dochuk, Matthew Avery Sutton, Dan Williams, Molly Worthen, and others, who have placed conservative Christian ideas, cultural commitments, and political activism firmly within a rich framework of social and political history.[58] As Sutton writes in regard to the mythical, multidecade flight of fundamentalists from the political arena after the Scopes trial, "They never retreated."[59]

Creationists' repeated denials of political activism are remarkably similar to recent claims by "denialists" of a different kind: anticommunist climate-change-denying scientists and their allies. In *Merchants of Doubt: How a Handful of Scientists Obscured the Truth on Issues from Tobacco Smoke to Global Warming* (2010), Naomi Oreskes and Erik M. Conway show how a handful of scientist activists were the key players behind campaigns to stop government action to reduce smoking, address a thinning ozone layer, and combat the effects of climate change. A superficial reading of these scientists' writings suggests that they were simply combating "bad science." But they were motivated, above all, by a "free-market fundamentalism" that expressed deep hostility to government regulation of the economy. In opposing action on climate change, the denialists ridiculed environmentalists as secret socialist "watermelons"—green on the outside and red on the inside.[60] While the history of the COVID-19 pandemic remains to be written, it would be hard to deny that conflicts over virus lethality, the wisdom of mask wearing, and testing data are not about science in any strict sense but rather rest on opposed political worldviews with deep historical roots.[61]

To reconstruct the century-long history of Red Dynamite politics, it is essential to begin by documenting the pre–Scopes trial reality of socialist and communist pro-evolutionism, which is the subject of chapter 1. Chapter 2 charts the early Christian evangelical response by centering on creationist geologist George McCready Price, a Seventh-day Adventist whose faith tradition uniquely encouraged a young-earth perspective and who began tying together the evils of socialism and evolution early in the twentieth century. In chapter 3 we follow the organizing activities during the 1920s of prominent creationists such as William Bell Riley, Gerald Winrod, J. Frank Norris, Mordecai Ham (who converted Billy Graham and preached that evolution was the result of a Jewish-Communist world conspiracy), and Catholic creationist George Barry O'Toole. In chapter 4, Christian anticommunism unfolds in a context highlighted by widespread labor struggle, political polarization, and the rise of Fascism on a global scale. Riley and Winrod both embraced the authority of the *Protocols of the Elders of Zion*, which explains that Jews are responsible for both Marxism and Darwinism. In the early years of the Cold War as narrated in chapter 5, conservative Christian leadership passed to a new generation who prominently included Baptist firebrand preacher John R. Rice. In the early 1960s, an organized antievolution movement reemerged under the leadership of "scientific" creationist Henry M. Morris, coauthor with John C. Whitcomb Jr. of the young-earth blockbuster *The Genesis Flood* (1961). Chapter 6 traces the continuation and transformation of the Red Dynamite theme in the writings of Morris and his allies in the Creation Research Society. In chapter 7, I bring the story through the end of the Cold War in the early 1990s. Chapter 8 traces echoes of the Red Dynamite theme into the twenty-first century.

The political connotation of the word "red" has been transformed into its opposite since George McCready Price coined his phrase nearly a century ago. Young people today associate "red" with the Republican Party, not communism. And yet, that political symbolism continues to evolve in surprising ways. When public school teachers launched a powerful wave of strikes in the spring of 2018, they wore red T-shirts. Teachers proudly proclaimed that they were "Red for Ed," drawing on labor movement traditions and calling attention to the sad state of state education budgets.[62] The shift in the political meaning of the color red may seem unrelated to the decades of political conflict unleashed at the Scopes trial, but it is a telling example of social evolution, which gets to the crux of the matter. The controversy over evolutionary science has never been primarily about science or religion, in a narrow sense, but about morality and power. Who will rule society and on what moral basis? Viewed in this light, the ongoing tensions over teaching Darwin and his ideas as they have evolved over 150 years are inseparable from the broader social and political conflicts of today. Not until those conflicts are resolved will we stop arguing about evolution.

1

LIGHTING THE DARWIN FUSE

In 1923, Rev. William Bell Riley painted a frightening picture for his congregation at the Minneapolis First Baptist Church. Delivered to some two thousand church members, Riley's sermon depicted the fruits of teaching atheistic evolution in Soviet Russia.[1] Inspired by Darwinism and a Marxist conception of social evolution, Bolshevik leaders had embarked on an immoral transformation of Russian society. "There isn't a single one of the civilities of the Christian civilization that this [Soviet] rule cares to retain," Riley charged. "They have deliberately attempted to destroy the family, to governmentize all women, and compel every babe that is born to be a bastard." This ruthless Soviet policy constituted Act One of the "Darwinian drama," with "survival of the fittest" taking center stage. Fueled by jungle ethics, the amoral Bolsheviks aimed to take over the world and had placed "secret agents of Lenine" in New York. They were planting "dynamite at many American points," saving their "largest charges" for the public schools. Those "charges" consisted of textbooks employing God-denying evolutionary arguments. Having planted such volatile materials, the clandestine communists would "light the Darwin fuse and witness the demolition."[2]

There was a large dose of fantasy in Riley's conspiratorial fears. But his sermon was correct in the claim that Marx and Engels supported Darwin. It was equally true that leaders of the pre–World War I Socialist Party of America promoted evolutionary science. They freely combined social and biological evolutionary arguments, mixing Marx and Herbert Spencer. After the Great War, the fledgling American communist parties took up the evolutionary cause. Their leaders made an uncompromising stand in defense of evolution, aiming to clarify for revolutionary workers what was at stake in the 1925 "Monkey Trial" of John Scopes. American communists were in step with their Russian counterparts, who promoted both atheism and "Marxist Darwinism" and paid close attention to American developments.[3] The Bolshevik regime did not "governmentize" women, but it did carry out an ambitious program of raising women's power and status. Over the next century, when American fundamentalists and anti-evolutionists linked evolution to communism and immorality, their claim was based not only on imagined conspiracies but on the real promotion of evolutionary science, atheism, and social change by socialists and communists both in the US and abroad.

The association between evolutionary science and Marxism began in the wake of Darwin's publication of *On the Origin of Species*. In December 1859, Friedrich Engels wrote to Karl Marx, "Darwin, whom I am just reading, is splendid."[4] A year later, after Marx finished the book himself, he wrote back to Engels, "This is the book which contains the basis in natural history for our view."[5] Marx and Darwin shared a basic materialist outlook. At the same time, Marx and Engels did fault Darwin for applying pro-capitalist economics to nature.[6] As Marx wrote in 1862, "It is remarkable how Darwin rediscovers, among the beasts and plants, the society of England with its division of labour, competition, opening up of new markets, 'inventions' and Malthusian 'struggle for existence.'"[7] By transposing mid-nineteenth-century capitalist England onto the natural world, Darwin was some variant of what came to be called a "social Darwinist."

The founders of the modern communist movement, although not unqualified endorsers of Darwin, continued to pay tribute to him.[8] In 1873,

Marx sent Darwin an inscribed copy of the second volume of *Capital* signed by a "sincere admirer."[9] In 1877, Engels published *Anti-Dühring*, whose chapter on "The Organic World" featured a spirited defense of Darwin.[10] As more and more German socialists promoted Darwin's ideas, the prominent anti-Darwinist scientist Rudolf Virchow publicly attacked Darwinism, as leading to socialism. German Darwinist Ernst Haeckel strenuously denied this, prompting Darwin himself to decry "the foolish idea . . . on the connection between Socialism and Evolution through Natural Selection." And yet at Marx's funeral in 1883, a year after Darwin's death, Engels eulogized, "Just as Darwin discovered the law of development of organic nature, so Marx discovered the law of development of human history."[11]

That law of development and its evolutionary character stood at the center of *The Origin of the Family, Private Property, and the State*, which Engels published the following year. Engels showed how these central institutions of modern capitalist society, far from being eternal or God-given, were of recent historical vintage. They emerged only as human society evolved from "savagery" (bands of hunters and gatherers) to "barbarism" (settled agricultural existence) to "civilization" (modern capitalist society). He concluded that in the earliest phase of this stage—which Engels called the "childhood of the human race"—our ancestors must have been tree-dwelling, apelike creatures who evolved the capacity of "articulate speech" over thousands of years.[12] Once modern humans had evolved, Engels argued, they lived in a condition of "primitive communism." Women held considerable social power through the institution of matrilineal descent. Humans lived in a relationship of rough equality. It was the emergence of class divisions—made possible by the production of a social surplus—that paved the way for the oppression of women, political tyranny, and economic exploitation. These evils were not a product of human nature, but of class society. Once modern class society was overthrown, humanity could restore the virtues of "primitive communism" but on a higher material level.

Along with Marx and Engels themselves, the other major influence on American socialists and reformers writing and speaking on evolutionary themes was Herbert Spencer. Author of the phrase "survival of the fittest," Spencer is best known today as an ultra-individualist who opposed government aid to the poor and thought socialism was tantamount to "slavery." But Socialists embraced not his reactionary book *Man against the State* (1894), written near the end of Spencer's life, but the younger, radical-minded opponent of the English landed aristocracy and established church.[13] Then a critic of individual landownership and supporter of women's suffrage, Spencer argued that, freed of government interference, and under Malthusian population pressure, society would evolve toward a perfect "equilibration" between the social organism and its environment.

Spencer drew heavily on a vision of evolution associated with Jean-Baptiste Lamarck that revolved around the inheritance of acquired characteristics. Lamarck thought that generations of giraffes gained longer necks by straining to reach leaves higher up in trees and passing on those longer necks to offspring. Spencer likewise believed that humans could improve the "fitness" of society by conscious effort during their lifetimes. In contrast, Darwin's proposed mechanism of natural selection did not depend on conscious striving. It only required that some members of a natural population were more successful than others in passing on their genetic material, thereby changing the profile of that population over long periods of time. More Lamarckian than Darwinian, Spencer's *First Principles* (1862) laid out the beginnings of his "synthetic" philosophy that combined physical, biological, social, mental, and political development into one grand evolutionary scheme. Concluding his chapter on "equilibration," Spencer averred that "evolution can end only in the establishment of the greatest perfection and the most complete happiness."[14]

It is appropriate that Spencer's ashes were deposited in London's Highgate Cemetery, facing the grave of Karl Marx, since early American socialist writers attempted to incorporate both men's work into their reckonings with evolutionary science. They were influenced by prominent Italian socialist and criminologist Enrico Ferri, who called for a "class struggle in the Darwinian sense," thus blurring the lines between social and biological evolution.[15] Ernest Untermann, a German-born American Socialist activist with university training in paleontology, geology, and biology, wrote widely on the relationship between socialist politics and science and was the translator of Engels's *Origin* into English. Identifying himself as a "socialist Darwinian," Untermann published *Science and Revolution* in 1905, one of the most notable attempts to wed Marxism and the

Darwinian science of the day. He was well aware that Marx and Engels themselves maintained a clear line between the two. But he also claimed that a "dialectic synthesis of Marxism, Darwinism and Spencerism" was allowed.[16]

More critical of Spencer but still enamored of the organic analogy was Chicago Socialist Arthur Morrow Lewis. Born in England, educated through the common schools, and trained as a molder, the largely self-taught Lewis became a highly effective popularizer of socialist evolutionism to an American working-class audience.[17] Lewis regularly delivered lectures on science to overflow crowds on Sunday mornings at Chicago's Garrick Theater. He debated in open-air meetings on a variety of topics, including science and socialism.[18] In 1908, Charles H. Kerr & Co., the best-known American publisher of socialist literature, issued a compilation of Lewis's lectures as *Evolution: Social and Organic*.[19] In his preface, Lewis observed that some Socialists had questioned the wisdom of lecturing on evolution, fearing that challenging workers' prevailing religious notions would result in "driving people away." In response, Lewis wrote that "I have yet to be convinced that there is any kind of knowledge which is good for university men, but unfit for workingmen."[20]

The knowledge Lewis sought to impart to Chicago's working people concerned both their past and their future. Lewis traced the origin of evolutionary ideas to the ancient Greek materialist philosophers—among them Thales, Heraclitus, and Empedocles—who provided, in Spencer's words, "vague adumbrations" of future evolutionary science. Noting that scientific knowledge was carried forward during the Middle Ages by "pagan Arabians" and not Europeans, he credited the rising capitalist class—in its progressive phase—as the "harbinger" of progress, bringing to the fore thinkers such as Linnaeus, Lyell, and Lamarck. But, echoing Marx, Lewis argued that by unleashing the forces of modern science, the capitalists planted the seed of their own destruction. Realizing this, today's plutocrats would rather "suppress science or at least prevent its reaching the proletarian brain." Yet, they rely on a mode of production that makes the education of workers "a relentless necessity" and thus are caught in a painful contradiction.[21]

One mark of the seriousness with which Lewis took his educational task is his engagement with recent trends in evolutionary science. Even as he trumpeted Darwin's central explanatory concept of natural selection, Lewis was well aware that Darwinism in the early twentieth century faced a range of challenges that evolutionists could not yet effectively meet. Fleeming Jenkin had posed problems with Darwin's concept of heredity, and Lord Kelvin's lowered estimates of the age of the earth seemed to make an imperceptibly slow process of natural selection impossible.[22] Meanwhile, the rediscovery of Gregor Mendel's work in genetics initially led pioneer geneticists Thomas Hunt Morgan, Hugo de Vries, and others to argue against gradual Darwinian natural selection. They preferred a theory of rapid evolutionary leaps, known as mutation theory.[23] Then there was August Weismann and his mice. A German zoologist who was determined to disprove the Lamarckian concept of the inheritance of acquired characteristics, Weismann cut off the tails of mice, bred five generations of offspring, and found that they all had intact tails, rather than the stubby ones as Lamarck would have predicted. Since Lamarck's ideas were still popular in the early twentieth century, Weissman's results seemed to confuse the evolutionary picture even further.[24]

Lewis took up these challenges and wove them into his vision of socialist politics. Dutch mutation theorist Hugo de Vries conducted studies of the evening primrose (*Oenothera lamarkiana*), which seemed, at random, to produce differently colored varieties from the original yellow stock, thus seemingly undermining the model of gradual evolution by natural selection. Mutation theorists used these results to attack Darwinians. But as Lewis pointed out, de Vries did not deny Darwinian natural selection; rather, he was making an argument about the cause of variation, on which natural selection could act. Lewis hypothesized that species might undergo alternate periods of stability and "mutability," which would shrink the total amount of time needed for natural selection to operate. Two happy results would be that Kelvin's new estimate of the age of the earth was no longer contradicted; and gaps in the fossil record could be explained. Lewis then derived the political lesson: just as natural evolution oscillates between stasis and rapid change, so does "social development" move between periods of "apparent social stability" and those of "social revolution when the entire social superstructure is transformed."[25] Some six decades later, paleontologist Stephen Jay Gould's Marxist-influenced theory of punctuated equilibrium would make similar connections (and draw similar attention from creationists).

Lewis also took up Weismann's Darwinian challenge. Weismann's experiments seemed to undermine a widespread optimistic belief, among reformers and many socialists, in the inevitability of a Spencerian-Lamarckian progress, based on continual efforts at human social betterment. If those improvements—the social equivalent of the stubby mouse tails—were not passed on to the next generation, did not the future look bleak? Not to Lewis. To the extent that the personalities of working people under capitalism were affected negatively by "degrading conditions," if their behavioral traits were passed on to future generations until they became "fixed characters," then a future socialist society that depended on the improving capacities of those same working people would be a utopian dream.[26] Thanks to Weismann, the future now looked brighter. Social evolution and biological evolution could proceed on parallel but separate paths.

The problem of the relationship between social and biological evolution was posed sharply for members of the fledgling Socialist Party of America who were women. When the first German-American socialists arrived in the US at midcentury, the "woman question" was barely asked. Holding to the prevailing patriarchal view of women's role in the family, male socialists dismissed demands for women's suffrage and opposed their entry into the paid workforce. It took decades of immigration, increasing entry of women into industry, and determined organizing by early feminist socialists to put the issue on the agenda. By the early twentieth century, a new generation of women, who had earned their spurs in suffrage, temperance, and labor activism, placed their feminist stamp on the young Socialist Party, through the Women's National Committee and the *Socialist Woman* magazine.[27]

Of considerable influence on American socialists, both male and female, was *Woman and Socialism* (1883) by German socialist August Bebel. Published the year before Engels's *Origin*, Bebel's work provided a similar account of the roots of women's second-class status, reaching back to the downfall of a primitive communist matriarchy. Departing from the view that most German socialists had taken, Bebel argued that the path forward was the "release of woman from her narrow sphere of domestic life, and her full participation in public life and the missions of civilization."[28] Not only did Bebel challenge the common view that women were ordained by nature to be intellectually inferior to men; he placed his argument in an evolutionary framework. While he agreed with Darwin that there were comparatively few known women of "genius," he disagreed that this stemmed from women's innate biological nature. Rather, Bebel argued, it was the "conditions of existence" that explained the differential social evolution of men and women. Only under socialism, he claimed, would the conditions be sufficiently changed to allow women to develop their full potential. Bebel was confident that under those conditions, which had never "existed in human evolution," "woman will rise to a height of perfection that we can hardly conceive to-day."[29] In the sense that social evolution allowed for the development of all individuals' potentiality, male or female, Bebel believed that "Darwinism" was an "eminently democratic science."[30]

A growing number of American Socialists were willing to consider the need for women's political and economic equality. But they were more resistant to challenging existing norms of women's sexuality. The Socialist Party did take part in anti-prostitution "social purity" campaigns in the early twentieth century and sharpened the traditional Progressive critique by focusing on the culpability of capitalism for the degradation of working-class women. But the bohemian "new intellectual" Socialists, around publications like the *Masses*, *New Review*, and New York *Call*, who pioneered the movement on behalf of birth control or "family limitation," were marginalized by the older, established party leaders. It was one thing to view woman as victim, another to see her as an active subject with sexual drives and interests equal to any man.[31] Intense pressures weighed on Socialists to play down aspects of their political and intellectual heritage (including Bebel's work) that might lead to charges of immorality or advocacy of "free love."

The perils of socialist evolutionism emerged in the activist life of Lena Morrow. An Illinois-born, college-educated veteran of the Woman's Christian Temperance Union (WCTU), and an energetic defender of women's rights, Morrow was a leading Socialist Party organizer who became the first woman ever elected to the party's National Executive Committee. Based in San Francisco, Morrow traveled throughout the West, visiting mining and lumber camps to boost the socialist cause. She was a fearless street speaker who in 1903 called her new husband, Socialist lecturer Arthur Morrow Lewis, to bail her out of jail. Though she supported women's suffrage, Lena Morrow Lewis rejected the idea that women were different from men in some fundamental way. Hewing to a Lamarckian evolutionary view, she believed that men's prejudice toward women had resulted from their "brain cells" adapting to lower economic forms, which were now being superseded.[32]

While Lena Morrow Lewis received socialist accolades, not all her comrades were comfortable with a woman in such a leadership role. Her short and apparently unhappy marriage to Lewis—they divorced in 1905—may have made things more difficult for her. Then there was the "scandal" that enveloped Lena Morrow Lewis in 1910–11. Articles in internal party bulletins charged that Lewis carried on an affair with Socialist Party national secretary J. Mahlon Barnes, who had himself been accused by a party rival of engaging in "free love" with office employees.[33] Writing in the *Masses* in 1911, and perhaps reflecting on these painful events, Lena Morrow Lewis noted that, historically, women had passed from the stage of primitive communism in which they could live "the life of a human being" to class society, in which they were defined by their "maternal functions." Hence we say "man and wife" and never "husband and woman." But with the prospect of socialism, once again woman could "live the life of a complete human being." In the new society, standards of "sex relations" and of "morality" would be determined by newly evolved conditions and social demands.

If there was one socialist who embodied the evils of evolutionist immorality in the eyes of creationists during the 1920s (and generated the "Red Dynamite" label), it was the Reverend Charles Browning "Bouck" White. Hailing from an old-line New York family, White graduated from Union Theological Seminary in 1902. He served as pastor at several Congregational churches, moved to New York City, and then, under the impact of the 1909–10 shirtwaist strike and subsequent Triangle Shirtwaist Fire of 1911, became a militant socialist preacher. White first attracted headlines (and courted arrest) in May 1914 when he led a labor protest at the Calvary Baptist Church in New York City. Its congregation included John D. Rockefeller Jr., primary owner of Colorado Fuel and Iron, where private police and state militia had just carried out what came be known as the Ludlow Massacre. On the eve of US entry into World War I, White was again arrested. He held a church service in which flags of combatant nations, as well as the American flag, were burned in a testament to "internationalism and universal brotherhood."[34]

White's writing also got him into trouble.[35] In 1911, White authored a popular biography of Jesus, *The Call of the Carpenter* (1911), which made a profound impact on Socialist Eugene Debs. Drawing on a nineteenth-century literary trend that portrayed Jesus as a real historical and non-divine figure, White was the first to create a book-length biography of Jesus as a proletarian revolutionary who had come of age as a young laborer in "working-class Galilee." While the book sold well, its radicalism led to White's dismissal from his position as head resident at the Holy Trinity Episcopal Church, prompting him to start his own unique denomination—the Church of Social Revolution.[36]

At White's new church, he attracted renewed attention by flouting prevailing notions of proper marriage. White favored a new set of vows that bound husband and wife not as long as they both shall live but "so long as love shall endure." Designed to accommodate a new ideal of "companionate marriage," the vows, said White, were for those who were "in a spirit of revolt against old customs."[37] In this spirit, he married Andree Emilie Simon, a wealthy young Frenchwoman. Upon their arrival in the US in 1921, the couple briefly lived in New York City, where White hired a tutor from the Intercollegiate Socialist Society to inculcate his young bride with radical ideas. When this educational experiment failed, they removed to White's run-down Ulster County country "estate," which quickly led Simon to file for an annulment of their marriage. Rumors of White's radical notions and Simon's charges of abuse led local vigilantes to abduct, tar and feather White, and dump him on the outskirts of town.[38]

For all his uniqueness, White was typical of his fellow socialists in one respect: his enthusiasm for evolutionary science. In *The Call of the Carpenter*, White hailed evolution's ability to undermine the idea of "God the father almighty"—which he thought had been used as a weapon of class exploitation. This explained why Darwinism had been hailed by the "proletariat" and the "democracy." To show that evolution aided the forces of social revolt, White quoted representatives of the forces of social order denouncing Darwin, drawing on evidence collected in Andrew Dickson

White's *A History of the Warfare of Science with Theology in Christendom* (1896). They included historian Thomas Carlyle, who called Darwin "the apostle of dirt worship," and a French Catholic apologist who claimed that the "offspring" of Darwinism were "revolutions."[39] Like his socialist comrades, Bouck White reminds us that a politicized left-wing "social Darwinism" was alive and well in early twentieth-century America.

In the half century before the Bolsheviks took power in Russia in 1917, reformers and revolutionaries there also embraced Darwin. The publication of Darwin's *On the Origin of Species* in 1859 coincided with a period of social and political ferment, symbolized by the freeing of Russian serfs in 1861. The close alliance between the conservative Russian Orthodox Church and the czarist regime, along with the relative lack of religious "modernists," promoted an identification of science and progress by nihilists and populists. Debates raged among scholars over the precise identification of the evolutionary mechanism. Czarist universities were filled with sparring neo-Lamarckians, vitalists (who believed that evolution sprang from an inner life-force within biological organisms), Mendelians, mutation theorists, and, due to the writings of Prince Kropotkin, those who believed that "mutual aid" by members of the animal and plant kingdoms was the primary vehicle for change.

But for the vast majority of Russian intellectuals, a common origin for all life on earth was not seriously in doubt. In 1909, they marked the centenary of Darwin's birth with conferences and a multi-authored volume of tribute, *In Memory of Darwin* (1910). Among them was animal physiologist and Nobel Prize winner Ivan Pavlov, who had planned to become a priest until he encountered Darwin's ideas as a seminary student. Prominent as well was Kliment Timiriazev, a highly esteemed plant physiologist who championed Darwin's ideas early on and who was widely known as "Darwin's Russian bulldog," a counterpart to British comparative anatomist T. H. Huxley.[40]

Russian Marxists—members of the Russian Social Democratic Labor Party—were proponents of evolutionary science as well. Among these, Georgi Plekhanov stood out in his belief that Darwinism and Marxism were allied for the cause of the working class. He echoed Engels's graveside eulogy in his description of the achievements of Marx and

Darwin: "Darwin succeeded in solving the problem of the origins of plant and animal species in the struggle for survival. Marx succeeded in solving the problem of the emergence of different types of social organization in the struggle of men for their existence. Logically, Marx's investigation begins precisely where Darwin's ends."[41] While Plekhanov insisted that different types of laws prevailed in social and biological evolution, other Russian Marxists attempted to creatively apply a Darwinian model to human society. Thus did physician and early Bolshevik A.A. Bogdanov argue that "social forms represent adaptations in the same sense and to the same degree as all biological forms."[42]

The Russian Bolsheviks whom American creationists were most likely to target in the 1920s were Lenin and Trotsky. Born Vladimir Ilyich Ulyanov in 1870, Lenin was baptized into the Russian Orthodox Church but grew up influenced by his mother's heterodox religious views and the revolutionary activism of his brother Alexander. The regime executed Alexander in 1887 for a failed attempt to assassinate Czar Alexander III. Around this time, Vladimir Ilyich renounced his belief in God and joined a series of revolutionary groups. Over the next decade, he was exiled to Siberia, and then traveled in Western Europe, debating strategy, exchanging views, writing, and gathering his forces for a coming Russian revolution. A published 1894 polemic against a leading Russian populist—N. Mikhailovskii—was one of the rare occasions during this period on which Lenin explicitly addressed the subject of evolutionary science. Mikhailovskii had compared Marx's *Capital* to Darwin's work and found the former wanting.[43]

In his response to this critique of Marx, Lenin chided Mikhailovskii and his fellow populists for their inconsistency in applauding Darwin's scientific achievement but denying that sociology could be scientific. In contrast with the Russian populists' utopian notions of the ideal society built upon their idealist conception of "human nature," Lenin argued that Marx had provided a much more objective (and materialist) grounding, and in this respect was similar to Darwin:

> Just as Darwin put an end to the view of animal and plant species being unconnected, fortuitous, "created by God" and immutable, and was the first to put biology on an absolutely scientific basis by establishing the mutability

and the succession of species, so Marx . . . was the first to put sociology on a scientific basis by establishing the concept of the economic formation of society as the sum-total of given production relations, by establishing the fact that the development of such formations is a process of natural history.[44]

Marx, like Darwin, had discovered a scientific "law of motion."

A debate between Bolsheviks led Lenin, fifteen years later, once again to visit the relationship between biological and social evolution. In *Materialism and Empirio-Criticism* (1909), Lenin squared off against A. A. Bogdanov, who was working to combine Marxist ideas with recent development in science, including Darwinian biology. Bogdanov had been influenced strongly by two German scientific thinkers: physicist Ernst Mach and chemist Friedrich Wilhelm Ostwald. Best known for his pathbreaking research into vision (including the physics of optical illusions) and acoustics, Mach wrote a number of works that sought to lay a foundation for unifying all the sciences on a solid empirical and non-metaphysical basis. In doing so, however, he adopted a "phenomenalist" stance that required scientists to base their conclusions exclusively on "sensations" and that rejected any attempt to establish a correspondence between such sensations and an "external" world.[45] For his part, Ostwald had developed a theory of "energetics," which held that energy, not matter, was the single unifying entity in nature. Not only did he argue that energetics could unite all the physical sciences, but Ostwald developed his ideas into a full-blown worldview that encompassed the humanities, social sciences, ethics, and morality.[46] Since all these ideas emerged when new developments in radioactivity and electrodynamics (and soon to include relativity theory) were exploding previous conceptions of "matter," Lenin was deeply concerned. "Machism" might open the door to a rejection of materialism and the fundamentals of Marxism, all in the name of science.[47]

Like his earlier polemical response to Mikhailovskii, Lenin's *Materialism* aimed to clarify the position of genuine Marxism. While most of his fire was directed at the fundamental philosophical issue of whether or not there was an external reality, Lenin also addressed the issue of evolution. Among the passages from Bogdanov's writing that Lenin singled out for scorn was the following, which he put in italics: "*Every act of social selection represents an increase or decrease of the energy of the social complex concerned. In the former case we have 'positive selection,' in the latter 'negative selection.'"* Lenin's reply: "And such unutterable trash is served out as Marxism! Can one imagine anything more sterile, lifeless and scholastic than this string of biological and energeticist terms that contribute nothing, and can contribute nothing, in the sphere of the social sciences?"[48] For Lenin, unlike many leading American socialists, the unique integrity of both Marx and Darwin forbade carelessly mixing them together in one analytical stew.

Lenin also took pains to clarify the Bolsheviks' perspective on religion. In a 1909 article, Lenin acknowledged that Marxism was atheistic; but he warned fellow Bolsheviks about the political dangers of "declaring war on religion." Traditionally, noted Lenin, "fear [of the natural elements] made the gods." The social roots of modern religion, he argued, lie in the feeling of helplessness on the part of working people, in face of "the blind forces of capitalism." The most effective way to reduce the power of religion is to increase workers' sense of power and control over social and political life. Rather than fervently and provocatively "preaching" atheism, Bolsheviks needed to "work patiently at the task of organizing and educating the proletariat."[49]

Lenin was picking up where Marx had left off in his contribution to *Critique of Hegel's Philosophy of Right* (1844). Marx wrote here that religion was "the opium of the people," which emphasizes the way in which it keeps working people passive and inert. But he prefaced this oft-quoted line with the following: "*Religious* suffering is, at one and the same time, the *expression* of real suffering and a *protest* against real suffering. Religion is the sigh of the oppressed creature, the heart of a heartless world, and the soul of soulless conditions."[50] Since oppressive conditions were the key to the existence of religious belief, in this view, they should be the focus for change. This sage advice would be pointedly ignored by later generations of Soviet leaders.

Nine years Lenin's junior, Lev Bronstein (1879–1940), later known as Leon Trotsky, was also a dedicated evolutionist. Born into a prosperous Jewish farming family in the Ukraine, Trotsky spent a short time at Hebrew school and then was sent to study in Odessa, which opened his eyes to the wider world. By 1896, he was organizing workers as a populist revolutionary. Arrested in 1898, Trotsky was to serve four years in exile in Siberia, but not before serving some time in an Odessa prison, from which he took both his name (after a jailer) and inspiration from reading books.[51] He initially had access only to prison-provided Russian Orthodox texts, but soon gained access to books from the outside. Among these were works by Mikhailovskii, Plekhanov, Italian Marxist Antonio Labriola, and Darwin, who made a special impact. When Trotsky entered the prison, he had been attracted to Marxism but had still resisted its lure. "Darwin," Trotsky recalled, "destroyed the last of my ideological prejudices." The young revolutionary left the Odessa prison for Siberia with a new sense of certainty. The "idea of evolution and determinism," he wrote, "took possession of me completely." Darwin, Trotsky told Max Eastman, "stood for me like a mighty doorkeeper at the entrance to the temple of the universe."[52]

Nearly two decades later, at the helm of the new Soviet republic, Trotsky and Lenin continued to promote Darwinism along with their Marxist politics and philosophy. The early Bolshevik regime sponsored a vast expansion of Russian science, including the new field of genetics. The People's Commissariat of Enlightenment (known as "Narkcompros") carried out a massive campaign to popularize Marxist ideas. Its Main Scientific Council included as a member Kliment Timiriazev, whose 1919 article "Darwin and Marx" emphasized the parallels between the two men who "marched under the banner of the natural sciences." In 1922, Engels's works on natural science—*Dialectics of Nature* and "The Role of Labor in the Origin of Man"—were for the first time translated into Russian.[53] Influential speeches and articles by both Lenin and Trotsky addressed the connections between Marxism, Darwin, and the other natural sciences. In *Under the Banner of Marxism*, Lenin called for fellow communists to forge an alliance with noncommunist but materialist Soviet scientists.[54]

More so than Lenin, Trotsky directly addressed himself to Darwinian evolutionary science in the early 1920s. In a 1923 article, Trotsky focused on the process by which young people became effective revolutionists. They faced not only external obstacles, but internal ones that inhibited their full commitment to changing the world. The potential revolutionist needed to shed any kind of "mysticism or religious sentimentality." Anyone who "believes in another world," wrote Trotsky, "is not capable of concentrating all his passion on the transformation of this one." Darwinism played an essential role in helping young people lose their belief in another world, argued Trotsky, and thus was "a forerunner, a preparation for Marxism." Taken together, Darwinism and Marxism could explain universal development of nature and society in their proper relationship, from "the living flow of being in its primeval connection with inorganic nature" to the modern class struggle.[55]

To those who claimed that Darwinian gradualism was incompatible with the Bolsheviks' revolutionary politics, Trotsky followed the logic laid out by Arthur Lewis fifteen years earlier. Trotsky acknowledged that natural history included long periods of "relative equilibrium" where species remain "relatively stable" and natural selection operates "almost imperceptibly." But then, on an evolutionary timescale, there are also periods of "geobiological crisis," during which natural selection works with "ferocity" to destroy whole species. Evolution, he concludes, is the "theory of critical epochs" in the natural world, just as Marxism is focused on such periods in the history of human society. Though they would soon be on opposite sides of the deadly factional politics of Stalinism later in the decade, Bolshevik leader Nicolai Bukharin agreed with Trotsky on this point. Contrary to the oft-quoted Latin aphorism, he wrote, in 1925, "Sudden leaps are often found in nature." Our failure to recognize this fact reflects our fear of sudden social shifts, that is, "fear of revolution."[56]

The Bolsheviks' embrace of evolutionary science was sincere, but it also served the practical purpose of undermining traditional religious belief, a process they viewed as essential to building a new socialist society. The early revolution's confrontation with organized religion took several forms, all of which would feature in anticommunist writings in the coming decades. The Russian Orthodox Church was its largest and most lasting target. It was not only tied organically to the czarist regime—the czar was head of state and church—but it permeated Russian culture. In the vast rural areas of Russia, the church, its icons, its rituals, and its network of local priests were deeply enmeshed in the rhythms of daily life. To be effective, a campaign against "religious belief" meant something approaching total cultural war.

A series of decrees issued after the Bolsheviks took power set the stage. They established legal separation of church and state, which meant that control of education as well as all church property now belonged to the Bolshevik regime. No minor could legally receive a religious education, except in a private home. Marriage became a civil relationship. State subsidies to any church institution were suspended. Military clergy were dismissed. At the same time, new positive rights were established. The state could pass no law that restricted freedom of conscience or privileged any particular religious belief. More straightforwardly, "Every citizen may confess any religion or profess none at all."[57]

It was one thing to make declarations, and quite another to enforce them. The ensuing civil war both enabled and hobbled enforcement. The decision of Tikhon, the Orthodox patriarch, to ally himself and the church hierarchy with the counterrevolutionary Whites gave the Bolsheviks license to use force to seize church buildings and land and to take harsh measures against bishops, priests, and believers who actively resisted. Bloodshed attended further attempts after the civil war to requisition church treasures to convert to badly needed hard currency. The very militancy of the Bolshevik response emboldened resisters, and led to further reprisals, which undermined any attempt to peacefully convince workers and peasants of the new materialist worldview.

A coercive, administrative, and in some cases extralegal antireligious campaign is not what Lenin himself had counseled in the years before the Bolsheviks took power. Along with their sometimes violent confrontations with church leaders, the Bolsheviks created a whole series of propaganda vehicles—films, traveling drama troupes, mass atheist organizations, and magazines aimed at enlightening the masses. Among the best known was *Science and Religion*, founded in 1922 and then renamed *Bezbozhnik* (Godless). A Society for the Friends of *Bezbozhnik* was soon formed, which, in the left-turn of the "Great Break" of 1928, became known as the League of the Militant Godless. The popular campaigns in support of Marxist-Darwinism fit into this broader antireligious campaign. So did plans to advance the application of science to the lives of Russian peasants. Pushed with special zeal by Trotsky, the materialist calculus suggested that the most powerful way to change thinking was to transform the material conditions of life. The Friends of *Bezbozhnik* in Samara illustrated this logic by confiscating church bells, selling them, and then purchasing tractors for local peasants.[58] Despite the wild exaggerations of later anticommunist conspiracy theorists, Bolshevik support for evolution and opposition to the organized power of religion were very real.

That support also inspired one truly bizarre venture that later produced fodder for creationists. The Bolshevik commitment to evolutionary science became international news in 1926 because of a controversial research project in Kindia, Guinea (then part of French West Africa), at a facility of the Louis Pasteur Institute of Paris. The lead researcher was Ilya Ivanovich Ivanov (1870–1932), an evolutionary zoologist who had pioneered the practice of large-scale artificial insemination with purebred horses. His project was to artificially hybridize humans and apes.[59] As strange as the scheme sounds today, the idea had been taken seriously by leading European scientists in France, Germany, the Netherlands, and Russia. Recent discoveries of hominid fossils, as well as living gorillas, fired a popular and scholarly interest in humanity's origins.[60] While Ivanov and the Bolsheviks did not motivate the project using racist terms, the colonization of West Africa and prevailing racist conceptions of a lower "African" race made the scheme sound reasonable to Europeans. Moreover, the preceding decades had seen a European vogue in the science of rejuvenation. The supposed virilizing powers of ape sexual glands fueled an interest in collecting specimens of live orangutans, gibbons, and chimpanzees.[61] Successfully appealing to the Bolshevik government for initial funding, Ivanov stressed the project's ability to aid the ideological campaign against organized religion and for Darwinism. In later discussions with the Academy of Sciences—which refused to support Ivanov's work—he stressed the scientific value of his research for human evolutionary studies.[62]

Once in Guinea, Ivanov did carry out at least part of the experiment—artificially inseminating several captive chimpanzees with the sperm of a local Guinean man. When the animals failed to become pregnant, the researchers sought to try their luck inseminating local African women with chimpanzee sperm (hoping to do so without the knowledge of the women, who were patients at a French colonial hospital). But the French authorities denied permission. When Ivanov complained about this to his Soviet sponsors, they ordered him not to attempt to impregnate women without their consent. One important legacy of the entire venture, however, was a primatological nursery in Sukhumi, in the Soviet Republic of Abkhazia (later Georgia), where Ivanov continued his work in the late 1920s, soliciting Soviet women volunteers for artificial insemination. Hybridization failed, but the population of chimpanzees gathered at Sukhumi would later produce the animals that rode *Sputnik* flights into outer space. Those voyages spurred Americans to strengthen scientific education, unintentionally inciting a backlash of creationist activism in the 1960s.[63]

Many Americans became aware of Ivanov's work because of prominent coverage in the US press. In June 1926, a *Time* magazine titled "Men
and Apes" reported that "Ivanoff," supported by Moscow, was headed to
Africa to "'support' Evolution by breeding apes with humans." Readers
also learned that the American Association for the Advancement of Atheism (AAAA), led by Charles Lee Smith, was publicizing the project and
actively raising funds for it, though Ivanov's staff in Moscow disclaimed
any connection with the group. That may well have been because leaders
of the AAAA had absorbed the "scientific" racist ideas of British anthropologist F. G. Cruikshank. His artificial breeding scheme recommended
the following pairings: orangutans with the "yellow race," gorillas with
the "black race," and chimpanzees with the "white race."[64] But the basic
story, as expressed in two June 1926 *New York Times* headlines, was
true: "Russian Admits Ape Experiments" and "Soviet Backs Plan to Test
Evolution."[65]

We do not know whether the refusal of Bolshevik authorities to sanc-
tion Ivanov's plans for the secret insemination of African women with
ape sperm was based on a principled feminist stand or a matter of realpo-
litik. We do know that the young Bolshevik regime took unprecedented
steps during these years to advance the status of women in revolutionary Russian society.[66] These controversial measures were reported—and
misreported—widely in the United States. The Bolsheviks' record—and
its impact in the US—became intertwined with the American debate over
evolution during the 1920s.

While the leaders of the Bolshevik Party were overwhelmingly male,
and hardly free of prejudice against women, they distinguished themselves
in the early years of the revolution by acting on the analysis of wom-
en's oppression developed by Engels and Bebel. Women workers, party
activists, and soldiers played a critical role in both phases of the 1917
revolution and the ensuing civil war. Once the Bolsheviks took power,
they did not hesitate to move forward on this front. Among those leading the charge was Alexandra Kollantai (1872–1952), who like American
socialist Lena Morrow Lewis believed that marriage and the patriarchal
family were products of an exploitative, class-based society, and that
under communism, new forms of human relations and social organization would evolve.[67] A member of the Bolshevik central committee during

the October revolution, Kollantai became the Bolshevik commissar for social welfare. Along with Inessa Armand, she founded the Zhenotdel, the Women's Department of the Russian Communist Party, from which she fought to change the conditions of Russian women's lives. In short order, divorce was made incomparably easier for women to obtain. Marriage became a civil legal relation, rather than one governed by the Russian Orthodox Church. Abortion was legalized, women gained the legal right to keep their maiden names, and the new regime offered a generous maternity leave policy. And it sought to combat what Lenin referred to as the "barbarously unproductive, petty, nerve-racking, stultifying and crushing drudgery" of the kitchen and nursery. Within the limits of a backward country devastated by civil war, the Bolsheviks pushed forward to build public laundries, cafeterias, and child care centers.[68]

These pioneering measures garnered close attention from both feminists and antifeminists in the United States.[69] The "information" that most Americans received, however, took the form of sensationalized newspaper stories about the "nationalization" of Soviet women.[70] The earliest report came in an Associated Press story printed around the United States on October 26, 1918. In Indiana, for instance, the *Huntington Press* bore the headline, "Decree Provides Maidens Become Property of State." A subheadline in the *New York Times* informed readers that "Decrees Compel Them to Register at 'Free Love Bureau' on Attaining 18 Years."[71]

The US Senate's Overman Committee amplified this coverage. Originally established to investigate pro-German propaganda in the US brewing industry during World War I and chaired by Senator Lee Slater Overman (D–NC), the committee targeted the American communist movement after the war ended.[72] Committee hearings held in February and March 1919 were sparked by a public meeting that featured Louise Bryant, radical journalist and wife of US Communist John Reed, who had spent time living in the new Soviet republic. She defended the Bolshevik Revolution, including its actions to liberate Soviet women. While Bryant denied the validity of the "free love" charges at the hearings, others deepened the accusations. A former US Commerce Department agent read into the record the text of several documents seemingly proving that the Bureau of Free Love and the nationalization of women were real. A proclamation of the "Anarchist Soviet" of the southern Russian city of Saratov lamented that the "best species of all the beautiful women had been the property of the bourgeoisie." To correct this, all women ages eighteen to thirty-two would become the "property of the whole nation." Any man bearing a certificate that he was a member of the working class was entitled to "use" one woman no more than three times a week. Such men were obligated to pay a monthly fee, out of which the women would receive a monthly salary. If a woman were to become pregnant as the result of her "use," the child would be given up to an "institution" at the tender age of one month.[73] This, then, was social evolution, Soviet-style.

The reality behind the Saratov decree and similar documents is difficult to discover. They emerged in the "fog" of a civil war between the Bolsheviks and the Whites, whose troops were joined by fourteen nations, including those of Britain and the United States. There was a rich history of fabricated documents in the drama of the Russian Revolution going back to the 1903 *Protocols of the Elders of Zion*.[74] There is evidence suggesting that the Saratov document was fabricated by a local monarchist to put the anarchists in a bad light. From there, the proclamation took on a life of its own, either in the hands of White generals looking to win peasants away from the Bolsheviks, or by anarchists who truly believed that the new socialist utopia included some version of "free love." Just as the Greenwich Village–based Socialist bohemians in the US placed more emphasis on women's sexual liberation than did the mainstream of the party, so did a small minority of the victorious revolutionaries in Russia aim to make dreams of a new sexual order a reality.[75] It is unlikely such schemes represented Bolshevik policy. In at least one case, when Lenin learned of a plan to "redistribute" women in a town in his native region of Simbursk, he sent an angry telegram, ordering the local Cheka to investigate and, if the rumors were true, to "arrest the guilty."[76]

Despite political factionalism, ongoing scientific debates, questionable hybridization schemes, and fabricated Free Love Bureaus, the authority of evolutionary science in the new Soviet Republic was largely untarnished. Debates over the mechanism of evolution presumed that evolution was a fact. In April 1932, on the fiftieth anniversary of Darwin's death, Soviet authorities carried out a broad campaign to celebrate Darwin's heritage. One prominent headline read, "The Working Class, Armed with Marxist-Leninist Theory, Takes Everything Truly Scientific from Darwinism for the Struggle to Build Socialism." According to a new popular slogan, "The Soviet Union is the second birthplace of Darwin." In contrast, as another headline pointed out, in the United States, or rather, "countries of dying capitalism and rotting bourgeois culture," Darwinism was "on trial as the accused."[77] In this early "cold war" over evolutionary science, the Soviets were clearly in the lead.

Back in the United States, the homeland of "dying" capitalism, a tiny but energetic communist movement also took up Darwin's banner. In 1919, former members of the left wing of the American Socialist Party, inspired by the Bolshevik Revolution, broke off to form two fledgling communist parties—the Communist Labor Party and the Communist Party of America. Joining the Communist International, the American organizations were fundamentally different from the former Socialist Party. Not only did they espouse a revolutionary outlook on how workers would reach a socialist future, but they built a new kind of party based on the Leninist model. This would be a disciplined, politically homogeneous organization that left behind the "big tent" approach of the Socialist Party.[78]

The Bolsheviks' American comrades were in a poor position in the mid-1920s to make an impact on the raging debate over evolution. The majority of them did not speak or publish in English. They tended to adopt an unrealistic "ultraleft" perspective that revolution was just around the corner. Owing to the post–World War I red scare, the parties remained underground for several years, emerging as an open, legal organization, the Workers Party, only in 1921. It took a determined fight by one faction, known as the "Liquidators," to convince Comintern leaders and the ranks of the party that it was time to start conducting politics openly. As soon as that question was settled in their favor (and against the "Goose" faction, who were fonder of the underground party), factional divisions quickly developed along other fronts. For the entire decade, the tiny American communist movement—numbering somewhere fewer than twenty thousand—was virtually at war with itself.[79] It was not until 1923 that a single united party emerged—awkwardly designated the Workers (Communist) Party.[80]

But once the nation's attention was riveted on *State of Tennessee v. John Thomas Scopes*, in the summer of 1925, the American communists jumped into the fray. The *Daily Worker* was full of coverage that clarified the stakes for workers and the communist movement in the battle in Dayton, Tennessee, and beyond. News articles closely followed daily developments in the trial, which lasted from July 10 to July 21, 1925. In the twenty-six days from July 3 to July 29, the *Daily Worker* ran at least one article on the trial on nineteen of these days. A typical specimen appeared on the front page of the July 14 issue and was titled "Anti-Evolution Law Branded Unconstitutional in Fight for Freedom of Education." It contained significant excerpts from a statement given by John Neal, the chief attorney for Scopes, who sought, unsuccessfully, to quash his client's indictment on the grounds that Tennessee's Butler Act was unconstitutional.[81] While no author's name appeared in the byline—only "Special to the Daily Worker"—the inclusion of visual and aural details such as "the judge looked worried" and "there was a gasp of surprise" suggests that the reporter sending in stories to the *Daily Worker* was on the scene.[82]

One of the "special" correspondents for the *Daily Worker* was John T. Scopes himself. On July 21, the day the jury convicted him of violating the Butler Act, Scopes filed a brief story about continuing the fight. He reviewed the factors militating against a victory in Dayton, but looked forward to winning the case on appeal. "We will stay by the ship and every point will be fought out bitterly," he wrote. "Success is ultimately with us." The same exact story appeared elsewhere as an "exclusive" for Hearst's International News Service—throwing into question what the appearance of the article in the *Daily Worker* signifies. But given what we know about Scopes's political upbringing, it is entirely possible that he would have had no objection to his words appearing under the Communist banner.[83] At the very least, John Scopes knew that the radical labor press took up a wide variety of social issues and related them to the struggles of working people.

The *Daily Worker* spelled out how evolution was relevant to workers in a variety of ways. In "Darwinism on Trial," published the day the trial opened in Dayton, the editors commented on what they viewed as a conflict between two wings of the American "bourgeoisie." Even though this was a fight within the ranks of the "class enemy," they wrote, "the working class cannot remain an idle onlooker." To the contrary, workers needed to realize that Darwinism was part of the Marxist worldview, the "Communist conception of the universe." Both Marx and Darwin had overturned the concept that the existing order, either social or natural, was immutable and had resulted from "eternal laws." When workers began to understand that capitalism was not eternal, they would move toward socialist revolution.

While the Communists defended evolutionary science, they also sought to expose unnecessary compromises made by pro-evolutionists. For the *Daily Worker*, the guilty parties included the Socialist Party, whose *Milwaukee Leader* ran an editorial in July 1925 titled "Evolutionists Defend Religion." "This paper is not irreligious," wrote the Milwaukee Socialist editors, "and it never attacks religion, either expressly or impliedly." The Socialists continued to hold a variety of views of religion, while the Communists, made up primarily of the former left wing of the Socialist movement, took a clearer atheist stand.

The compromisers also included Clarence Darrow. While the Communists supported Darrow as against Bryan, another editorial portrayed both as representatives of the "middle class." Darrow was the "middle class intellectual," and Bryan was "the leader of the well-to-do farmers of the middle west and south" whose interests were opposed to big capital. As a middle-class figure who found himself hemmed in by opposing classes, Darrow consequently waffled on evolution by refusing to take a clear atheist stance. As the editors put it, "Only the Communists stand squarely against religion as 'the opium of the people.'"[84] This distinctive Communist critique of middle-class pro-evolutionism was reflected in *Monkey or Man?* (1925), a satirical play composed by Mike Gold. Gold was a Communist activist and well-known exponent of "proletarian literature," a Bolshevik-inspired genre of Marxist-infused fictional writing based on the lives of working-class people.[85] According to the *Daily Worker*, Gold's play was "a characterization of the struggle that is taking place between the Fundamentalists and the Modernists. It will show them both up."[86]

Despite the way in which the Communists seemed to say "A plague on both your houses," they reserved a special level of venom for William Jennings Bryan. Not only was Bryan the leading voice of the fundamentalists on the issue of evolution, but he had falsely portrayed himself for decades as the champion of workers and farmers. One line of attack portrayed Bryan as an ally of the Ku Klux Klan in its campaign against evolution and for fundamentalism. Bryan was, one article claimed, "the most perfect type of kluxer." In the wake of Bryan's death, another article argued that Bryan had literally been "a member of the

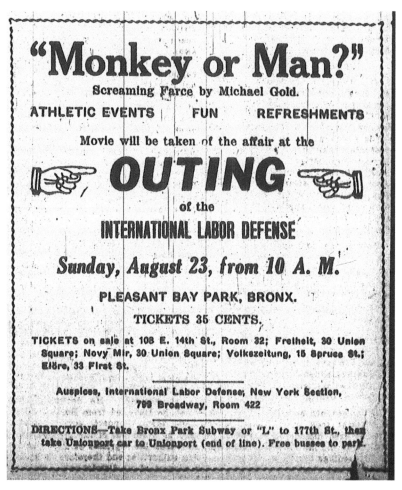

Figure 3. Advertisement for an International Labor Defense picnic, 1925.
Just weeks after the Scopes trial, a performance of the satirical evolution-themed
play *Monkey or Man?* formed the centerpiece of this event sponsored by the
International Labor Defense, a Communist Party–organized united-front group.
The early Communists gave an uncompromising defense of evolutionary science. *Daily
Worker*, August 20, 1925.

ku klux klan." The evidence was drawn from publicity for an upcoming
Klan gathering, the stated purpose of which was to seize "the torch of
fundamentalism from the falling hand of Bryan and carry it aloft in national conflict."[87] While this was hardly conclusive evidence, Bryan knew

that the Klan played a powerful role in pushing forward antievolution laws. Whatever his qualms about Klan tactics, he led a successful fight at the marathon 1924 Democratic National Convention against including a plank in the party platform explicitly denouncing the hooded order. And like most leaders of the Democratic Party, Bryan publicly proclaimed the idea—and defended it against prominent African American critics—that whites were "the advanced race."[88]

Another contribution of the Communists was their confidence that evolutionary science would prevail. That attitude stemmed from their Marxist standpoint on the relationship between capitalism, scientific progress, and the working class. In a *Daily Worker* editorial reiterating the point Arthur M. Lewis had made years earlier, party leader William Schneiderman argued that those fighting against the advance of science are doomed to fail. The editorial was illustrated by a Soviet political cartoon mocking Americans. Reprinted from *Komsomolskaya Pravda*, it depicted a Tennessee legislator as a "monkey" and was captioned, "Living Proof of Darwinism." Such characters could not succeed, Schneiderman explained, because the capitalists themselves need science in order to revolutionize production. By promoting scientific thinking among the workers, the capitalists inevitably and unwittingly erode "the superstition and ignorance upon which the bourgeoisie depend to maintain their strangle hold on the toilers."[89]

Promotion of scientific thinking among workers—deliberately by communists or unwittingly by capitalists—was all but impossible, according to the best-known journalistic commentator on the Scopes trial, H.L. Mencken. Known for his original style and biting satirical wit, Mencken was also deeply elitist and antidemocratic, as Daytonians quickly discovered. In a *Baltimore Sun* column titled "Homo Neanderthalis" published on the eve of the trial, the journalist insulted the intelligence of Dayton "yokels" by claiming that they would and could never understand scientific ideas. "It would be as vain to try to teach to peasants or to the city proletariat," Mencken wrote, "as it would be to try to teach them to streptococci." The working people of East Tennessee, that is, were no brighter than bacteria.[90]

In contrast to Mencken, and armed with their revolutionary confidence in the capacities of ordinary working people, the Communists took to the streets to reach them with a pro-evolution message. Readers of the

Daily Worker learned that Communists planned an "open air meeting" in Minneapolis, William Bell Riley's home turf. Workers (Communist) Party organizer John Gabriel Soltis was projected to speak on the "Principles of Evolution and the Working Class." At a Chicago Communist meeting, party members voted to "organize as soon as possible a mass meeting on the Scopes trial, in order to bring out the attitude of the Communists toward religion and science."[91] Area communists were experienced street speakers. Two weeks earlier, Communists J.K. Dante and Irving Search had been speaking in neighboring Cicero, where the Workers Party held weekly street meetings. This time, however, "a gang of sluggers, led by a priest, attacked the meeting and tried to break it up." Rather than corral the attackers, police arrested Dante and Search. At their trial, the city prosecutor mentioned the Scopes trial and proclaimed that he would protect the residents of Cicero from "heresy."[92] The judge dismissed the charges, but the dangerous mixture of evolution and communism was in the Chicago air.

Two very different American communists, a preacher and an editor, are worthy of note for their sustained focus on evolutionary science. In late June 1925, on the eve of the Scopes trial, when Chicago Communists hosted a meeting launching the International Labor Defense, the speakers' rostrum featured not only central party leaders and left-wing Socialist veterans such as Benjamin Gitlow and James P. Cannon, but a relative newcomer to the party who had spoken out in support of both evolution and communism: William Montgomery Brown (1855–1937).[93] Serving as bishop of Arkansas in the Episcopal Church, Brown had developed unorthodox views under the impact of books by Darwin, Haeckel, various Christian Socialists, and the German monist philosopher Paul Carus. In 1920, he self-published a quirky procommunist tract, *Communism and Christianism: Analyzed and Contrasted from the Marxian and Darwinian Points of View*. Brown soon joined the Workers Party. In 1924, using quotations from Brown's book, the Episcopal Church tried the "Bad Bishop" for heresy and officially deposed him the following year.[94]

The motto of *Communism and Christianism*, printed on the book's cover, called on its readers to "Banish Gods from Skies and Capitalists from Earth." Featuring portraits of Marx, Engels, Darwin, and Lenin,

Brown's book included generous quotations from the *Communist Manifesto*, Charles Kerr's *Scientific Socialism Study Course* (presented as a catechism), and Socialist Party platforms, along with his distinctive prose. Brown focused on what he called the "twofold revelation" of Marxism and Darwinism, which together constituted the truth (referred to in John 8:32) that "shall make you free" from "commercial imperialism" and its attendant "unnecessary suffering." As in this passage, the book used religious language and biblical quotations in the service of proletarian revolution and modern science. "Darwinism and Marxism," wrote Brown, "constitute . . . the only, true, comprehensive, and sufficient gospel" that would lead to the "salvation" of humanity through Bolshevism. The Soviet regime translated Brown's book into Russian and used it during the 1920s to erode the authority of the Russian Orthodox Church.[95]

Last but hardly least in the efforts of early Communists to spread the gospel of evolution was the indefatigable Ludwig Katterfeld (1881–1974). He edited and published *Evolution: A Journal of Nature*, the very first popular-oriented monthly magazine in the United States to promote the cause of evolutionary science.[96] Though Katterfeld and his little magazine have been largely forgotten, he was a central figure in the radical movement. As a member of the National Executive Committee of the Socialist Party, he helped lead the party's opposition to US entry into World War I. In 1919, Katterfeld became a founding member of the new Communist Labor Party. As the result of the Palmer Raids early the next year, Katterfeld was arrested and stood trial in July 1920 in Chicago for violating the Illinois criminal syndicalism law. Though he and his comrades were defended ably by none other than Clarence Darrow, Katterfeld was sentenced to one to five years in prison and fined $2,000. While he appealed the conviction, Katterfeld was in the inner circles of the new united Communist Party of America, which twice sent him to Moscow. There he conferred with Lenin, Trotsky, and others about pressing political questions, including the "liquidation" of the underground American party. Upon his return to the US in 1923, Katterfeld served a year in the Illinois state prison at Joliet. When he emerged from behind bars, he moved to New York and managed the East Coast distribution of the *Daily Worker*.[97]

In 1927, some two years after Katterfeld arrived in New York, he took the initiative to launch *Evolution*. Present at the creation of this publication was a young Communist Party member who later became world famous for his role in the Alger Hiss espionage case: Whittaker Chambers. According to Chambers, Katterfeld started up his magazine when the party faction he supported—the Foster-Cannon caucus—lost out. "When the Lovestoneites took power in the party," recalled Chambers, "Katterfeld withdrew from it. He began to publish a magazine called *Evolution*, for his mind was in many ways a petrifact of 19th-century radicalism, and among its oddments of conviction was militant Darwinism."[98] Factional shifts may help explain the timing of Katterfeld's decision.[99] But the suggestion by Chambers that, in the eyes of proper Communists, "militant Darwinism" was a quaint artifact ignored recent Communist history. *Evolution* began as a monthly, but by the early 1930s Katterfeld had to suspend publication, only to revive it later in the decade. When he did so in 1937, however, no less than *Time* magazine ran a substantial story on the reappearance of *Evolution*, its "mild-mannered" former Socialist editor, and his "crusade for scientific truth."[100]

As the *Time* article indicated, *Evolution* was a remarkable publication on a number of counts. It adopted an in-your-face graphic strategy: the cover of the first issue sported a photo of a gorilla, whose caption read, "Man's Blood Cousin: The Gorilla." As historian Constance Clark has noted of the use of the gorilla in evolution cartoons, these depictions drew upon long-standing traditions in which "these gentle animals often carried sensational, even salacious, metaphoric freight—and racial connotations—in European popular culture."[101] Knowing this, Katterfeld seems to have chosen this provocative image for precisely this reason—it would get people talking. Talk they did. As the editor reported in the next issue, the cover image "caused comment all over the continent." In his opinion, at least, nearly all of it was "very favorable," although some less radical-minded pro-evolutionists found Katterfeld's methods counterproductive.[102]

Despite its provocative covers, *Evolution* was avowedly "non-political," meaning, nonpartisan. The magazine's proclaimed purpose was "to combat bigotry and superstition and develop the open mind by popularizing natural science." Regardless of party affiliation, all who embraced this goal could use the information in the magazine for their own purposes. At the same time, Katterfeld specified that the publication, though being "non-religious," would also not be atheist, knowing full well that atheism had explosive political implications.[103] While Katterfeld's was an inherently political venture, there was little indication that the publication's editor was, or ever had been, a central leader of the Communist and Socialist parties. (One clue was the regular appearance in "Some Good Books" offered by the Evolution Book Service of Bishop William Montgomery Brown's *My Heresy*.)[104]

Given Katterfeld's lack of scientific credentials—and his past conviction for sedition—another striking thing about *Evolution* is the impressive roster of contributing writers. They included David Starr Jordan, the famed ichthyologist, Darwinist, and emeritus president of Stanford University, who wrote an introductory piece that ran in the magazine's first two issues; William King Gregory, a professor of zoology at Columbia University and staffer at the American Museum of Natural History who was regarded as the world's leading authority on the evolution of human teeth; Hermann J. Muller, who had recently demonstrated the ability of radiation to induce genetic mutations (for which he won the Nobel Prize in 1946); and Harry Elmer Barnes, a Columbia historian deeply influenced by Darwin's ideas in the realm of social evolution. Early praise for the new magazine came from Columbia faculty as well. Philosopher John Dewey congratulated Katterfeld for "enlisting as writers persons of unquestioned competency and having a clear style." "I am impressed," Dewey wrote, "with the fact that the Journal is scientific as well as popular. You are rendering a public service and I wish you every success."[105]

As much as *Evolution* popularized evolutionary ideas, the magazine also helped evolutionists to become better acquainted with their enemies. Each issue featured a "Funnymentals" column that quoted from the likes of Gerald Winrod, William Bell Riley, George McCready Price, and others. Katterfeld was providing a measure of "comic relief" for his readers, who were likely to view as bordering on lunacy this statement from Winrod: "I would rather my babies' eyes be gouged out at this minute than to have them taught this blatant atheism."[106]

But in other respects, this "humor" column was deadly serious. It provided a revealing look at the arguments fundamentalists were making, thus helping to arm and encourage pro-evolutionists to fight back. In the August 1928 issue, as a battle raged over teaching evolution in Arkansas, Katterfeld printed the full text of Riley's "The Fundamentalist Challenge." He then printed, at the bottom of the page, a call to "ANSWER THIS

FUNDAMENTALIST CHALLENGE by sending copy of EVOLUTION with Ward article to each of FOUR HUNDRED THOUSAND FAMILIES IN ARKANSAS." (Charles Henshaw Ward, one of the most successful popularizers of evolutionary science, had already contributed several articles to Katterfeld's journal.)[107] "Funny" fundamentalist excerpts also included several that pointed at the confluence of communist politics and evolution. Such was the reprinted *Defender* cartoon of a bearded, sneaky-looking man holding a bomb labeled "Evolution," above the caption "Red Russian Ravages."[108] Whether or not Whittaker Chambers knew it, Katterfeld and his fellow communists were aware that antievolutionism and anticommunism were closely allied.

Of the evolutionary biologists on the roster of Katterfeld's *Evolution*, the one most likely to appreciate this brand of communist humor was probably Hermann J. Muller (1890–1967), whose story wraps up this chapter. From 1914 to 1918, Muller taught and conducted research at Houston's Rice Institute, where Julian Huxley, famed evolutionist and grandson of T. H. Huxley, had founded the biology program. Recruited by Huxley, Muller continued his work on *Drosophila* that he had started as a student in the fly lab of famed geneticist Thomas Hunt Morgan at Columbia. But he also developed a new theoretical explanation—based on "lethal mutations"—for the observations of rapid evolutionary leaps in generations of *Oenethera lamarkiana* made by Hugo de Vries that was consonant with Morgan's Mendelian ideas.[109] This put him on the road to his work on X-rays and made a significant contribution to what became the modern evolutionary synthesis. But it was the political events of these years—most notably the Bolshevik Revolution—that changed the course of Muller's life. He was already a socialist by 1917, and he would soon become deeply enmeshed in the politics of science in the Soviet Union.

In 1922, then teaching at the University of Texas, Muller visited the young Soviet republic for the first time. Recalling his visit, Muller appreciated the degree to which evolutionary science was endorsed by the Bolshevik government, providing a "curious commentary" on the state of affairs in the US. Muller's observation emerged from a conversation with a Professor Berg, who conducted research on plant genetics. In his studies of cereal plant evolution, Berg had put forth a theory known as orthogenesis, which meant that evolution moved in a predictable, predetermined linear direction. When Muller mentioned that some American states were moving in the direction of banning the teaching of evolution, Berg related that his own book was currently under review by Bolshevik authorities who regarded his idea as "pernicious and subversive." There was potential censorship in both cases. But Berg was in hot water not because he was supporting evolution, but because he was perceived to be undermining it. His book stood outside the new political orthodoxy of Marxist-Darwinism, based on the central idea of natural selection.[110]

Muller's conversation captures the sharp contrast between the situation facing evolutionary biologists in the US and USSR in the 1920s. It points as well to the political pressures bearing down on Soviet researchers, which later would prove dangerous for not only their academic futures but their physical existence. Muller found himself on the wrong side of the Stalinist divide in the 1930s when he was living and working in Moscow.[111] But the later Lysenkoist campaigns against genetics in the Soviet Union have obscured a fact that was evident to observers, including Muller, in the early years of the Bolshevik Revolution: the Communists were outstanding promoters of evolutionary science. Their comrades in the United States were as well. As the next chapter documents, American antievolutionists were paying attention.

THE LAMB-DRAGON AND
THE DEVIL'S POISON

"You don't think much of scientists, do you?" So Clarence Darrow asked William Jennings Bryan, as he cross-examined the Great Commoner on the porch of the Rhea County Courthouse, in Dayton, Tennessee, on July 20, 1925, the seventh day of the Scopes trial. Knowing that Darrow was attempting to paint him as an unlettered ignoramus, Bryan insisted on naming specific examples of scientists he respected. "I will give you George M. Price," he replied. But Darrow was not impressed. Bryan had cited "a man that every scientist in this country knows is a mountebank and a pretender, and not a geologist at all."[1] It is true that George McCready Price (1870–1963) had no fans among American scientists. A Canadian-born Seventh-day Adventist writer, teacher, and self-trained geologist, Price was best known to them for his pioneering books in the field of what has come to be called "creation science." For nearly two decades, starting in 1906, Price had published antievolutionary books and articles and taught science at a string of Seventh-day Adventist colleges. In 1923, Price published *The New Geology*, a college-level textbook that

denounced evolution as a scientific fraud. The book argued that a universal Noachian flood, and not eons of evolution, explained the geological
features of the earth, which he estimated to be six thousand years old. Scientists knew Price, but they knew him, in the words of *Science* magazine,
as the "principal scientific authority of the fundamentalists," and not as a
credentialed geologist.[2]

But there is a deeper sense in which Darrow's characterization of Price
as "not a geologist at all" can tell us more about Price than his academic
marginalization. While Price published thousands of pages analyzing the
conclusions of geologists, subscribed to scientific journals, and regularly
corresponded with eminent researchers,[3] his main objection to evolution—
as he admitted—had nothing to do with the veracity of scientific claims.
Rather, it was the "philosophical and moral" consequences of evolution—
the "evil fruits" of the "corrupt tree," in the words of Jesus's Sermon
on the Mount—that turned Price into a creationist. In a series of works
published both before and after the Bolsheviks took power in Russia,
Price made it clear that socialism and communism were among those
evil fruits.[4] A strange duality thus pervades the work of Price—who
coined the phrase "Red Dynamite"—and that of his creationist successors. They seem to focus on the scientific evidence, or the lack thereof,
for evolution. But what really troubles them are the alleged sociopolitical
consequences of evolutionary belief. As Price's geological work formed
the intellectual basis, decades later, of John Whitcomb Jr. and Henry
Morris's highly influential young-earth creationist *Genesis Flood* (1961),
Price is rightly viewed as the godfather of the modern creation science
movement. But he was a creationist pioneer on both the geological and
political frontiers.

It is impossible to understand George McCready Price and his creationist
ideas without some appreciation of his Adventist theological perspective.
Seventh-day Adventism originated as an offshoot of the millenarian movement led by farmer and lay preacher William Miller. He had fixed the
date for Christ's Second Coming as October 22, 1844, based on prophetic
passages in the book of Daniel that foretold a "cleansing of the sanctuary."[5] In the wake of the "Great Disappointment" that followed the Messiah's failure to appear, a group coalesced around the idea that Miller had
not erred about the date, only about the nature of what had taken place.

Figure 4. George McCready Price, c. 1930. A devout Seventh-day Adventist and amateur geologist, Price was a pioneer in promoting young-earth creationism and linking the perceived dangers of evolution and communism. Courtesy of Office of Archives, Statistics, and Research, General Conference of Seventh-day Adventists.

Christ was indeed cleansing the sanctuary, but the event was taking place in the heavenly realm instead of on earth. By 1863, one "remnant" of the Millerite movement formally constituted itself as the Seventh-day Adventist Church, established its own weekly newspaper, and set up headquarters in Battle Creek, Michigan. Ellen G. White, née Harmon (1827–1915), who grew up in a Millerite family in Portland, Maine, began to have waking visions as a teenager. She married church cofounder James White and became the central seer and prophet of Seventh-day Adventism. Her writings are second only to the Bible as authority.[6]

Adventist theology developed into a variant of premillennialism. Christ was coming to establish his reign of a thousand years of heaven on earth (the millennium), but humanity would first pass through a terrible period—the tribulation—in which the Antichrist would gather strength and cause horrible suffering. Although those who accepted Christ would ultimately triumph, humanity was headed for disaster. For historicist premillennialists—which included Adventists, as well as most Protestants before the mid-nineteenth century—the biblical prophecies foretold history from ancient times through their own time. In regard to Revelation 13, the Adventists agreed with Protestant tradition that the first "beast," with seven heads and ten horns, represented the Catholic Church.[7]

But the Adventists added a unique feature—their interpretation of the second beast of Revelation 13, which had "two horns like a lamb" and "spake as a dragon." This hypocritical creature, who appeared Christlike but was later revealed to be Satan, was none other than the United States of America. The horns were, respectively, the republic and Protestantism, standing for civil and religious liberty. Despite their premillennialism and inclination toward political quietism, many founding Adventists were abolitionists and felt that the US was betraying its founding republican ideals. The early Adventists also felt betrayed by Protestantism because of the scorn they had suffered as they focused on the expected Advent. Their minority Sabbatarian beliefs made them sensitive to the movement for Sunday laws. While it was fellow Protestants leading these campaigns, the historic identification of the Antichrist with the pope, and the association of the Catholic Church with political tyranny, led Adventists to believe that the threat of a papal "despotism" was always imminent.[8]

The Adventist focus on the Saturday Sabbath day not only set the church apart in terms of liturgical practice, but also provided the framework for its distinctive position on evolution. By the late nineteenth century, many Protestant evangelical leaders had accepted the latest scientific discoveries that pointed to an ancient earth. They hewed either to the day/age theory, in which each biblical day of creation represented an indefinite period, or the gap theory, which postulated an unaccountably long delay between Genesis 1, the creation of the earth, and Genesis 2, the creation of Adam and Eve.[9] But few Adventists took either of these positions, since in *Spiritual Gifts* (1864), Ellen White rejected both. Stating that she had been transported during a vision back to the time of creation, White reported that the week of Genesis was "just like every other week." Genesis days meant "literal days." To deny this fact was to launch a direct attack on the Fourth Commandment—"Remember the Sabbath day, to keep it holy." White's claim that the earth "is now only about six thousand years old" became part of bedrock Adventist doctrine.[10]

Adventist responses to evolution also were informed by White's explanation for biological diversity in the aftermath of the Noachian flood, which hinged on the concept of "amalgamation." According to White, "every species of animal which God had created were preserved in the ark. The confused species which God did not create, which were the result of amalgamation, were destroyed by the flood. Since the flood there has been *amalgamation of man and beast*, as may be seen in the almost endless varieties of species of animals, and in certain races of men."[11] Adventists have argued over the proper interpretation of this passage, which seems to imply that humans and animals mated and produced offspring, and that some racial groups were less than fully human. Such implications not only violated existing scientific knowledge, but also cast doubt on the egalitarian values that the Adventist founders had embraced. Some Adventists contended that White meant, in effect, "of man and *of* beast," but the evidence supports a plain reading of her words. Adventists understood her to mean that "certain races" included Africans, Native Americans, and others who were commonly classed as inferior.[12] As for the cause of amalgamation and the proliferation of new "confused" non-godly species after the flood, White never explicitly identified it, but Adventist commentators commonly assumed that it was Satan.[13] Ellen White thus laid a rich foundation for Price's antievolutionary thinking.

George Edward Price was born in 1870 on a farm in Havelock, New Brunswick, Canada. His father, George Marshall Price, farmed seven hundred acres. Susan McCready, Price's mother, came from a more educated family. Her brother J.E.B. McCready was editor of the *Daily Telegraph* in Saint John.[14] Because of the strength of young George Price's literary ambitions—"I cannot remember a time in my early youth and young manhood when I did not aspire to be a writer"—he adopted his mother's maiden name as his own.[15] Soon after his father's death, Susan McCready joined the Seventh-day Adventist Church and George took up a new occupation—selling Adventist books.

His stock-in-trade included Ellen G. White's *The Great Controversy*, which focused on the contest between Lucifer and Jesus Christ.[16] Part history, part prophecy, White's book traced this struggle by following the fortunes of "God's children." They included early Christian martyrs, European Protestant reformers, William Miller, and the early leaders of the Seventh-day Adventist Church. On the side of Satan stood the false Catholic Church, which presided over a long period of "spiritual darkness." Moreover, readers were reminded about the various "snares" that Satan had planted among well-meaning but easily fooled Christians: "He is intruding his presence in every department of the household, in every street of our cities, in the churches, in the national councils, in the courts of justice, perplexing, deceiving, seducing, everywhere ruining the souls and bodies of men, women and children, breaking up families, sowing hatred, emulation, strife, sedition, murder." To this familiar litany of Satan's activities, White added the distinctive Adventist apocalyptic vision of how the "great controversy" would be resolved in favor of Christ. While Price would soon have his own experiences of battling demonic forces in the big city, this book, more than any other, convinced him that he should spend his life spreading God's word.[17]

For the next several years, George shared the experience of spreading this stormy but ultimately hopeful vision with his bookselling partner Amelia Anna Nason, a fellow native of New Brunswick. They developed a mutual affection, and in 1887 they were married.[18] Both George and

Amelia attended Battle Creek College, an Adventist institution in Battle Creek, Michigan.[19] Neither finished college, but both would enter the teaching profession. The couple had three children and would remain married for sixty-seven years.[20] But for the first decade of the new century, they spent much of their time apart. George struggled to make ends meet in a succession of jobs as a bookseller, school administrator, teacher, preacher, writer, and handyman. One of these jobs landed him in Tracadie, New Brunswick, where he wrote his first book: *Outlines of Modern Christianity and Modern Science* (1902).[21]

This book set the mold for Price in two fundamental ways. First, he focused his fire on evolutionary geologists' alleged circular reasoning when determining the age of rock layers. Geologists assign dates to strata in the geologic column based on the types of creatures and plants fossilized therein. The simpler types of fossils are found in the lower layers. The contents of these layers are roughly consistent around the world. Evolutionists conclude that the lower strata must be older. But, Price wrote, "it is nothing but a pure assumption, utterly incapable of any rational proof."[22] Price argued instead that the specific gravity of different living creatures during the flood determined their place in the geologic column. (Later, in *Illogical Geology* [1906], after he had discovered that in some mountainous regions, the layers were out of expected order, with a "newer" stratum on top of an "older" one, Price attacked the idea that there was even a truly uniform geologic column. He scoffed at the commonly accepted geologic concept of thrust faults, tremendous pressures that, geologists believed, could accomplish this feat.)[23]

Second, the book made clear his concern with the "political" consequences of evolutionary science. Price demonstrated how Adventist eschatology was intertwined with his developing moral and political critique of evolution. He prefaced the argument by invoking Christ's teaching on false prophets in Matthew 7:15, using language creationists would repeatedly invoke: "It is rightly considered that the supreme test of any doctrine, religious, social, or scientific, is its bearing upon life and human action. 'Ye shall know them by their fruits.' What are the fruits of the evolution theory?" According to Price, evolution was "utterly subversive of civil and religious liberty for the individual."[24]

Evolution led to tyranny by accelerating social disorder. According the Price, acceptance of evolutionary ideas—the survival of the fittest—had caused "the increase of crime and lawlessness of every kind, the increased lack of self-government on the part of the individual."[25] In associating evolution with lawlessness, Price may well have been influenced by discussions in the Adventist *Review and Herald*. Less than a month after the assassination of President McKinley the previous year, the editors opined that "every seed of evolution planted is also a seed of anarchy."[26]

To bolster his case in *Outlines* that evolution had caused lawlessness and thus drove society toward despotism, Price pointed to two "signs." One was imperialism. "By our taking up the 'white man's burden' of governing what we are pleased to call half-civilized peoples beyond the seas," Price wrote, "we shall end up finding a similar state of things requiring attention at home." Price echoed the concerns of other Seventh-day Adventists. Their peculiar concern with liberty had led the church to denounce the annexation of the Philippines.[27] Percy Magan, who had taught Price Roman history at Battle Creek College, published a church-endorsed book on the subject in 1899 with a telling title: *Imperialism versus the Bible, the Constitution, and the Declaration of Independence; or, The Peril of the Republic of the United States.*[28]

Price's implicit criticism of imperialist racism also drew on Adventist traditions reaching back decades. Because of the abolitionist sympathies of William Miller and prominent Millerite abolitionists such as Joshua Himes, antislavery feeling and even belief in racial equality found a relatively accommodating home in the early Adventist movement.[29] While Adventists were conflicted about performing military service, Ellen White and other Adventists publicly supported the Union side in the Civil War, viewing the slaveholders' rebellion as satanically inspired. During Reconstruction, Adventist publications gave voice to Radical Republican views in favor of racial equality, though violence directed at interracial Adventist missions in the South led the church to modify its stance.[30]

At the same time, Price was hardly a champion of full racial equality. He attributed human racial variety to three factors: God's dispersal of humanity in punishment for the Tower of Babel; the changing environment; and the process of racial amalgamation, taught to him by Ellen White. In a poem penned in 1910, Price focused on the first two in explaining the origins of the allegedly inferior Negro race. According to Price, "the poor little fellow" who fled Babel to Africa "got lost in the forest dank," acquired dark skin from the "fierce sun," and "his mind became a blank."[31] In a later work, Price argued that the distinct human races "greatly resemble true species" and that "natural instincts," aided by God's providential action at Babel, should have kept them separate. Contrary to nature and God's will, however, a mixing of the races or "amalgamation" had taken place.[32] Price acknowledged that he joined with other Adventists in identifying "the great primal hybridizer" of human races, plants, and animals as Satan.[33] This claim anticipated mainstream evangelical arguments made decades later against the presumably satanic desegregation of the races in America.

Imperialism was one sign of the growing danger of despotism; another, closer to home, was the amassing of collective power by large corporations and by workers. "What with the labor unions, and what with the trusts," wrote Price, "we are certainly beholding the fast passing of individualism." In capital letters, Price warned of "THIS HEAVEN-DARKENING DESPOTISM OVER THE GRAVE OF LIBERTY."[34] Premillennialists expressed an evenhandedness on the subject of class conflict, viewing its very existence as a sign of the end times. But Adventists were not neutral on the subject of labor unions. In the wake of the 1902 anthracite coal strike, Ellen White had made the position of the church clear. "Unionism," she wrote, "is controlled by the cruel power of Satan. Those who refuse to join the unions formed are made to feel this power." The next year, the *Review and Herald* called labor unions "a dragon voice which is heard speaking in the nation to-day," a clear reference to the second beast of Revelation 13.[35] Not only were unions a threat to individual liberty; they were inextricably tied to the city and the "snares" set there by Satan.[36] The satanic snares of the big city, including modern labor unions and their rebellious politics, played a key role in creationist and anticommunist thinking for decades to come.

Despite Ellen White's warning, Price spent a short but eventful six months in the nation's biggest city. In September 1904, Price arrived in New York, hoping to make a living as a writer. He moved into a room in a four-story brick apartment building at 95 Christopher Street, in the heart of the West Village.[37] While the neighborhood's bohemian days still lay ahead, it already had a reputation as a literary enclave. Herman Melville, Henry James, Mark Twain, Edgar Allan Poe, and other leading lights had lived and worked just blocks from Price's temporary quarters.[38] The trip was an economic and spiritual gamble. Not only was Price betting that he could obtain work writing for cosmopolitan, secular publications— something he had never done—but he was also directly disobeying prophetess Ellen White's injunction to avoid the big city in order to follow his quixotic dream of literary success. On both counts, the New York sojourn was a profound failure. As he informed SDA church elder William Guthrie in late December 1904, "experience has made me a wiser and sadder man." After nearly four months in the city, Price had worked only about one-third of the time. Knowing that his own family back in New Brunswick was "destitute and almost starving" drove Price to thoughts of suicide and damnation. "Heaven only knows what privations I have gone through and what torment of soul I have suffered," he told Guthrie.[39] In early 1905, he moved out of the Village and into an apartment building on the edges of Hell's Kitchen. He got steadier work, laboring sometimes up to fourteen hours a day, but worried about his "present associations and oc-cupation, which are not right." Feeling the lure of those satanic snares, Price wrote that my "eternal welfare is at stake in making a change and cutting away" from New York.[40]

One of the temptations that New York City offered George McCready Price was the young but growing socialist movement. As a regular reader of the Adventist *Review and Herald*, Price would have encountered fairly regular discussions of the new party. Consistent with the approach that Price later took in his own writings, Adventist editors expressed sympathy with socialist aims, but they rejected collective political action. In a 1905 article commenting on the gains of the Socialists in the 1904 elections, church leader Leon A. Smith commented that "from a political stand-point, much may be said in favor of socialism as compared with other political systems"; and yet, the only solution to humanity's problems was "the coming kingdom of Christ."[41]

Early 1905 was a heady time for New York City's socialists. On Janu-ary 22, which became known as "Bloody Sunday," the troops of Czar Nicholas II fired rifles into a crowd of workers and peasants who had trav-eled to the Winter Palace to present a petition to their ruler. "Civil War Threatened, Workman Have Lost Faith in the Czar, and Now Mean to Fight," read one headline in the *New York Times* the next day. As strikes quickly spread through St. Petersburg and beyond, Russian-Jewish social-ists on the Lower East Side of New York, just across town from Price's former digs, were electrified. "In that part of the city," the paper reported, "thousands of men and women who have cared and suffered for the cause of Russian freedom, have found a haven, and there was not one of these who did not feel a personal share in the events."[42] Socialist intellectuals from more privileged backgrounds were also inspired by the scale and depth of the Russian revolt. Soon after "Bloody Sunday," New York socialist William English Walling headed off to Russia to cover events for the socialist press.[43]

In the spring of 1905, well before the Russian Revolution reached its climax that fall, George McCready Price left New York City. He worked a succession of jobs for the Seventh-day Adventist Church, landing eventually in Loma Linda, California, a budding Adventist settlement sixty miles east of Los Angeles. Starting in 1907, he began teaching at the Loma Linda College of Evangelists, and he would spend most of the remainder of his life living and teaching at Adventist institutions.[44] As one of Price's students recalled, "He opened every class with prayer. He always had a twinkle in his eyes." As for his teaching methods, "I can't remember his just telling us things," she noted. "We found them out by experimentation."[45]

For Price, there was no contradiction between opening class with prayer and then jumping into scientific investigation. Loma Linda students were required to take a course in "Spirit of Prophecy" (on the life and writings of Ellen White).[46] The idea that nature and revelation were mutually reinforcing sources of truth drew on a centuries-old Christian apologetic tradition reaching back to the writings of the early church fathers. In North America, it appeared as early as 1721 with Cotton Mather's *Christian Philosopher*, which referred to the "Book of Nature" along with the Bible as proof of God's glory.[47] These two sources of truth—and the dangers of straying from them—were the primary focus of the book Price published in 1911: *God's Two Books: Or Plain Facts about Evolution, Geology and the Bible*. Published by the Adventist Review and Herald Publishing Association, this work was the first one in which Price explicitly addressed socialism and the labor movement. The intellectual framework for his critique of evolution was the by-now familiar "fruits" argument, but this time draped in a more rigorous scientific guise. Perhaps owing to his newfound academic authority in a college classroom, Price paid more serious attention in this work to the question of scientific method.[48]

He did so by joining Protestant theology with Baconian empiricism.[49] A wide range of American Protestant thinkers had embraced Sir Francis Bacon's inductive method. Originally a weapon wielded against medieval scholasticism, Baconian ideas were now enlisted as a defense of the existing order against what were viewed as dangerously speculative hypotheses arising from the French Revolution. Pure facts, unadulterated by any (false) assumptions, came first; only then could conclusions follow. Without acknowledging the deductive character of their own theistic worldviews, Protestant leaders fixed on the words of scripture as the essential objective facts to be collected, classified, and organized. As American evangelical leader Reuben Torrey put it in his contribution to the *Fundamentals* on the subject of the Resurrection of Jesus Christ, "We shall not assume anything whatever."[50]

The focus on collecting facts had a democratic flavor: any literate person with access to the Bible and a dose of what Price called "enlightened common sense" could use these facts to reach conclusions about both spiritual and earthly matters. A popularized version of Baconianism had become so firmly entrenched in England by the mid-nineteenth century that even Charles Darwin, who was putting forth the audacious hypothesis of natural selection, clothed his effort in proper Baconian garb on the frontispiece of the first edition of *On the Origin of Species*. The selection he chose from Bacon's 1603 work, *The Advancement of Learning*, trumpeted the value of studying both "the book of God's word" and "the book of God's works."[51] For exactly opposite purposes, Price included on his title page of *God's Two Books* a similar quotation from Bacon's *De Augmentis Scientiarum*.[52]

In Price's exploration of the facts of geology that fills most of the book's pages, the speculative, nonfactual theory of evolution comes up wanting. Acknowledging that a certain amount of common descent must have taken place, Price derides evolutionists for their unwarranted "assumptions," which get them into trouble.[53] But Price's own assumptions are revealed in his opening chapter, titled "Moral and Social Aspects of the Evolution Theory." His argument that we can know the scientific theory of evolution by its "fruits" is by now familiar: "And surely the moral issue, as set forth above, is a surer way of gauging the truth or falsity of the Evolution theory than the long, complicated methods connected with 'variation' and 'selection,' 'heredity' and 'environment,' and the other biological problems, *even supposing the theory apparently capable of the most exact proof*. In short, we need offer no apology for thus measuring this scientific hypothesis by other and far more certain standards of proof."[54] Price makes here a remarkably un-Baconian and anti-intellectual argument. The validity of a biological scientific idea has little to do with the status of the facts drawn from the natural world that support or refute that idea. To rescue his Baconianism, he suggests that the facts that truly matter are "moral" ones—the societal consequences of adopting evolutionary logic.

In *God's Two Books*, Price for the first time made an explicit connection between the "moral" fruits of evolution and socialism. The basic framework—that evolutionary theory threatens liberty—harked back to his discussion in *Outlines* nine years earlier. But Price now seems more alarmed at the potential consequences of the "ceaseless struggle for existence and survival at the expense of others." He points to the danger of "the grim, Red terror loading its pistol and sharpening its dirk while awaiting the opportune time to strike." Price argues that evolutionary "ethics" are the primary cause of "firing the blood and quickening the pace of the present strenuous age, until the only apparent outcome will be the wreck and anarchy of Revolution."[55] Evolution was not only unscientific but politically dangerous.

Price's growing concern with the Socialists had everything to do with the real political gains they had made since 1902, covered amply in the Adventist press. Since Price was based near Los Angeles, it is likely that his thinking about the fruits of evolutionary science was affected by a literal explosion that took place there shortly before *God's Two Books* was completed. On October 1, 1910, a dynamite bomb ripped through the *Los Angeles Times* building, setting it on fire, killing twenty-one people, and injuring one hundred. The blast took place during a strike by unionized ironworkers against the city's iron manufacturers. *Times* publisher Harrison Gray Otis was bitterly antiunion and ran the city's employer association, which aimed to break the strike. Otis promptly accused unionists—whom he called "anarchist scum"—of setting the bombs. In April 1911, authorities arrested and charged ironworker union leaders J. B. and J. J. McNamara for the crime, to which they pleaded not guilty. The American Federation of Labor rallied to their defense, as did the Socialist Party. Job Harriman, Socialist front-runner in the Los Angeles mayoral race, joined the McNamaras' defense team, which was headed by Clarence Darrow. But soon after the trial opened in October, the brothers changed their pleas and admitted to carrying out the bombing.[56]

George McCready Price and other readers of the Adventist press received a steady stream of commentary on the McNamara case. Upon the arrest of the two brothers, an article in the Adventist magazine *Signs of the Times* noted that the Socialists "propose to make 'California a battleground.'" Observing that opinion was deeply divided over the McNamaras' guilt, the *Signs* editors placed the conflict in prophetic perspective. "Strifes of this kind are growing both in frequency and in bitterness," they wrote.[57] In the spring of 1911, *Signs* reported that William "Big Bill" Haywood had proposed a general strike to protest the "capitalistic conspiracy" against the McNamaras, a move the editors opposed.[58] *Signs* also ran a lengthy article on the rise of socialism—"the world-wide spirit of revolution"—that focused on the McNamara case. It featured a substantial excerpt of a piece by Socialist Eugene Debs, who also called for a general strike and a massive Socialist election day turnout.[59]

The optimism of Debs and fellow Socialists in early 1912 was bolstered by the growing Socialist vote. In November 1912, Debs received some nine hundred thousand votes in his presidential bid. The Adventist press followed these events and provided readers with ample coverage of Socialist proposals.[60] A 1912 *Review and Herald* article quoted a recent *Outlook* article that noted with alarm the election of Socialist mayors in a number of industrial cities and towns, as well as the election in New York and Rhode Island of Socialist state legislators. That article also drew from a speech given by the president of Cornell University Jacob Gould Schurman, who stated that "the spirit of discontent is far more widely diffused than ever before, and the causes are at once more fundamental and more permanent." His assessment, wrote the Adventist editors, "is worthy of serious consideration."[61]

As the *Review and Herald* educated Adventist readers about how socialists fit into the "signs of the times," Price published *Back to the Bible* (1916), in which he once again addressed socialism and its evolutionary connections. Price warned of the danger inherent in humanity trying to organize on a global scale to improve the world. Whether such efforts were led by "the capitalistic classes" or "the proletariat," they both fell prey to same evil: the "deification of man." Rather than accept that the ultimate cause of misery is "man's evil nature," such schemes of world federation rested on the false idea that an "evil environment" was to blame. This idea, in turn, derived from the "Evolution doctrine," which argued that "all things relating to human life are equally and entirely mere matters of convention, matters of expediency; that morality is only petrified custom."[62] The concept of social evolution was dangerous.

While Price claimed that either capitalists or workers could push such a scheme, his discussion focused on the latter. "The radicals among the Socialists, the labor-unionists, the I. W. W.,—in a word, the whole of the proletariat,—are raising issues which they consider are the real first steps toward the goal of their ambitions," warned Price. But following the lead of the *Review and Herald*, Price added a disclaimer, saying that he was merely studying the subject in the "impartial spirit of science." He went even further, saying he wanted to clarify his position: "All honor to those who are trying to secure by every righteous means a greater degree of 'social justice' for the oppressed and downtrodden."[63]

The degree to which Price salutes the socialists is striking. It may be that his own struggle for survival on the margins of academic respectability and economic security made him more sympathetic to the socialist message. Price's career trajectory fluctuated in the period after 1912. His position at the renamed College of Medical Evangelists in Loma Linda ended in that year. Over the following six years, he taught at two Adventist secondary institutions—Fernando Academy in Los Angeles County, from 1912 to 1914, and then Lodi Academy from 1914 until 1920, when he once again obtained a college-level position.

Not only does Price concede socialists their good intentions, but he provides an analysis that mirrors the *Communist Manifesto*. Price attributes the trend toward world federation to several "material factors," including the railroad, steamship, automobile, telephone, and telegraph. They have converted the world into "one vast community with common interests, common aspirations, and a unified self-consciousness." Moreover, writes Price, corporations that no longer are confined within national borders are also contributing to this growing sense of "internationalism." Whether one seeks to build a global capitalist empire or an international labor movement, "consolidated humanity" is essential.[64] Price even shares some of Marx and Engels's revolutionary optimism. Witness Price's identification of the radical section of the labor movement with "the whole of the proletariat."

Whether or not Price sympathized with secular rebels, he was certain that the socialist quest for international proletarian brotherhood would end in disaster. In *Back to the Bible*, Price concluded that in comparison with the looming threats to American liberty, "the Roman Empire . . . was a mere baby." And yet, for premillennialists, there is always a silver lining in bad news. Price predicted that this latest drive for "federation of the world" by the socialists could lead God to end the "long reign of sin."[65] In a perverse way, the socialists might speed the Second Coming of Christ.

When Price returned to the topic of socialism five years later, there was no mistaking his negative tone. In the aftermath of World War I and the Bolshevik Revolution, his attitude had hardened. Socialism now represented more than meetings, agitators, and subversive books—it meant a revolutionary government in power. It also appeared ever more closely intertwined with evolution. *Socialism in the Test-Tube* (1921) warned of the dangers of the Bolshevik government and its evolutionary philosophy. Richly illustrated and coauthored with Seventh-day Adventist missionary Robert B. Thurber, the book was aimed at a popular audience. Its argument takes the form of a fictional conversation between Gordon, a young American soldier on leave from the fighting in France, and some of his friends and neighbors.

The most influential of those friends was Colonel Newcome, a well-traveled former army surgeon with an encyclopedic knowledge of the history of socialism from Marx to Lenin. On the good colonel's veranda, Gordon learns how Marx's materialist interpretation of history embraces an evolutionary conception of morality. "All man's notions of right and wrong, all of his habits of thought, his ideals, and also his religion," explains the colonel, "are in the final analysis wholly the product of his economic life." The colonel quotes Engels from *The Origin of the Family, Private Property, and the State* on the passing away of the "monogamous family." The book, says the colonel, "gives an economic twist to the ordinary Darwinian theory." Without mentioning the "Bureau of Free Love," Newcome attributes to Karl Kautsky the notion that women, under socialism, are common property of men. "Say, Colonel, that's abominable!" responds Gordon.

Thurber and Price reinforce this image of evolutionary abomination with a "free-love caricature": a hand-drawn illustration of a "Soviet Russian marriage" that can be broken off as soon as one member of the couple loses sexual interest. The betrothed couple stand in a city office surrounded by communist bureaucrats. Husband and wife grasp the red flag and gaze at portraits of Lenin, Trotsky, and Marx. Summing up the connection between the Soviet republic and evolutionary thought, another neighbor tells Gordon, "The ethics of the jungle and the cave, inspired by Darwinism, and the doctrine of the class war and dictatorship of the proletariat, taught by Socialism, may be trusted to evolve the vulgar tyranny of Bolshevism, but never the orderly democracy of America."[66]

Price's *Poisoning Democracy* (1921) sounded similar themes. Published by Fleming H. Revell, whose imprint included a wide swath of American fundamentalist authors, the book made it clear how closely allied evolution and socialism were. The "Evolution doctrine," wrote Price, "develops logically and inevitably into Socialism and Bolshevism as its natural expression in the department of social and civil life."[67] In comparison with the vague discussion in *Outlines*, Price clearly explained why socialism could also be viewed as a form of religious despotism. He drew an intriguing parallel between his own eschatology and Marxism: "The picturesque stories of Darwin's struggle for existence and the ape origin of man constitute the Genesis and Exodus of the socialist Bible; the economic interpretation of history makes up the rest of its Old Testament; while the cheerful doctrine of the class struggle is its Apocalypse, with its prophecy of a coming Armageddon, followed by a socialist new heaven and new earth." Socialism was a religious faith. It was the "*devil's poison for democracy*,—a poison for the working classes who accept it as their religion."[68] Casting socialism as religion made it doubly dangerous in Adventist terms—it encompassed both horns of the lamb-dragon. And yet, danger offered hope. The *Review and Herald* captured the paradox in a headline: "Bolshevism as a Sign of Christ's Coming."[69]

Elaborating on the "devil's poison," Price pointed to Russia, where, he said, "these doctrines have been carried to their logical results." Reflecting Price's concern with the moral impact of communist evolutionism, his examples revolved around the family, gender, and sexuality. Price reported that under the new legal regime, in order to obtain a marriage or divorce, Russians needed only to walk to city hall and sign a register. Drawing from an account published in the *Literary Digest*, Price informed his readers that Russian children were subject to a "fiendish" and "Satanic" scheme of public school indoctrination directed by Anatoly Lunacharsky,

the Bolshevik commissar of public education. Children attended dancing sessions into the wee hours of the morning, without parents present.
"Last winter," an eyewitness reported, "it was painful to see miserable
mothers waiting all night in the snow outside of brilliantly illuminated
school buildings, where the boys and girls were dancing the tango and
foxtrot." The original *Literary Digest* article added the following salacious commentary, withheld by Price: "All the children's time is taken up
with flirtation and dancing-lessons. In the state boarding-schools boys and
girls are quartered in the same dormitory."[70] Perhaps with these words in
mind, Price asked, "What normal individual, whose mind has not been
perverted and depraved through worshipping the false gods of an unnatural and irrational philosophy, desires these experiences to be repeated in
America?"[71]

The Adventist concern with the immoral "fruit" of Russian communism dovetailed with a broader critique of "companionate marriage" in
the US during the post–World War I period on the grounds that it was
"Bolshevik" and "Anti-Christian."[72] As one 1919 article quoted a witness
testifying before a US Senate investigating committee, "They are aiming
at free love and hope to do away with marriage; to make marriage a con-
tract for a term of years, so to speak."[73] Price provided "damning" quotations from socialists in *Poisoning Democracy*, such as one from August
Bebel's *Woman and Socialism*, in which the author wrote that "the con-
tract between the two lovers is of a private nature, as in primitive times,
without the intervention of any functionary."[74] And what of the impact
on children? Price also quoted John Spargo, a leading Socialist Party intellectual who blended Marx, Darwin, and Spencer for a popular audience.[75]
Spargo eagerly anticipated that a socialist regime would prohibit religious
education for children until they reached an age where they could exercise "independence of thought." For Price, Spargo was foreshadowing a
nightmarish time when children of the new socialist marriage would be
"the property of the State."[76]

Poisoning Democracy garnered some high-profile reviews. An anonymous reviewer for the *Literary Digest* wrote that Price's latest work was
"truly a remarkable little book entitled to more consideration than it is
likely to get." Price received plaudits for an argument that was "inexo-
rably logical" and "ingenious" and a writing style that showed "admirable lucidity." At the same time, the reviewer did find Price's eschatology

"curious," coming as it did from a college geology professor.[77] Price must have been pleased with the review, especially its comment on his literary prowess.

The reviewer's reaction was not shared by Bryn Mawr College geologist Malcolm Bissell. He sent off a blistering attack on Price to the *Digest*. Far from being worthy of "'more serious consideration,'" Bissell wrote, *Poisoning Democracy* is "not worth noticing at all." Citing Price's dismissal of thrust faults, Bissell thought the book showed an "astonishing ignorance" of science. When Price wrote to the geologist in defense of his work, Bissell's response was blunt: "There is something wrong with your mental processes." Bissell also questioned the "evil fruits" argument that Price had made central to *Poisoning Democracy*, since it implied that all evolutionists were immoral. "This is absurd," wrote Bissell.[78]

Price's Adventist flood geology continued to have limited appeal, but in the post–World War I years, his twin indictment of Bolshevism and evolutionism struck a chord with some secular conservatives. In early 1922, *Poisoning Democracy* received favorable coverage in the *Constitutional Review*, published by the National Association for Constitutional Government. Price, wrote a reviewer, provided a "scathing indictment" of "socialism's shuddering aversion from religious beliefs and observances, its degrading attitude toward the relation of the sexes and family life, and its fluctuating and opportunist standards of right and wrong."[79]

Poisoning Democracy also drew praise from conservative evangelicals who shared Price's apprehension about the moral and political fruits of evolutionary thought. A reviewer for *Sunday School Times*, a longstanding independent tabloid with a circulation of some eighty thousand, noted the "cogency" of Price's contention that socialism was the "economical aspect" of evolution. He called for the book to be "widely circulated, especially among young men and women who have been attracted by the glamour of Socialism."[80] Price supporters included Virginia educator and conservative Presbyterian Joseph D. Eggleston. He had sent Price a string of friendly missives during the World War I years, while serving as president of Virginia Polytechnic Institute. In 1921, as newly appointed president of his alma mater, the Presbyterian-affiliated Hampton-Sydney College, Eggleston wrote to congratulate Price on *Poisoning Democracy*, which he termed a "smashing indictment." Writing in the early days of Prohibition and facing unruly students at his beloved school, Eggleston shared the news with Price that fellow Presbyterian crusader William Jennings Bryan had come through town speaking on the "menace" to young people of "Evolution and the Higher Criticism."[81] Price must have been thrilled to be placed in company with the "Great Commoner," the most prominent opponent of evolutionary thought.

Bryan had been making his own fruitistic arguments about evolution. He famously attributed the carnage of World War I to the spread of evolutionary ideas among the German General Staff.[82] He also held Darwin's ideas responsible for the crime at the center of the first "trial of the century." In 1924, Clarence Darrow defended the Chicago teenagers Nathan Leopold and Richard Loeb, who had killed a young classmate in order to prove they could commit the perfect crime. Darrow saved them from the death penalty by arguing that Leopold and Loeb were victims of their social environment, both the evolution-tinged ideas of Friedrich Nietzsche and, even more powerfully, the example and patriotic glorification of mass murder on European battlefields. Bryan likened Nietzsche's rejection of moral codes to a bottle of "poison" infusing "the souls of our boys." Society needed to accurately label the Nietzschean and Darwinian bottles before they did any more damage.[83]

Bryan also joined Price in identifying communism as one of evolution's evil fruits. In a 1921 sermon "The Bible and Its Enemies," Bryan focused on the dire practical consequences of Darwinism. After making his customary link to the Great War, Bryan drew from a book published by popular English writer Harold Begbie. Like Bryan, Begbie had been an advocate of Christian-based social reform and a pacifist. After the war, his politics turned sharply right. In a passage Bryan quoted in his sermon, Begbie expanded the list of evolution's evil fruits. "Darwinism," he wrote, "not only justifies the Sensualist at the trough and Fashion at her glass; it justifies Prussianism at the cannon, and Bolshevism at the prison door."[84] Driving home the latter point, Bryan wrapped up his sermon as follows: Darwinism undermined Christian belief, promoted world war, and, in his own words, "is dividing society into classes that fight each other on a brute basis."[85]

In the wake of World War I, as Bryan, Baptist William Bell Riley, and other antievolution activists went into action, Price and Adventist church leaders increasingly found common ground with fundamentalists.

In a 1925 issue of the *Review and Herald*, an advertisement for all of Price's books appeared under the heading "Fundamentalist Literature."[86] *Signs of the Times* even ran a seven-part series of articles on evolution and Christianity by Riley.[87] That fall, after Riley had spoken in Portland, Oregon, Price received an encouraging letter from the president of Adventist-affiliated Union College. "The manager of our Pacific Press branch in Portland told me," he reported to Price, that "he sold sixty-six books mostly on writings of yours in one evening on the occasion of a lecture by Dr. Riley on evolution."[88] Price was moving closer to Protestant fundamentalist respectability.

It is likely that one of the books sold at Riley's talk was Price's latest work to engage the topic of evolution and socialism, *The Predicament of Evolution* (1925).[89] While *Predicament* introduced no new arguments, its relative brevity (128 pages) and its ninety illustrations may have made it a more powerful vehicle for Price's message than either *Poisoning* or *Test-Tube*. *Signs of the Times* advertised the book as a "little volume" written in "popular style." Compared to *Poisoning*, which sold for $1.40, *Predicament* cost only 50 cents.[90] Another Adventist publication promoted *Predicament* with a full-page ad featuring an orangutan-looking creature seated on a chair, alongside the heading, "'Gorilla Sermons': Did your ancestors originate in the Garden of Eden or the Zoological Gardens?" The Scopes trial had broadened the appeal of Price's subject, which, the ad noted, was "now being discussed in the newspapers."[91]

"Red Dynamite" was the title of Price's chapter on socialism and its connection to evolution.[92] Price played on a quotation from a 1914 interview the *New York Sun* had conducted with *Call of the Carpenter* author Bouck White, a month before the Church of the Social Revolution's "invasion" of Calvary Baptist. White had been speaking about his alma mater, the Union Theological Seminary, and aimed to counter criticism from fellow Socialist Party members that modern biblical scholars were disconnected from the class struggle. To the contrary, argued White, liberal seminary teachers, by revealing Jesus as a social rebel, aided the cause of socialist revolution. As quoted by Price, White approvingly described their teachings—which included an openness to theistic evolution—as "social dynamite" that will "blow up the whole apparatus" of capitalist civilization.[93] To bring home the point about literal and figurative dynamite,

Price included a photo of the aftermath of the September 16, 1920, bombing on Wall Street, just outside the banking house of J.P. Morgan. The blast killed thirty-eight and injured hundreds.[94] Although a culprit was never identified, the event provided evidence for Price that "'Red' influence is wide in America."[95]

To develop his argument about the "Red" connection to evolution, Price asked readers to imagine the following scenario: You have a million dollars that you need to transport by car at night down a "long, lonely road." You need to enlist the help of an armed guard who will sit in the backseat with the treasure. There are two candidates for the job. One is a Bible-believing Christian raised in a "Puritan" home. He lives to repay a debt to God, his Creator, for giving him life on earth, and to Christ, for giving him the promise of eternal life. He feels indebted to "all his fellow men" for they were, like him, created beings. The other man views the Bible as a "collection of myths," believes that humans descend from "brute ancestors," and feels no obligation to anyone. He wants to get the most he can out of life and, first and foremost, to take care of himself. Whom would you pick? As Price informs us, the treasure represents modern culture and civilization. Do we want to entrust it to those whose moral code is based on belief in a Creator, or to those who believe that our morality is "only what developing anthropoids" have agreed is best for our stage in the evolutionary process?[96]

Taking on this materialist tenet of Marxist thought—that the morality of a given society is a product of its mode of production and ruling class—Price argued the idealist opposite. Civilization, he writes, is *not a cause, but a consequence*" of ideas. Only religious faith can produce morality, and the two of them generate civilization. Since evolutionary ideas pull the rug out from under the foundations of faith and God-given morality, civilization—defined as the family, the "sacredness of human life," and "the rights of private property"—is in trouble. Under the heading "Evolution and Socialism One," illustrated by portraits of Marx and Lenin, Price spelled out the nature of the threat: "Marxian Socialism and the radical criticism of the Bible, though arising first in point of time, are now proceeding hand in hand with the doctrine of organic evolution to break down all those ideas of morality, all those concepts of the sacredness of marriage and of private property, upon which Occidental civilization has been built during the past thousand years."[97] Focusing his attention on the danger of teaching evolution in the schools, Price urged his readers not to become complacent. The post–World War I wave of radicalism had ebbed. But since "*Marxian socialism and the dictatorship of the proletariat are merely the economic aspects of the doctrine of organic evolution,*" and since evolution continues to be taught to schoolchildren, a resurgence of a movement for "Social Revolution" is "inevitable."[98]

George McCready Price was outside the evangelical mainstream. His Adventist theology and his insistence on a literal six-day creation set him apart. But under the impact of the Bolshevik Revolution, the birth of an American communist movement that openly championed evolutionary science, and the increasingly public battle over evolution in the schools, Price's equation of socialism, Bolshevism, evolution, and immorality gained currency. In the years surrounding the Scopes trial of 1925, a range of figures from William Bell Riley to Gerald Winrod to J. Frank Norris to Mordecai Ham all gave voice to this antievolutionist theme in sermons and writings. The influence of Price's flood geology on the content of "creation science" would not reappear in a significant way until the 1960s. But his view of evolution as "Red Dynamite" would prove to be tailor-made for the 1920s.

Blood Relationship, Bolshevism, and Whoopie Parties

Gerald Winrod was worried. After establishing Defenders of the Faith in Salina, Kansas, in 1925, the fundamentalist preacher had seen several state antievolution campaigns go down to defeat. His pamphlet *The Red Horse* (1932) analyzed the nature and scope of the problem. Knowing that his readers were familiar with the apocalyptic imagery of the book of Revelation, Winrod argued that the red steed—traditionally representing war—was none other than Soviet Russia.[1] What made Russia especially significant to Winrod was communism's intimate connection with evolution. Since belief in evolution—or "animalistic ancestry"—led to atheism, and atheism led to communism, teaching evolutionary science had "opened the door" to the acceptance of communist ideas in the United States. After all, wrote Winrod, "Every leader of the Russian Revolution was, and is, an evolutionist." To illustrate the connection between communism and evolution, Winrod pointed to the "moral collapse" evident in Russia. Because of Soviet animalism, "A man in Russia may live with several women the same day, all as his wives, because

Figure 5. "EVOLUTION is the Bunk!" cartoon, 1930. Appearing in the July 1930 issue of Gerald Winrod's *Defender* magazine and exemplifying the "monkey-superior" genre that inverted the human-ape hierarchy, this cartoon conveyed the range of purported social and political evils Winrod attributed to evolution. Courtesy of Wichita State University Libraries, Special Collections and University Archives.

marriage and divorce are practically abolished." This resulted from an evolutionary logic: "Our ancestors observed no marriage laws in the jungles, so why should we?" As evidence that American schoolchildren were imbibing this "godless, beastly, antichrist thing," Winrod summarized a passage on the Darwinian concept of sexual selection in a popular high school psychology textbook. "Do not," he warned, "teach the children of America that they have the bodies, minds and sex-impulses of the beast and expect them to live clean, Godly lives."[2]

By 1932, *Defender* readers were familiar with the themes Winrod raised in *Red Horse*: the apocalyptic role of Russia; the attribution of evolutionary thinking to Satan; the secret, manipulative methods of Moscow; the association of "free love" and immorality with the young Bolshevik Revolution; the allegations of moral decline in Russia and the United States, particularly linked to youth and sex; and the claims that revolutionary Russian leaders were evolutionists. *Defender* cartoons reinforced the message. One featured a monkey disclaiming any "blood relationship" with human beings, whose unsavory deeds, pictured in newspaper headlines, included "murder," "divorce," "Bolshevism," and "whoopee parties."[3]

Other leading fundamentalists during the 1920s—most notably William Bell Riley and J. Frank Norris, but also Mordecai Ham—were delivering a similar message. Unlike George McCready Price, these figures were not primarily identified as "creationists," but an anticommunist-inflected version of creation*ism* was key to their fundamentalist activism. They reinforced each other's ideas and worked together in the World's Christian Fundamentals Association, which played a leading role in mobilizing the forces to ban the teaching of evolution. They preached in each other's pulpits and wrote for common publications. Their cohort included the theologically marginal Price, who was a contributing editor to the *Defender* and a guest speaker at a Defenders conference. More distantly, they were joined by George Barry O'Toole, a prominent Catholic professor of philosophy and biology and author of *The Case against Evolution* (1925).[4] O'Toole also tied evolution to the dangers of socialist revolution. Together these thinkers made clear for conservative Christians that evolution posed a deadly danger to their worldly and eternal welfare.

It is appropriate that William Bell Riley, a man who figured so prominently in the conflicts within the Baptist Church, was born in deeply divided southern Indiana in 1861 exactly three weeks before the American Civil War began.[5] Since his father, Branson Radish Riley, was a Kentuckian by birth and, by his son's account, a "slavery sympathizer," the family moved from Bloomfield, Indiana, in Greene County, back to the Bluegrass State within a month of William's birth.[6] While Riley would come to be identified with his newly adopted home of Minneapolis, his second wife and biographer, Marie Acomb Riley, suggested that her "Kentuckian" husband retained his flair for southern "hospitality." He loved to tell jokes, "especially Negro stories," which contributed "spice" to conversations with friends.[7] As this description of Riley appeared in a biography published in 1938 by Eerdmans, a prominent evangelical publisher based in Grand Rapids, Michigan, it speaks not so much to Riley's distinctive southern character as it does to the national reach of casual racism.

Looking back on his childhood, Riley wondered what effect the timing of his birth might have exerted on his personality. He thought the question was best left to psychologists, but there is no doubt that he enjoyed the art of intellectual brinksmanship. Riley traveled back into Indiana to earn a teaching certificate, hoping to become a lawyer. Compelled to return to his family farm, he felt the pull of the ministry, and stayed the course for the rest of his life. He graduated near the top of his class from orthodox Presbyterian-affiliated Hanover College in Hanover, Indiana, in 1885. Back across the border in Louisville, he earned a doctorate in divinity from Southern Baptist Theological Seminary in 1888.[8] Though never a lawyer by profession, Riley became a formidable advocate for the fundamentalist cause.

Riley's early assignments took him to southern Indiana, central Illinois, and then, from 1893 to 1897, the South Shore neighborhood of Chicago, where he was pastor of the newly formed Calvary Baptist Church. He preached blocks from the newly established University of Chicago, a bastion of Baptist modernism founded by John D. Rockefeller Jr. and William Rainey Harper. Riley's success in growing his young congregation and his ambition to pastor a large, centrally located urban congregation won him a prize position in 1897 at the blue-blooded First Baptist Church in downtown Minneapolis. In his new position, Riley remade the church in his own image. He took on the elite lay leadership of First Baptist and pushed for an open membership policy, regardless of social class, and for the abolition of the elitist practice of pew rentals. He cracked down on dancing, card playing, and theater. And the masses of Minneapolis Baptist faithful responded. Church membership nearly doubled the first year. Now ensconced in his pulpit, Riley founded the Northwestern Bible and Missionary Training School in 1902.[9]

In his subsequent travels around the Midwest, Riley preached, as William Trollinger has noted, "standard revival fare." His messages on sin and salvation were laced with a strong dose of premillennial dispensationalism, which stressed the imminent Second Coming of Christ, a heavy emphasis on "signs" of that imminent return, and the inerrant truth of the Bible. He also took on the Socialists. Initially, his approach was to co-opt and channel the potentially radical sentiment. In one of his first sermons delivered in Chicago in September 1893, titled "Christ and Laboring Men," he explicitly addressed and praised "those who are counted among the laboring class," in company with Christ, the "lowly Nazarene" carpenter.[10]

By 1912, he was going for the political jugular. His chief text was *Christianity vs. Socialism*.[11] Speaking in nearby Duluth, Minnesota, Riley zeroed in on the supposed socialist philosophy of replacing marriage with "mating." Without using the word "evolution," Riley complained about the socialist view that past societal norms needed to change—that the "social codes of the past have no more sacredness or binding authority for the modern man than the customs of the anthropoid ape should have for his human and educated descendant." He then charged socialists with proposing a "collectivist" form of marriage, which meant "free love" and the conversion of children into "wards of the state." As evidence, he quoted Bebel on the need to abolish both marriage and private property. In the booklet, Riley clarified for his readers that if they are attracted to the idea of the equal division of property, they must realize that this inevitably means "free love." This "doctrine of devils," writes Riley, will "degrade men and women to a level with the beasts of the field." In the Duluth speech, Riley described the socialist dystopia this way: "Men would fight for the possession of the most beautiful women and the streets of the civilized world would run with blood."[12] Without characterizing socialist ideas about "free love" as "evolutionary," the language Riley used— "beasts," "savagery," and "blood"—resonated well with his developing critique of evolution.

As early as 1910, the decision of the Northern Baptist Convention to hold its national convention at the University of Chicago got Riley thinking more seriously about organizing the Baptist fundamentalists. The outbreak of World War I accelerated their growth. The capture of Jerusalem by the British and the prospects for a Jewish return to Palestine seemed to bear out biblical prophecy about the end times. By 1918, Riley was meeting with Reuben Torrey of the Bible Institute of Los Angeles to plan a prophecy conference for the next year, which Riley engineered into the founding meeting of the World's Christian Fundamentals Association (WCFA). In 1922, Riley convinced the WCFA that it should concentrate on evolution. The group provided the organizational base for successful passage of the antievolution Butler Act in Tennessee three years later. The WCFA flourished only through the late 1920s, but as its president and head of the Committee on Conferences, Riley kicked off its inauguration with great fanfare, leading a six-week tour of North America.[13]

Among the Bible institutes present at the 1919 Philadelphia conference was Riley's own Northwestern Bible Training School. It had grown steadily since its founding, owing largely to Riley's superb administrative and fund-raising skill. While he boasted that the school survived thanks to a multitude of small contributions, Riley also managed to enlist the help of wealthy supporters.[14] His curriculum was distinguished from the education Riley had received at the Southern Baptist seminary, which had included a large degree of classical literature and biblical commentary. At Northwestern Bible, the focus was on reading and studying the plain "facts" of the Bible, with the clear fundamentalist guidance of superintendent William Bell Riley.[15]

Riley first addressed the Red connection to evolution in the early 1920s in a pair of sermons that he delivered at First Baptist: "The Theory of Evolution—Does It Tend to Anarchy?" (1922) and "Evolution or Sovietizing the State through Its Schools" (c. 1923). Delivered at the outset of the WCFA campaign against the teaching of evolution, "Anarchy" was based on Second Peter 2:1–12, in which the author warns of false prophets who would spout "damnable heresies" and lead the people astray with their "pernicious ways." Riley first claims that the "evolutionary hypothesis" propounded by such modern-day false prophets is "unproven." Then he argues that it was built on pure speculation, which has turned out to be "utterly false." Evolution is therefore "a lie," and as such, it cannot produce "desirable fruits." Rather it produces "social putridity," "social slavery," total absence of government, and more concretely, "BESTIAL BOLSHEVISM!"[16]

Freely mixing anarchism and socialism, Riley quoted from anarchist Mikhail Bakunin to illustrate the "dominant" interpretation of socialism, but then provided as one of the few concrete examples of evolution's evil

fruits the situation prevailing in the Soviet Union. Just as George Mc-
Cready Price asked whom his readers would trust with a large sum of
money, Riley wondered if any of his parishioners would feel safe bringing
up their children in any country that had "tried out Evolutionary Socialism." Surely, answered Riley, they would not want to be in Russia. He
borrowed from Price, retelling the horror story of children dancing all
night in "brightly illuminated school buildings." Nor would they want to
be in Germany, which, under the spell of Darwinism, had become "brazen
and defiant as a nation—ravishing, maiming, poisoning, burning, suffo-
cating, deporting, enslaving." In either case, the proponents of Darwinism
were like the false prophets of Peter's time who bring about their own
destruction. They were like "natural brute beasts, made to be taken and
destroyed."[17]

Even more dangerous were promoters of evolutionary thinking right
here in America who menaced the country. Borrowing again from Price,
Riley gave the example of Bouck White, who used spiritual "dynamite"
to attack the established capitalist order.[18] Riley and Price agreed that at
stake for American Christians in the battle over evolution was nothing
less than morality and civilization. In a "godless world," warned Riley in
the sermon, children would no longer honor their parents; parents would
abandon their children, kill with impunity, steal without guilt, and "commit adultery without conscience." On this last point, comparing this socialist nightmare to current-day America, Riley explained that the law
against adultery "was only made sacred by capitalism."[19] Riley's rhetori-
cal shift here from anticommunism to pro-capitalism anticipated a theme
that became increasingly visible in later decades.

By 1926, Riley joined forces with Gerald Winrod, under the auspices
of the Minnesota Anti-Evolution League, and both men were barnstorm-
ing the state. Riley gave sixty-five speeches. But in a stinging rebuke to
Riley in March of that year, the University of Minnesota refused to give
permission for the First Baptist pastor to speak on campus. Angry but
undaunted, Riley spoke in Minneapolis's Kenwood Armory to more than
five thousand people on the "white-hot" issue of evolution. He focused
on natural and social science textbooks used in Minnesota public schools
that contained evolutionary material. As before, he tied evolution to German philosophy, and at one point associated one of the textbook authors
with defenders of the Bolsheviks. He wisely ended his talk with a populist

appeal to the "God-fearing majority": "Whose university is it over there, will you tell me? Does it belong to a dozen regents? (A voice: No sir!) Does it belong to fifty or seventy-five professors? Does it? (No!)."[20]

Riley's "Sovietizing" sermon conveyed a more detailed and politically pointed message. To preempt any suspicion that his anticommunism might be masking a defense of wealth and privilege, Riley began by establishing his credentials as a "friend . . . of the poor and oppressed." Despite his identification with the downtrodden, Riley warned his flock about the "menace of the Soviet." Knowing that his congregation might have heard that the word "Bolshevik" meant rule by the majority, Riley argued that a small minority now ruled Soviet Russia. They were motivated by "class hatred." As did Winrod and Norris, Riley granted that there was ample reason for such feelings to exist under Russian czarism. Nonetheless, he contended, such a movement was inevitably "dangerous," because whenever "ignorant masses" overthrow a ruling power, they risk coming under the rule of an even worse tyranny. This had now happened, Riley said, quoting Lenin to prove his point, as follows: "Today the revolution in the interest of socialism demands the absolute submission of the masses to the single will of those who direct the labor process."[21] Not only were the Bolsheviks tyrannical, but they were brutal. After offering a catalog of their actions, Riley quoted "one writer" who described the peril of Russian communism as "a brutal savagery which, like a wild beast, tortures and kills to vent its bestiality."[22]

That wild beast had already appeared in two arenas of American life. One was the labor movement. Led by the Industrial Workers of the World and the radical socialists, workers were pressing the US government to extend diplomatic recognition to Bolshevik Russia. Citing the Lusk Committee, a state legislative body investigating seditious activities, Riley claimed that there were more than five hundred thousand "red agitators" in New York alone. They were circulating literature in "shops and factories" and had even held public rallies to popularize their cause. The second manifestation of Sovietism was more subtle and insidious: the influence of Darwinism in high schools and universities, including at the University of Minnesota. The "outstanding leaders" of "Sovietism" in America, declared Riley, were the "professors in our modern universities who are naturally materialistic in their conception of the universe."[23]

As he did at the Kenwood Armory, Riley highlighted a handful of evolutionary social science textbooks currently in use at the University of Minnesota and the city's North High School. Not a single one of those books taught biology: Maurice Parmelee, *Criminology*; Charles A. Ellwood, *Sociology and Modern Social Problems*; Edward A. Ross, *Social Psychology*; F. Stuart Chapin, *Social Evolution*; and Henry Reed Burch and S. Howard Patterson, *American Social Problems*. As Riley summarized their approach, morality was a creation of human beings in specific historical circumstances. So, too, the human mind evolved from animal intelligence; the family developed from "the very conditions of life itself" and was not created by divine authority; and Christianity evolved out of older religions and retained elements of "wizardry" and "magic." The subjective religious conversion experience could be explained by the phenomena of hypnosis, the power of suggestion, and the subconscious mind. In short, the young people of Minnesota were learning from "God-denying, Christ-repudiating, Bible-scorning" textbooks.[24]

Evolutionary teachings were so dangerous, Riley explained to his flock, because they built a bridge that enabled an invasion of "Soviet propaganda." The textbook authors were not socialists, but their materialistic ideas, fueled by the scholarship of higher criticism, paved the way for Marxism. With swarms of paid agitators already active in America, this infusion of communist ideas would lead to revolution. America would become like Russia: "infidelity, mental and moral; rapine, plunder, robbery— these will be universal." Riley concluded his sermon with the following plea: "God forbid that we should be silent while America is thus being menaced and the immortal souls of all men are being thus imperiled."[25]

For Riley and his supporters around the country, the importance of the debate over teaching evolution went far beyond the classroom and even the church. It was not only that the salvation of individual Christians was imperiled. Nothing less than the future of civilized life on earth was at stake. Did you want to live in a world in which rape, marital infidelity, robbery, and all forms of immorality were rife? That is the world that evolutionary teachings—starting in biology and moving into the social sciences—were threatening to create. They were doing so by opening the door to the atheistic Communists in Soviet Russia, who claimed to be fighting for a world run in the interest of working people, but who were actually a band of infidel criminals who needed to be fought at all costs. In delivering his message in Minneapolis to the Baptist faithful in 1923, Riley was pulling out all the stops.

As a crack organizer and administrator, Riley was not working alone. Having established the Northwestern Bible school two decades earlier, he now had a growing institutional base of operations. Published by Northwestern and edited by Riley, *Christian Fundamentals in School and Church* (*CFSC*) projected his message on a regular basis. In the April-May-June 1923 issue, readers learned about the fifth annual conference of the WCFA, held in Fort Worth, Texas, at J. Frank Norris's First Baptist Church. Expected speakers included William Jennings Bryan, Norris, Riley, and T. T. Shields, a prominent Baptist conservative in Toronto. It was also reported that a week later, Shields, Riley, and others presided over the founding conference of the Baptist Bible Union (BBU) in Kansas City, Missouri.[26]

Fueled by resistance to the growth of modernism, including theistic evolution, the BBU was an early incarnation of the separatist fundamentalist impulse within the northern Baptist church. As the convention call in the *CFSC* put it—drawing on Jesus's parable of the tares secretly planted in the wheat fields—"The tares of rationalism have been sown in both our schools and churches, and they have not only taken root but are now developing to such proportions as to demand attention."[27] As an example of such "attention," T. T. Shields, who served as a member of the board of trustees of McMaster University in Toronto, led a protest campaign the following year when the school awarded an honorary degree to modernist Baptist William H. P. Faunce, then president of Brown University. While Shields did not accuse Faunce of communist leanings, the militancy of Shields's fundamentalism was reflected in the title of his 1924 article on Faunce in the *Gospel Witness*: "Why Some Individuals and Institutions Need to Be Blown Up with Dynamite."[28]

The Red connection to evolution emerged several times in the spring 1923 issue of Riley's paper. One item, based on a letter sent to editor Riley, reported on a "blasphemous" ceremony in Moscow on the previous Christmas, in which Komsomol members burned the "great dignitaries of the religions of the world" and mocked the biblical story of Christ's birth. In his letter, Riley's friend asked rhetorically whether such action was any different from the rejection of traditional Christianity by American modernists. Was it not just a different expression of the same "anti-Christ" spirit? "The question is so pertinent that it requires no discussion," was Riley's reply. While the young communists seem to have condemned world religions in an equal opportunity fashion, the *CFSC* headline was more partial: "Burning the Image of Christ in Effigy."[29]

If that article did not make the connection between evolution (as a component of modernism) and communism explicit, an extensive piece by Baptist fundamentalist missionary T. A. Blalock did.[30] It concerned an arena of conflict over communism and Christianity of increasing interest to fundamentalists as the twentieth century wore on: China. Based in the western Chinese city of Tai'an, Blalock sought to explain what he termed China's political "confusion"—the seemingly endless political instability that had prevailed since the reform period of the 1890s and then since the republican revolution of 1911. They key "fact" according to Blalock was that China was a "nation of liars," thus undermining any basis for trust and stable rule. This racial flaw, moreover, was accentuated by the Western education received by Chinese abroad. There they learned about the animal origins of humanity and the lack of an eternal moral code. This was especially true for students who attended Columbia University, which Blalock labeled "one of the most notorious hot beds of evolution and modernism." If Blalock had read the list of endorsements for Ludwig Katterfeld's *Evolution* magazine from that university later in the decade, he would not have been shocked.[31]

Within China, added Blalock, the situation was possibly worse. Since 1890, Chinese reform intellectuals had taken up the cudgel of Darwin in the battle to modernize China. They stressed its eugenic side, focusing on the dominant Han nationality and on the survival of the fittest, a lesson that China's humiliating defeat by Japan in 1905 emphasized with great clarity. They also tended toward neo-Lamarckianism, looking for a new Chinese who would thrive in modern conditions and pass racial improvements to the next generation.[32] Regardless of these grand ambitions, what Blalock stressed about the "incoming tide of new thought" was the simple fact that in government schools, as in those in Tai'an, where he lived and worked, students were learning about "the ape origin of man."

These evolutionary teachings had alarming political implications. Just as the "evolution seminars" in the US were paving the way for communism, so in China, wrote Blalock, was the "false science of evolution . . . making this land ready for a hot bed of Socialism, Anarchy and Bolshevism."[33] With evolution running rampant, Blalock warned, things were about to get much worse. Evolution-crazed revolutionaries would turn the Middle Kingdom upside down. Whether it was China, Russia, New York, or Minneapolis, the message from Riley and his counterparts on the subject of evolution was consistent. They pointed not only to the Bible and otherworldly salvation, but to the common moral and political dangers facing Christians in the here and now.

If Riley was the preeminent fundamentalist based north of the Mason-Dixon Line in the first half of the twentieth century, his southern counterpart would have to be J. Frank Norris (1877–1952). Norris and Riley were remarkably similar. They grew up in the rural South, earned degrees from the Southern Baptist seminary in Louisville (Norris was first in his class), pastored big-city First Baptist churches for decades (Norris in Fort Worth), founded independent Bible schools, edited fundamentalist publications, and campaigned against both evolution and communism. In some respects, however, Norris's early life and personality were sharply different. Born in 1877 in Dadeville, Alabama, of Warner and Mary (Davis) Norris, young Frank experienced the terror of an alcoholic father. Earning his living as a steelworker in Alabama and later as a cotton sharecropper in Arkansas and Texas, Warner Norris drank much of his meager earnings away. He was physically abusive, once beating Frank when the boy dared to pour out his precious bottles of liquor in a vain attempt to protect his mother. When Frank was fifteen, he and his father were nearly killed when some local cattle rustlers, hearing that Warner was prepared to testify against them, appeared on their farm and started shooting. In a reflection of his combative spirit, Frank came at them with a knife and lived, barely, to tell the tale.[34]

The violent confrontations of Frank Norris's childhood continued in adult life. In 1912, Norris was tried for arson, when the First Baptist Church of Fort Worth, which he began pastoring in 1909, burned to the ground. It seems that his aggressive campaigning for prohibition had provoked local liquor interests to frame him. Hired gunmen twice shot at Norris on the street. But the attempt to frame Norris for the arson failed, and he was acquitted in a jury trial. Norris became best known for a second trial, in 1926, in which he was again the criminal defendant. Sitting in

Figure 6. Rev. J. Frank Norris, c. 1920–25. The foremost southern fundamentalist of the pre–World War II era, Norris repeatedly warned his flock about the twin threats of evolution and communism. George Grantham Bain Collection, Library of Congress.

his office, Norris had shot and killed D. E. Chipps, a Fort Worth business-man. Norris claimed that Chipps had physically threatened him and that he shot in self-defense. In a 1927 trial in Austin, Texas, the jury agreed, and Norris was exonerated.[35] But his reputation as the "shooting parson of Texas" was secure.[36] As biographer Barry Hankins has suggested, "It seems that people were attracted to Norris not in spite of his brushes with the law but in part because of them."[37]

They also liked his dramatic flair in the pulpit. After having great success preaching at a Kentucky revival in 1911, Norris adopted a more expressive, emotional style at First Baptist. One observer described the high point of a Norris sermon: "His movements were jerky and impetuous. His arms at high points wave like flails. He would yank out his handkerchief, fitfully mop his face with it and thrust it back in one pocket and then in another. Once for emphasis and to drive home [his point] . . . he seized the Bible and pounded it with the palm of his hand."[38]

His performance won over working-class Fort Worth Baptists and drove away wealthier parishioners, just as Riley had in Minneapolis. As Norris told one of his flock, "I would rather have my church filled with the poor, the halt, the lame, the sinning . . . than to have it filled and run by a high-browed bunch." For the ever-growing congregation—which reached some three thousand by the early 1920s—Frank Norris was one of them.[39]

His flock also liked the fact that he did not back down in the face of his perceived enemies. Before the evolution wars of the 1920s, those foes included liquor dealers, corrupt politicians, profiteering employers, and the Catholic Church. Norris fought them with the weapon of the *Searchlight* newspaper, which he founded in 1917. From the beginning, Norris made it clear that it would not be a narrow church publication. Rather, it was "an independent weekly, which will deal with all matters social, political and religious." Initially considering the *Searchlight* an open forum, he soon was describing it as the "official organ" of the First Baptist Church. Starting in the early 1920s, the front-page masthead featured a drawing of Norris, standing at left, with a Bible in one hand and a searchlight in the other. Its light rays traversed the title of the newspaper and hit their target at the far right—a cape-wearing, tail-spouting, and pointy-hat-bedecked devil, who was cowering and seemed dismayed that Norris had exposed him.[40] For many Texas Baptists in the 1920s confronting satanic forces in the world, J. Frank Norris was their man.

During the 1920s, Norris was sure that the devil was in league with the communists and evolutionists. In 1923, he delivered an address to the Texas state legislature in support of a bill to ban the teaching of evolution

in the public schools of the Lone Star State. After Representative J. P. Lane
delivered the invocation to the assembled crowd of politicos, a fellow legislator introduced Norris as the pastor with the largest Sunday school in
the entire country. Having lobbied at the state capitol against organized
gambling and for prohibition, Norris was a familiar figure among Austin lawmakers, and he began his long speech with joking reminiscences
of those early days. His easy rapport with the crowd is reflected in the
fact that his speech was interrupted repeatedly with applause and laughter.[41] Those audience interruptions point to the mobilizing function of
such speeches. Norris did present ideas about evolutionary science, but his
main aim was not to intellectually persuade. He aimed rather to move legislators into action. As did Riley, Norris accomplished this goal by focusing on the consequences of evolution for people's ordinary lives.[42]

After disposing with the preliminaries, Norris began his defense of the
bill. To the accusation that the bill was anti-science, Norris held up the
Bible as a source of true science. To the claim that the bill violated academic freedom, Norris relied on the majoritarian argument that Bryan
would use to good effect in Dayton. Since the public believed in creation
according to Genesis, and since they paid teachers' salaries, they were
entitled to classroom content that did not undermine their children's precious faith. To the argument that the bill violated constitutional guaran-
tees embodied in the First Amendment, Norris readily agreed that "no
creed, no dogma, no tenet of faith" should be taught in the public schools.
But all major Protestant denominations—representing the majority of
Texans—agreed on the fundamental idea of creation, which meant that it
was not a sectarian creed but rather good Christian common sense. To be
sure, Norris allowed, if Buddhists or "Mohammedans" demanded mandatory teaching of their faiths, if Catholics insisted that papal infallibility
were to be taught to all Texas children (or even if Baptists were to insist on
requiring instruction in the theology of adult baptism), that would violate
the Bill of Rights. But teaching run-of-the-mill Christian creationism in
public school posed no constitutional issues.

In spelling out these points, Norris shifted back and forth between
straightforward exposition and homey stories and amusing anecdotes to
keep his audience's attention and to engage their emotions. In ridiculing
the speculative and "unproven" nature of evolutionary science, Norris
summarized humanity's evolutionary origins: "Away back yonder some

time, nobody knows when, six million, six hundred million, six hundred billion (one fellow put it at a quadrillion, on the matter of time these fellows are very extravagant and a few billion years is immaterial with brains that deal in wild guesses)—away back yonder some time—when, nobody knows, something happened away back yonder somewhere—where, nobody knows, something happened away back yonder somehow, something happened, nobody knows how." After matters proceeded a bit further from protoplasm to tadpole to fish to nameless cave-dwelling animals, who grew hair and "began to devour each other for breakfast and dinner," human beings finally appeared. His account continued: "Those living in caves kept on growing and being in the shade had no need for the hair and that accounts for so many baldheaded men. (Laughter.) One day one of these hairless, tailless animals ran off and stole a suit of clothes and became professor of biology out here at the State University. (Applause. Applause. Applause.)"[43] Evolutionary science is full of partial evidence and unanswered questions. But the remainder of the speech made it clear that such evidence was irrelevant to the argument Norris was making.

The question that framed the last third of the speech, and the key to its central meaning and impact, was "What will evolution do for us?" What Norris really meant is, "What will evolution do *to* us?" The picture painted by Norris indicated that God-fearing Christians were the victims. Norris introduced this portion of his speech by answering the hypothetical question, "Why is it that you insist on the Genesis account of creation?" His revealing answer is that since the Bible is the foundation of society, and since evolution attacks that foundation, he *must* reject evolution. It is not a matter of inadequate evidence for an academic claim but rather a matter of moral, social, and, one might say, political obligation. "Destroy the foundation," Norris explains, "and the whole superstructure of society will give away [sic]." Under the sway of evolutionary ideas, Norris told legislators, "man is reduced to a mere machine." He no longer had any personal obligation to his Creator.

The consequences, as Norris then proceeded to illustrate, were dire. "My friends, we are in a terrible hour," he began. "Wave after wave of crime is sweeping over the land, and the reign of lawlessness is engaging the best thought of our engaging statesmen." Norris tipped his hat to Governor Patrick Neff, who had sent troops to quell an epidemic of "bootlegging, gambling, prostitution and robbery" in the oil boom town of Mexia, Texas.[44] And yet, in the next breath, Norris shifted focus to "lawlessness" of a very different kind: "Last year I saw in New York a mob of 15,000 or 20,000 people. They had the red flag on the lapel of their coats. They waved them in their hands. A man would harangue a while and then a woman. They were ready to overthrow this government." His next example, concerning a group of defendants in Illinois who had been tried and acquitted of murder, also had strong political overtones: "Here a few days ago in Herrin, Illinois, a jury of twelve men lifted their hands to high Heaven and swore they would enforce the law. Red-handed murderers sat before them. There was no doubt about the murder. Nobody denied it. And yet these murderers were turned loose." The perpetrators of the 1922 "Herrin Massacre" were coal-mining union activists retaliating against company violence against their members. They received support from labor activists nationwide and were acquitted owing to the widespread sympathy with the United Mine Workers of America in the southern Illinois coalfields.[45]

After denouncing the Herrin verdict, Norris segued from the "crime wave" to the "wave of liberalism" sweeping the United States on the subject of gender relations and sexual morality. Using that redolent phrase made famous by newspapers attacking the Bolshevik regime, Norris proclaimed that "we are in the days of free-loveism." Whereas in days gone by, a divorce would ruin a woman's reputation, Norris explained, now a series of divorces was a ticket to "moving picture" stardom. Men who abandoned a wife and children no longer risked absolute disrepute. Women and girls were now even rivaling men in their criminal capacities. "I could tell you story after story tonight," said Norris, "of the shame and disgrace of 12 and 14-year old girls and high school girls who have gone to drinking and smoking cigarettes and who are throwing to the winds the priceless jewel of womanhood."

Finally, there was Russia. The threat was spiritual and military. Norris offered his own version of the effigy-burning in Red Square. It was not "Jesus" but "God" who was destroyed by the flames. The culprits were "students" led by their "professors." Even if this was one crowd in one city, America confronted a nation with 170 million atheists. But this was not all. German rationalism was the seed that had bloomed into full-blown evolutionary thought. Now that Russia and Germany were allied—through the Treaty of Brest-Litovsk—it was only a matter of time before the "hordes" reached France. America, foretold Norris, "is going in again." From world war to violent crime to sexual immorality to socialist revolution, the consequences of evolutionary thinking were monstrous.

If there was one prominent Christian fundamentalist during the 1920s who made it easy for readers and listeners to grasp the connections between evolution and communism, it was Gerald Winrod. Born in 1900 the son of a former saloonkeeper turned upstanding citizen and preacher, Gerald Winrod may have been particularly aware of the pressures weighing on Americans during the Jazz Age to not live "clean, Godly lives." It was not only his father's spiritual journey but his mother's experience of breast cancer and subsequent battle with morphine addiction that made an impact on him. According to family lore, prayer saved her life. The Winrods then shunned medicine and relied on God for their bodily and spiritual health. Gerald was even more devout than his parents, making him a "battering ram of righteousness," in the words of his father. The young Winrod started preaching at the age of eleven and began editing his own paper at twenty-one. He founded Defenders of the Faith in the wake of the Scopes trial, angered by the evolutionary implication that Jesus himself was descended from the animal kingdom. In 1927, his fledgling *Defender* magazine reached sixteen hundred subscribers; by 1934, that number had grown to sixty thousand.[46]

Even before launching the *Defender*, Winrod was spelling out the moral, social, and political consequences of evolutionary teachings. He began with *Christ Within* (1925), a self-published collection of "lectures" he had been delivering around the Midwest. Like Riley, Winrod delivered carefully prepared lecture-like sermons, in contrast to the hellfire style of J. Frank Norris.[47] Issued in four editions by 1932—with substantial revisions along the way—the book presented Winrod's defense of Christian fundamentalism and his attack on modernism and, more specifically, evolution.[48]

In the opening chapter, Winrod contrasted the sinful "Adamic" man and the regenerate "Christ" man. The former had a heart filled with sin. From it issued all manner of modern evils: world war; the "crime-wave"; mass viewing of motion pictures that featured sex and vice; rising divorce rates; "rivers" of bootleg liquor; public dance halls as recruitment grounds for white slavers; and women who wore revealing clothing and who "smoke, drink, swear, and carouse as a man."[49] Winrod emphasized the noxious character of immorality by using olfactory language: garbage "rotting," "putrescence," "stench," and "putrid."[50] In later editions of the book, Winrod offered a vivid picture of Satan, the force battling "the work of Christ in human lives." He stood atop a vast hierarchy of evil, with "millions" of demons at his disposal. He ran the equivalent of "a world-wide secret service department." And he was real—a figure with "huge, bat-like wings."[51]

Winrod attacked evolution in his next chapter, "Mark of the Beast." Noting that premillennial dispensationalists used this phrase from the book of Revelation to forecast events during the period of the Tribulation, Winrod told readers he was using it instead as a synonym for evolution. It was a "diabolical doctrine" teaching that humans were descended from lower forms of life, thus giving "man" the "mark of the beast in his heredity." Winrod sidestepped debates about the details of premillennial dispensationalism but harnessed the emotional power of Revelation to identify evolutionary thinking with the Antichrist and the end times. He then proceeded in a populist vein to attack evolutionary scientists for veiling their fraudulent ideas in fancy language. "History discloses," wrote Winrod, "that all truly worthwhile knowledge has been within the reach of the common people." Quoting George McCready Price for support, Winrod concluded that evolution was a "guess" with no basis in fact.[52]

While calling out evolutionists for lack of evidence, Winrod also sought to ground his own argument in the philosophy of science. For Winrod, God's spiritual power was a physical force. Now that physicists had identified elementary particles as essentially consisting of energy, rather than mass, and energy was invisible—which Winrod defined as the "spiritual realm"—it was clear that spirit, not matter, was the fundamental basis of "reality." "Murky materialism," in Winrod's words, had failed. Ideas, which were based on human beings' unique God-given ability to reason abstractly, were the moving force of the universe. In a subsequent article, Winrod referred to our thoughts as "superphysical energies" and as "mental electrons."[53] Here was a bold declaration of philosophical idealism based on the "new physics"—just what Lenin had in mind when he warned fellow Bolsheviks about Mach's ideas more than a decade earlier.

But just as Price did not rest content to attack evolution on the scientific front, Winrod followed his review of the literature with a blatant political point. Not only was opposing evolution a matter of defending religion, he told readers, but "it is a patriotic duty also." According to the "wicked jungle doctrine," explained Winrod, morality is viewed as a mere convention; religion is a human-created phenomenon based on "hallucinations"; marriage was instituted by men to control women; and property rights were created by those who had property to hold on to it. Social evolutionary ideas constituted a threat to capitalist civilization. They spawn "anarchy and extreme radicalism" and were allied with the "defenders of the Red Flag." Summing up, Winrod echoed Bouck White's explosive metaphor. Evolution, he wrote, "is intellectual T.N.T."[54]

Within the next year, as Winrod began to publish the *Defender*, his association of the moral and political consequences of evolution with the Bolshevik Revolution became more explicit and extensive. "Evolution Wrecks Youth" paraded a series of horror stories drawn from "Satan's cess pool" illustrating how teaching evolution in public schools created a "psychology of lust." He told tales of high school students participating in alcohol-soaked orgies ("whoopee parties") and male college students fighting over women at fraternity fests. In this early article, Russia came into the picture only as a joke. Students at the University of Kansas saw what they believed to be the work of student left-wing radicals—the Red Flag flying from an ROTC flagpole—which turned out to be a pair of "flaming bloomers."[55]

One year later when Winrod reported with alarm on the radical student newspaper at Kansas called the *Dove*, the Bolshevik Revolution assumed a more prominent place. According to Winrod, the paper had mocked Christian ethics and the Defenders. In response, one Major M. A. Palen of the American Legion had pointed to the *Dove* as evidence of communism on campus and demanded the university ban it. Winrod supported Palen's demand and offered the supporting evidence, apparently with a straight face, that "the magazine always appears printed on bright red paper." Winrod was surely serious when he informed readers that "the principle of Evolution applied in the national life of Russia, poisoning its heart-springs, is largely responsible for the present chaotic condition existing there."[56] The title of Winrod's article, "'Russia' in Universities," was doubly significant. It conveyed how he factually connected "Red Russia" to evolution. But the quotation marks around Russia and suggested how the Bolshevik Revolution had come to symbolize, for Winrod, a range of evolution-caused evils.

The fuller story on Russia and evolution emerged in Winrod's October 1926 *Defender* article titled "Russia's Mistake." Winrod opened by posing the question of whether or not it was "safe" to regard fellow humans as animals. The consequences of following the "beast" theory were the familiar litany of sexual perversion, crime, and moral degradation. Giving the example of a young woman who was learning about Haeckel's theory of recapitulation at a "certain state college," Winrod argued that the claim was factually false and was morally damaging to students. If such material had to be taught, students should be separated by sex. Simply by learning about the similarities between the physiology and embryology of humans and other animals in "mixed" company was dangerous. To these "fruits" of evolution, Winrod added one other that would later become central to the antievolutionist case: abortion. "To take the life of the unborn, according to the beast doctrine," Winrod wrote, "would be simply to destroy a forming animal."

Moving to the international stage, Winrod pointed to evolution's responsibility for war, immorality, and anarchy. It was, he said, "the devil's wrecking crew to the twentieth century." The signal case was revolutionary Russia. The Bolshevik masters had abolished organized religion, destroyed the churches, and made it illegal to teach children religion. As the "jungle consciousness" spread through the masses, marriage was widely spurned in favor of "free love." Men and women were pursuing sex with a single-minded focus and forgetting about their children. The result was that "335,000 waifs" were running in the streets of Russia like "little animals," riding the rails, and sleeping in abandoned freight cars. They were "victims of Evolution."[57]

As further evidence of the disastrous effects of evolutionary thinking, Winrod quoted Clarence Darrow as favoring liberalized divorce laws; a psychology textbook that discussed "animal tendencies" in humans; and a University of Chicago professor who favored eugenics. To wind up his tour of evolutionist horrors, Winrod once more brought Russia back into the picture, for he was aware of the widely reported Bolshevik-funded project to interbreed humans with apes in French West Africa. He accurately cited Charles Smith of the AAAA as his source for the information that the Pasteur Institute would house the experiments, with $10,000 contributed by the Russian government. Having now established that evolutionists were funded by communists to breed apes with humans, Winrod declared that "the great middle classes of people want nothing to do with such a putrid system of thought." Ever the populist, Winrod charged that an elite of evolutionists in the US were prevailing over the sensible "masses of American people."[58]

Over the next five years, Winrod popularized the dangers of evolution in booklets such as *Red Horse*, as well as *3 Modern Evils: Modernism, Atheism, Bolshevism* (1932).[59] To help readers grasp the Russian connection, *Red Horse* provided a political cartoon. It pictured Russia—personified as a bearded, barefoot peasant—standing on one side of a wall representing "U.S. Refusal to Deal with the Soviet Government." In the distance, beyond the wall, stood an American schoolhouse emanating, like the rays of the sun, a panoply of evil "communist" notions. They included evolution, animalism, disregard for God and the Bible, atheism, and immorality. As if to emphasize its importance, the word "evolution" appeared twice, once with and once without exclamation points. "'Taint Right!" complained the figure representing Russia. "Their government refuses to recognize me, but their schools steal all my ideas!"[60]

Like Riley, Winrod also kept his eyes on China. In 1927, Dr. Henry G. C. Hallock, an American missionary based in Shanghai, wrote to an American friend and Winrod supporter in Burton, Kansas, about the dramatic events there. Conservative warlord forces holding the city had just been defeated by an armed uprising of workers led by the Chinese Communist Party and the left wing of the Kuomintang Nationalists, opposed to the right-wing faction led by Chiang Kai-shek. It was the greatest victory up to that point in the long Chinese civil war, which would last, in fits and starts, another twenty-two years. "Wild Joy in Moscow on Fall of Shanghai," read the headline in the *New York Times* on the morning Hallock wrote his letter. "Tens of Thousands March to the Comintern Offices, Singing and Cheering in Streets," added the subhead.[61] But Hallock was not cheering for the Nationalists. Led by the Communists, they had deceived the gullible Chinese masses with cries of "Down with imperialists! Give the people freedom!"

Hallock expressed this idea with an animalistic analogy. The people, Hallock wrote, were "like a flock of sheep." They are pursued by "mad—dogs or wolves—by men in the pay of Bolshevists." And if these agitators, these "beasts of men," show too much concern for the hardships of the people, the Bolshevists just find others to do the dirty work of stirring up the workers to go on strike and terrorize those who refuse to do so. As they yell out their false slogans, the Chinese Bolshevists are roaming the country "like fierce, wild animals."[62] In Winrod's terms, they bore the evolutionary mark of the beast. In a subsequent letter to the *Defender*, Hallock was more explicit about the thinking of the "terrible Bolshevists." In his view, they were "just Modernists gone to seed."[63] If Winrod's readers needed it, here was yet another testimony to the poisonous effects of the beast doctrine.

Gerald Winrod, Frank Norris, and William Bell Riley were the most prominent members of the cohort of antievolution activists who led the charge during the 1920s and made a clear connection between the dual threats of evolution and socialism/communism. A lesser known but influential member of this group was Mordecai F. Ham Jr. (1877–1961), best known for winning a young Billy Graham for Christ at a Charlotte, North Carolina, revival in 1934.[64] Born in rural Kentucky, Ham was the son and grandson of Baptist ministers. He preached widely throughout the American South, especially in Texas, and reached an even wider congregation through his weekly radio sermon from his eventual base in Louisville, Kentucky.[65]

Despite living nearly all his life in the South, Ham spent a brief sojourn in Chicago, from 1896 to 1900, working in sales. Then came an offer to join an acting company—his good looks and deep voice were useful in both sales and the theater. According to one account, "his father intervened to save him from such satanic influences." Just as George Mc-Cready Price discovered, it was difficult to avoid the snares of Satan in the big city. Perhaps it was his struggle with those temptations that led Ham to become an uncompromising moralist. He returned home from Chicago, spent some time in prayer and contemplation, and soon hit the revival road. In the next two decades, Ham became known for his hell-fire preaching and soul-winning ability.[66] By the 1920s, Ham also became known for an intense anti-Catholicism, which he shared with Norris, and for a deep belief, which Riley and Winrod would also soon adopt, in a satanic international Jewish conspiracy.[67]

Having spent much of his energies preaching against the demon rum over the previous two decades, Ham joined the campaign against evolution. From 1921 through 1924 he crisscrossed the state of North Carolina in service of that cause. While activists failed to secure passage of an antievolution bill the next year, Ham did reach thousands of North Carolinians with his fiery message. He also won the support of Governor Cameron Morrison, a conservative Presbyterian who attended a number of Ham's revival sermons. In January 1924, in a move of questionable legality, Morrison blocked the adoption of two biology textbooks for North Carolina public schools on the grounds that they taught that humans were descended from a "monkey." The illustrations of one of these drew special attention from the governor, who commented that "I don't want my daughter or any body's daughter to have to study a book that prints pictures of a monkey and a man on the same page."[68]

The revival sermons Ham subsequently delivered in Raleigh were no-holds-barred attacks on modernism. In response to the contention that humanity could work together to improve the world—the preeminent message of the Social Gospel—Ham took the position that all such efforts were pointless. There was no hope, except regeneration through Christ. "Show me one word in the New Testament that exhorts Christians to make the world a better place to live in," he dared his audience, "and you may hang me to a telephone pole."[69] In another revival session, Ham again offered to submit himself to hanging, this time if anyone could identify a single fact that supported the theory of evolution. He blamed German rationalism for modernism, evolution, and worldliness, and he offered an appropriately violent Old Testament solution to the problem. If only the prophet Elijah had arrived in Germany thirty-five years earlier, lamented Ham, and "cut the heads off two thousand college professors"—who were presumably tribunes of modernism—that would have saved the lives of thirty-five million, the casualties of the Great War, caused by evolutionary thinking. He also used the story of Elijah challenging apostate Israelites who had gone back to worshipping Baal. Ham pointed to the false Jazz Age gods that modernists were now supposedly worshipping when things turned against them. "Call upon your picture shows," mocked Ham. "Call upon your boxing matches, call upon your Venus, your Charlie Chaplin, call upon your evolution, call upon your modernist who makes my Christ the illegitimate son of an impure woman."[70]

Finally, Ham addressed Russia. His sermon featured the story of Jehoshaphat, who nearly lost his life in a battle after allying with Ahab, who was under the influence of false prophets. For fundamentalists, the modernists were just such deceivers, and Ham reminded his audience that the devil prefers a "Christian to pull off his devilment." The stakes were nothing less than the souls of Christian children "who will be damned by the thousands" if good Christian parents did not take responsibility. The cautionary tale was Russia, "swept by bolshevism," and even the United States, to which the "disruption" had spread. In a signature line often repeated at the end of his evolution sermon, Ham warned, "The day is not far distant when you will be in the grip of the Red Terror and your children will be taught free love by that damnable theory of evolution."[71] Evolution appears as a satanic force that breeds sexual immorality, violence, and political oppression.

Not all North Carolina residents shared Governor Morrison's enthusiasm for Mordecai Ham. After the *Raleigh News and Observer* criticized Ham for calling modernists "damned infidels"—a violation of the Third Commandment not to take the Lord's name in vain—Ham's choir director lashed out at *News and Observer* editor Josephus Daniels for having misquoted what Ham had actually said: "damnable infidels."[72] But this spat paled in comparison with the war of words that Ham sparked during his visit that November to the small coastal town of Elizabeth City. There he butted heads with W. O. Saunders, the modernist editor of the local *Independent*, who took an unusually dim view of Ham's religion and politics.[73]

Ham's Elizabeth City revival sermons included the usual fare on sin and salvation, modernism and fundamentalism that he had served up in Raleigh earlier that year. But this time he added an ingredient that increasingly became part of the fundamentalist and creationist arsenal in the 1920s and 1930s: the Jewish conspiracy, which he tied to both evolution and communism. In a series of sermons in Elizabeth City—publicly rebutted by Saunders—Ham targeted the Jewish Chicago-based philanthropist and owner of Sears, Roebuck and Company, Julius Rosenwald. In Chicago, Ham charged, Rosenwald used his money and influence to promote prostitution, gambling, and the morally degenerate music of jazz nightclubs. Rosenwald had served on the Chicago Vice Commission, which turned a light on the racially segregated "vice" districts of the city. Though they recommended that the city act against all such activity, Ham led his listeners to believe that Rosenwald was in favor of racially integrating the vice industry. Houses of prostitution serving Black men would employ white women, thus raising the explosive prospect of "social equality."

Regardless of the details Ham cited about Rosenwald's activities in Chicago, his main point was that Rosenwald was part of a vast conspiracy, documented in the *Protocols of the Elders of Zion*, to corrupt and destroy the Christian world. The original version of the *Protocols* was drafted by ultra-nationalist journalist Pavel Krushevan, from the Russian province of Bessarabia (now Moldova). His Jew-hating articles that spread rumors of the Jewish blood libel helped spark the murderous 1903 Kishinev, Bessarabia, pogrom.[74] Morphing into a tool employed by czarist police agents in the years approaching the 1905 revolution, the *Protocols* defended czarism by discrediting the revolutionary movement as essentially Jewish. The document's authors observed that Jews played a disproportionate part in Russian revolutionary movements, and in various sectors of the economy. To explain these undeniable facts, they fabricated a story about a secret world council of wealthy Jews—the Elders of Zion—who were quietly and invisibly manipulating world events. The heart of their strategy was to create disorder, demoralization, and desperation—"chaos"—which would render the world's peoples willing to yield control to a dictator supplied by the Jewish elders. One reason that journalists and scholars long ago concluded that the *Protocols* were a forgery—that the Elders of Zion did not exist—is that much of book is copied word for word from a novel written in the mid-nineteenth century by French political satirist Maurice Joly. Published in 1864, *The Dialogue in Hell between Machiavelli and Montesquieu* is a fictional conversation between the two diabolical plotters, which Joly intended as a veiled political attack on the Emperor Napoleon III. The true origin of the *Protocols* explains its literally and figuratively Machiavellian language.[75]

In explaining to the attendees at his revival tent meetings the cosmic stakes in the Rosenwald affair, Ham leaned heavily on Henry Ford. The auto industry mogul was the most prominent promoter of the *Protocols* in the form of a four-volume group of pamphlets called *The International Jew*, selected from articles run in Ford's company-funded newspaper, the *Dearborn Independent*, in 1920–21.[76] According to *The International Jew*—which quoted from the *Protocols* extensively—a "cabal" of wealthy Jewish bankers was behind the plan Ford's editors called the "Jewish World Program." In Ford's eyes, not only was Julius Rosenwald a typical member of this evil group, but he had sinned further by financing an ex-posé of Ford's anti-Semitic writings. The author who unmasked Ford was John Spargo, a former Socialist leader and devotee of the Darwin-Spencer synthesis. Spargo skewered Ford and opened his book by reprinting a statement against anti-Semitism signed by a long list of notables "of Gen-tile birth and Christian faith," including former presidents Woodrow Wil-son and William Howard Taft.[77]

Readers of *The International Jew* learn that the Jewish plotters have in-troduced ideas to weaken and eventually enslave the "Gentiles." In Proto-col 2, quoted in Ford's "'Jewish' Estimate of Gentile Human Nature," the Jewish masters brag about the "successes we have arranged in Darwin-ism, Marxism, and Nietzscheism" and the "demoralizing effect of these doctrines upon the minds of the Gentiles." Moreover, they explain, we will use our "Press" to generate "blind confidence in those theories."[78] In "Does Jewish Power Control the World Press?" the editors underlined the importance of these systems of thought, which they called "the three most revolutionary theories in the physical, economic, and moral realms."[79]

Ford and his associates followed the *Protocols* closely in identifying Jews as the source both of economic oppression from above and proletar-ian revolution from below. But the Bolsheviks took center stage. Not only did *The International Jew* present a table listing the percentage of the staff of Soviet government agencies who were Jewish, in order to demonstrate the numerical superiority of Jews in the Bolshevik Party. It identified ex-actly what was most alarming about the "Jewish" Bolshevik Revolution. It was not state control of industry, but rather Jewish control of educa-tion, whereby they introduced "sex knowledge" to the minds of Gentile children in order to corrupt and weaken them. In other words, "It is the downright dirty immorality, the brutish nastiness of it all; and the line which the brutish nastiness draws between Jew and Gentile."[80]

While not drawing in Darwin here explicitly, the image of the "brute" plays a role similar to that of the "beast" in Riley's, Norris's, and Win-rod's accounts. The Bolsheviks were reprehensible because they acted like animals. Sex, animalistic imagery, and children stand at the center of *The International Jew*'s claims about Jewish influence on American popular music. Bolstering their claims with the fact that Jews were dis-proportionately prominent in the music industry, the editors described "Jewish Jazz" as "Monkey talk, jungle squeals, grunts and squeaks and gasps suggestive of cave love."[81] Their use of "monkey" created strong evolutionary overtones.

In his sermons denouncing Julius Rosenwald, Ham drew freely on the *Protocols* tradition that Ford had amplified for millions of Americans. According to Ham, Satan was behind a shadowy, secret organization with a "tremendous banking connection" that acted through the Soviet government. His ultimate goal was to overthrow Christian civilization and install the rule of the Antichrist. Dutiful tools of the Jewish conspiracy and ultimately of Satan, the Soviets were seeking to undermine America with a revolution that would equalize all wealth. To "demoralize" Americans and make that revolution possible, Moscow was promoting the liquor industry, prostitution, "corrupt literature," dancing, jazz, Charlie Chaplin movies, universal suffrage (regardless of race, sex, or "intelligence"), and the "false philosophy" of evolution. Posing as "friends of the negro," the communist plotters aimed to draw African Americans away from the church and toward their own immoral communist schemes. They would stop at nothing to attain their goals. Moscow had assassinated President Harding and was infiltrating workers and farmers to bring down the American economy. The "modernists" in American churches were collaborating with the conspiracy, as shown by the support given by the pro-evolutionary "Bad Bishop" Brown to the Communist Party. "Darwinism is not a science," declared Ham, "it is a propaganda."[82] Since that propaganda formed a key part of the diabolical Jewish-communist conspiracy, it had to be stopped.

Consistent with the conspiratorial connections that Riley, Norris, Winrod, and Ham had made by 1924 between evolution and communism, it comes as no surprise that the following summer, Rev. Timothy Walton Callaway chose to share damning facts "not generally known to the public" about the socialist-tainted background of defendant John Scopes.[83] Callaway's own background went back to Americus, Georgia, where his parents were Sunday school teachers at the First Baptist Church. After a brief career working for the railroad and as a local grocer, young Callaway felt the pull of the ministry, and soon was ordained as a Baptist preacher. Several years later in Macon, a growing regional market and manufacturing center, he founded the Baptist Tabernacle in a working-class section of the city to minister to the needs of newly arrived migrants from the Georgia

FATHER OF SCOPES
RENOUNCED CHURCH

Was Staunch Socialist and Follower of Debs.

Born in England and in Early Life Held to the Presbyterian Faith.

BY THE REV. T. W. CALLAWAY,
(Pastor, Baptist Tabernacle, Chattanooga.)
Chattanooga Times Special.

DAYTON, July 9. — The arrival in Dayton of Thomas Scopes, the father of our young professor, naturally causes interest to center around the early environment of the defendant in the evolution trial.

It is not generally known to the public that the father of young Scopes was born and reared to manhood in England—afterwards coming to America, where the professor was born whose name has now become famous. It is to be sincerely hoped that the glory thus suddenly thrust upon the young man will not soon be forgotten, nor will he be left alone in his new-found glory, unsung and unhonored.

An acquaintance of the Scopes family for seventeen years in Paducah, Ky., is authority for the statement that the elder Scopes was a machinist as far back as 1904 in the Illinois Central shops in Paducah. In the earlier years he was a professed Christian and a member of the Presbyterian church. Later he renounced his religion, became a stanch socialist and a faithful follower of Eugene Debs. It is claimed that Thomas Scopes was a man of strong personality, a fluent speaker, and could talk long and loud against the political and religious systems of America. It is also stated that he had quite an influence among his fellow workmen

Figure 7. "Father of Scopes Renounced Church," 1925. Rev. T.W. Callaway's July 10, 1925, article in the *Chattanooga Times* red-baits defendant John Scopes as the son of a dangerous Socialist on the opening day of Scopes's trial for teaching evolution in Dayton, Tennessee.

countryside. Perhaps this experience sensitized Callaway to the potential pull of socialist politics. By 1920, he had moved to take up the pastorate at the Baptist Tabernacle in Chattanooga, the city where he spent much of the rest of his ministerial career.[84]

Tarring Dayton mine manager and ACLU member George Rappleyea with the red brush, Callaway posed as a source of objective information for "the uninformed well-meaning public." Quoting from R.M. Whitney's *Reds in America* and the report of the New York State Lusk Committee that investigated labor radicals, Callaway argued that the ACLU was linked with communism through "interlocking directorates" with the aim of "penetrating" a variety of organizations to prepare for a violent general strike to overthrow the government. The ACLU supported "all subversive movements" and protected criminals. ACLU chairman Harry F. Ward, a Methodist minister, cooperated with the Socialist Party, the Industrial Workers of the World, and other "anti-American" movements.[85] Since Callaway viewed the Scopeses and Rappleyea as part of the Communist conspiracy, he might not have been surprised a few weeks later to see John Scopes's "special" in the *Daily Worker*. Certainly there were other fundamentalists who noticed the attention that left-wingers were giving to the trial. As Indiana-born and New York–based fundamentalist John Roach Straton commented, "A large group of outside agnostics, atheists, Unitarian preachers, skeptical scientists and *political revolutionaries* . . . swarmed to Dayton."[86]

Callaway's fundamentalist political activism went beyond the Scopes trial. In *Romanism vs. Americanism* (1923), he had already warned about the threat to the separation of church and state posed by the "religiopolitico organization" of the Catholic Church.[87] In 1926, Callaway cosponsored a resolution aiming to place the Southern Baptist Convention on record as opposing any theory that "declares or implies that man has evolved to the present state from some lower form of life."[88] In 1934, when the first graduating class walked across the stage at fundamentalist William Jennings Bryan University in Dayton, Tennessee, Callaway gave the commencement address.[89] Thus did the red-baiting of John Scopes become tied to a pioneer institution of American creationism and fundamentalism.

For all the connections that antievolutionists made between evolution and politics, they continued to insist that "true" science offered the most effective refutation of Darwin's ideas. In September 1925, just months after the Scopes trial ended, John Roach Straton paid tribute to the latest attempts by scientists of Christian faith, including George McCready Price, to refute evolutionary arguments. He was especially impressed with *The Case against Evolution* (1925). It was unusual in two respects: its publisher, Macmillan, was a major trade press, and its author, George Barry O'Toole, was Catholic. Straton thought that O'Toole's "logical grasp of the subject" and use of "scientific nomenclature" enabled him to strike a "deadly blow" to evolutionists.[90] O'Toole did review the scientific shortcomings of evolution. But he ended the book by arguing that evolution was dangerous to people of faith because it stirred social and political revolt.

George Barry O'Toole (1886–1952) was born in an Irish working-class neighborhood of Toledo, Ohio. He attended a local Jesuit college and then spent from 1906 to 1912 in Rome, earning his PhD and STD (doctor of sacred theology) from the Urban University. After doing a stint as an army chaplain during World War I, O'Toole took a position teaching theology and philosophy at the Benedictine Saint Vincent Archabbey in Latrobe, Pennsylvania. Having taken a few science courses at Columbia University, O'Toole was hired to teach an animal biology course at nearby Seton Hill College, enabling him to describe himself as a "biology professor."[91] In an age that lacked for creationist PhD holders, this designation, however specious, lent legitimacy to the cause. Citing O'Toole's work in the late 1920s, Gerald Winrod referred to him as a "scientist of recognized standing."[92]

Initial responses to Darwin among Catholics had included some prominent voices favoring theistic evolution, most notably English biologist and Catholic convert St. George Jackson Mivart (1827–1900).[93] Mivart's *On the Genesis of Species* (1871), published the same year as Darwin's *Descent of Man*, defended the general idea of evolution—including the evolutionary origins of the human body—while arguing that Darwinian natural selection alone was incapable of explaining natural history. A God-infused process of "individuation" was necessary to explain embryological development, as well as the presence of the distinctly human soul. Mivart believed that a God-directed evolution was "perfectly consistent with strictest and most orthodox Christian theology."[94] Illustrating the relatively noncontroversial nature of Mivart's conclusions for the Catholic hierarchy, the "liberal" Pope Pius IX granted Mivart a pontifical doctorate in 1876.[95]

Many American Catholics agreed. Mivart's American counterpart was Notre Dame's John Zahm, a physics professor and priest in the Brotherhood of the Holy Cross. Zahm championed a theistic non-Darwinist evolutionism in his *Evolution and Dogma* (1896). He was initially tolerated by Rome and even secured a private audience with Pope Leo XIII.[96] In the US, early responses to evolution suggested that Catholics did not feel threatened by Darwin. A number of prominent Catholic thinkers, including professor of theology John Gmeiner, claimed none other than Saint Augustine as an early exponent of evolutionary ideas. They found ways to reconcile their theistic evolutionism with a revival of Aristotelian neo-Scholasticism, promoted by Leo XIII starting in 1879, with its distinctive conceptions of the unity of form and matter. They also benefited from the absence of any papal pronouncements against evolutionary teachings.[97]

But the toleration of the likes of John Zahm did not last. Catholic pro-evolutionists soon became casualties of the Roman hierarchy's campaign against "modernism." The Roman Congregation of the Index banned Zahm's book in 1897. In 1899, Rome excommunicated Mivart for his increasingly heterodox views, particularly his rejection of the traditional view of hell. Perhaps the clearest statement from Rome of the antimodernist impulse was *Pascendi Dominici Gregis* (Feeding the Lord's Flock) issued by incoming conservative Pope Pius X in 1907. Pointing to the satanically inspired "enemies" of the church, Pius focused on those who are "hid" within the "very bosom and heart" of Catholicism, members of the laity and especially the priesthood who were spreading "poisonous doctrines." Because they attacked from within, these crafty characters were the most "pernicious" of all the church's enemies. Using a favored analogy from Jesus's Sermon on the Mount, the pope warned that such foes had struck at the root of Catholic truth and were spreading the "poison through the whole tree." *Pascendi* did not specify biological evolution as one of these poisons, but it did attack the concept that Catholic doctrine itself could "evolve."[98]

The Catholic campaign against modernism also took aim at socialism. Chief among the ideological weapons employed by Catholic clergy were papal pronouncements beginning with Leo XIII's encyclical *Rerum Novarum* of May 1891, which was subtitled "On the Condition of Labor." The pope endorsed labor unions but also upheld the sanctity of private property. Furthermore, the pope instructed that "the main tenet of socialism, community of goods, must be utterly rejected, since it only injures those whom it would seem meant to benefit, is directly contrary to the natural rights of mankind, and would introduce confusion and disorder into the commonweal." His successor Pius X issued similar declarations in 1903 and then again in 1912.[99] American Catholic lay activists took those declarations and ran with them. Among the most effective were Bostonians David Goldstein and Martha Moore Avery, both Catholic converts and former Socialist Party leaders, who hounded Socialist Party activists in the Bay State for decades. Their coauthored antisocialist diatribe, *Socialism: The Nation of Fatherless Children* (1903), flayed socialists for promoting "free love" and other forms of immorality. Boston's Cardinal William O'Connell, a militant antisocialist and strong ally of Rome's antimodernist campaign, gave the book his imprimatur.[100]

In *The Case against Evolution*, O'Toole spent nearly the entire length of his book reviewing the evidence and arguments for biological evolution. He began by describing the "crisis" in evolutionary science, citing William Bateson (a mutation theorist) on the "death" of Darwinism. He then devoted a chapter each to comparative anatomy and fossil evidence. For the former, he made good use of comments from T. H. Morgan about mutations in *Drosophila melanogaster* and how a series of intergraded species might seem to provide proof of historical development, but could instead be nothing of the kind. For the latter, he generously showcased George McCready Price's critique of the geological column and concluded that the argument for evolution from paleontology was "simply a theoretical construction which presupposes evolution instead of proving it."[101]

O'Toole devoted the remainder of his biological discussion to the origins of life, the human soul, and the human body. Life, according to O'Toole, did not arise by spontaneous generation, but rather by divine action. But since the elements in organic creatures were already present in organic matter, explained O'Toole, this was not a supernatural act, nor was it a "miracle." Just as Winrod had described God as a physical force, O'Toole defined creation as a natural act of God. As for the human soul, O'Toole compared humans to animals with respect to sensation, instinct, and intelligence. He concluded that humans uniquely possessed the latter, defined as "the power of abstract thought," which "cannot be evolved from matter." Finally, O'Toole reviewed in some detail the considerable evidence from embryology, anthropology, and recently discovered fossils that suggested that humans and living apes shared a common ancestor. Not surprisingly, he concluded that the "connecting links between men and apes are found, on careful examination, to be illusory."[102]

One might be forgiven, at this point in O'Toole's book—96 percent of the way through—for thinking that it was the evaluation of such evidence that guided the author's conclusions. But the key to O'Toole's thinking—and the dimension of his work that joined him to Riley, Norris, Ham, and Winrod—resided in his twelve-page afterword. Here he took up the cudgel against those who argued for evolution on the grounds of "materialistic" and "metaphysical" "monism." These were partisans of evolution who were convinced of its truth because of its simplicity or based on the principle of Occam's razor. O'Toole objected to this logic on the grounds that "*simple* explanations are not necessarily *true* explanations." He also indicted such thinkers for adopting an "attitude" based on something other than "the actual results of research." And yet, O'Toole proceeded to do just this when he argued against evolution from the "standpoint of moral and sociological consequences." From this angle, O'Toole contended, the "gravest count against evolution is the seeming support which this theory has given to the monistic conception of an animalistic man." If man is a mere "brute," if free will does not exist, O'Toole continued, then there is no basis for morality, and all hell breaks loose in a "wake of destruction." In ordinary social life, this translated into "undermined convictions, blasted lives, crimes, misery, despair, and suicide."

There was also a political aspect. If there is no reward in heaven, then wrongs must be righted here on earth. It is time for the "proletariat" to take back the world from the "coupon-holding capitalists." In a nice anticipation of Orwell's *Animal Farm*, O'Toole used a porcine analogy. If we are just animals fighting for enjoyment, then "the starving swine must hurl their bloated brethren from the trough that the latter have heretofore reserved for themselves." This struggle portended the "disintegration" of civilized life. Such was already happening in the Russian revolutionary "reign of terror," added O'Toole, with the application of "Marxian Socialism," which is called scientific because "it is based on materialistic evolution." Summing up this analysis, O'Toole ended the book with an extended quotation from noted Italian Jesuit scholar and anti-Darwinist Giuseppe Tuccimei, who bemoaned the baleful effects of evolutionary thinking, its link with socialism and anarchy, and its growing acceptance among the "ignorant and turbulent masses."[103]

O'Toole was hardly the first antievolutionist to deploy both scientific-sounding as well as "fruitistic" arguments. But the degree of disproportion in space allotted to each argument was striking in *The Case against Evolution*, so much so that one reasonably wonders whether anyone bothered to read O'Toole's afterword. Reviews in Catholic journals do not mention it.[104] But in at least one very prominent review in the *Brooklyn Eagle*, which dealt with O'Toole's arguments about science, the editors created a text box, designed to draw readers' attention to O'Toole's political argument. Headlined "What Marxian Scientific Socialism Has Done," it quoted a passage from his afterword text regarding Marxism's relation to evolution, the Bolshevik and French Revolutions, "free love," and suicide. The review's strategy paralleled O'Toole's in terms of quantitative coverage, but in contrast with the cleric, the newspaper's editors made sure their readers got the political message.[105]

For all of their differences in style, region, age, and religious affiliation, Riley, Norris, Ham, O'Toole, and Winrod agreed to a remarkable degree. Evolution was not only scientifically false and in contradiction to the truth of Genesis. It was a mortal threat to the future of human civilized society. It undercut any basis for a reliable moral code and so encouraged human beings to act as if there were no consequences for their actions. This explained why women were now smoking, drinking, cutting their hair short, and divorcing their husbands; why crime seemed to be sweeping the country; and why the socialist and communist movements, with their rejection of private property, the traditional family, and established sexual mores, were gaining ground. With its open embrace of evolutionary science and sociology, its aggressive actions to liberate Soviet women, its rejection of the value of the church, and its promotion of social and political revolt, the Bolshevik Revolution seemed to epitomize the satanic mark of the beast.

4

THE WOLF PACK AND THE UPAS TREE

On August 22, 1934, striking truck drivers celebrated a historic victory in their fight to make Minneapolis a union town. Carrying out three militant walkouts that year, led by a dedicated core of local Trotskyists, Local 574 of the International Brotherhood of Teamsters (IBT) won recognition at the majority of the city's trucking companies. Along with the San Francisco general strike and the Toledo Auto-Lite strike, the Minneapolis Teamsters' battle paved the way for the creation of the Congress of Industrial Organizations (CIO) and industry-wide unions in auto and steel.[1] On Sunday, September 16, less than a month after workers voted in the Teamsters, William Bell Riley delivered a sermon to his Minneapolis First Baptist flock that portrayed the dark side of the strike. It was, for Riley, a menacing development that showed the evil confluence of communism, evolutionism, and the Jewish conspiracy.

Riley's animalistic theme was "The Russian Boll-Weevil—Bolshevism." He tied the strike to communism by quoting IBT president Daniel Tobin, who opposed the local Trotskyist leaders.[2] "The purpose of the Communists," Tobin had said, "is to overthrow American institutions." Not only had "hundreds of Communists" invaded Minneapolis to "irritate and agitate" during the Teamsters strikes, Riley explained, but they were criminals and noncitizens. Using another animalistic analogy, Riley added that these "agitators" were drawn to strikes in the same way that "green bottle flies" were attracted to a rotting corpse, where they lay their eggs and produce more of their own kind.

The spreading communist invasion, Riley told his flock, was part of a diabolical plan, outlined in the *Protocols of the Elders of Zion*. The agents carrying out the "Protocol plan" were the criminal and Jewish Bolshevik leaders. These "inhuman beasts," said Riley, were using "terrorism" to maintain control over the Russian people. In the United States, as specified in Protocol No. 2, they were spreading the ideas of Darwin, Marx, and Nietzsche. In the name of science, the Protocol plotters were deliberately confusing and demoralizing the American people, who would eventually call for a ruler to bring order from the communist-caused chaos. Finally, the Elders of Zion, prophesied by John in Revelation 13, would crown their "sovereign lord"—the Antichrist—as "king of the whole world."[3]

In the year following the Teamsters strikes, Riley would flesh out the horrors of the Protocol plan in sermons drawing on the work of his protégé Dan Gilbert, who would soon join Riley in his exposure of the CIO's industrial unionism as a satanic plot. But in his early works Gilbert explained how communist influence and evolutionary teachings at the nation's colleges were sending young people down a hellhole of sexual immorality, meaninglessness, and crime.[4] Riley also liked Gilbert's *Evolution: The Root of All Isms* (1935), which taught that evolution was the source of Nietzschean amorality, Marxian communism, determinism, atheism, and various "Free-Love-Isms."[5] Giving a new twist to the horticultural analogy, Gilbert dubbed evolution the "upas tree of atheist-communism," referring to the legendarily poisonous *Antiaris toxicaria* of Indonesia.[6] Whereas Nietzsche promoted an amorality based on the individual "superman," explained Gilbert, Marx pushed a collective proletarian morality based on lust, hate, class-consciousness, and brutality, or, as Gilbert put it, the ethics of the "wolf pack."[7]

The integrated web of evils represented by communism, evolution, free love, the hidden hand of the international Jewish conspiracy, and labor revolt—this was the target for Riley, Gilbert, and their fundamentalist

Figure 8. Rev. William Bell Riley, c. 1930. Known as the "Grand Old Man of Fundamentalism," Riley (1861–1947) led the fight to ban evolution in the 1920s, and by the 1930s attributed evolution to a Jewish-Communist conspiracy. Courtesy of Berntsen Library, University of Northwestern–St. Paul.

counterparts from the early 1930s through the end of World War II and beyond. They were joined by veterans of the fight, including J. Frank Norris and Gerald Winrod. For Riley and Winrod, anticommunism led to public support for Adolf Hitler and the Nazi Party. The decade also witnessed the arrival and amplification of new voices making connections between evolution, communism, and moral decline: Elizabeth Knauss, premillennialist and anticommunist, and Aimee Semple McPherson, the monumentally popular founder of the Los Angeles–based Church of the Foursquare Gospel.

The appointment of Adolf Hitler as *Reichskanzler* of Germany on January 30, 1933, and the inauguration of Franklin D. Roosevelt as president of the US on March 4 of the following year were two critical signs of the times that set off the 1930s from the previous decade. They signaled a new context in which the battle against evolution would be fought: economic depression, labor revolt, political polarization, and world war. William Bell Riley's earliest commentary on the Nazi regime came five months after Hitler took power. Riley knew about protests in the US against the Nazi regime but had little sympathy with them. The German state had a right to defend itself, and the main threat it faced was communism. As Riley wrote, "The question involved here is not a question of Jewry at all; it is a question of Communism, and nothing more." Riley hated hypocritical liberal preachers who called on the US government to recognize Soviet Russia and yet spoke at meetings denouncing Hitler. Millions perished in Stalin's Russia, and they said nothing. But "the moment a Jew-Communist in Germany had his store closed, they were tearing their hair," wrote Riley. To deepen the point that Jews embodied the communist threat, Riley claimed that when he debated evolution before student groups, "the most vicious atheists and the most intolerable Communists that I have met, have been Jews from Russia, and other countries."

In light of subsequent events, the last paragraph of Riley's unapologetic anti-Jewish and pro-Nazi article is worth quoting at length:

> Jewry, from the day that she crucified Jesus Christ until the present time, has given many occasions for her own rejection and for that opposition which she has politically pronounced persecution. Hear Hitler, who speaks from

first-hand knowledge: "The Jew is the cause and beneficiary of our slavery. The Jew has caused our misery, and today he lives on our troubles. That is the reason that as Nationalists, we are enemies of the Jew. He has ruined our race, rotted our morals, corrupted our traditions, and broken our power."[8]

The "rotting" of morals spoke powerfully to Riley's readers, who were familiar with the charge. It reinforced Riley's point that Jewish Communists had assailed him when he critiqued evolutionary science. The attribution to "Jews" of everything that had gone wrong provided a handy way to explain the real troubles facing the masses of German people. As suggested by Riley's invocation of the "Jew-Communist" store owner, the Jew as enemy was a remarkably shape-shifting character.

Riley gave more positive publicity to the "Vienna painter" in a January 1934 *Pilot* article. He began by flaying liberals for hypocrisy in the face of starvation, murder, and religious persecution in Soviet Russia, which he called the "beast of materialism." Playing on the historicist premillennial idea that the beast had made an earlier appearance in the French Revolution, Riley described the decline of morality in revolutionary France in familiar animalistic terms: "Citizens had to awaken in a moral pigsty, and recognize the swinish level to which they had come." Turning now to Germany, Riley explained that Hitler's "anti-semitism has some just basis." He cited an article that praised Hitler for reducing unemployment, restoring law and order, and promoting a unified nationalistic spirit. Riley even suggested that Hitler was a divine agent. "To me, at least," Riley wrote, "it was nothing short of help from on high that enabled him to snatch Germany from the very jaws of atheistic Communism." If he had to choose between Germany and Russia, Riley concluded, he would choose Hitler's Germany "a thousand-fold."[9]

By early 1934, no one attending Riley's sermons or reading the *Pilot* could doubt his support for Hitler. But he went further in *The Protocols and Communism* (1934), which tied the evildoings of Jew-Communists to a worldwide conspiracy.[10] Riley knew that the *Protocols* might be a forgery. A number of prominent fundamentalists—including William Jennings Bryan—had condemned it as such.[11] But his "concession" on this point—"Jews in general deny that their elders wrote it"—implied that such "elders" existed, thus underlining the document's authenticity.[12] Riley repeatedly quoted from the *Protocols* to prove Jews' evil intentions.

World War I, for example, was a Jewish plot to create "pandemonium and destruction." Jews also created the motion picture industry, which Riley identified as "the most vicious of all immoral, educational, and communistic influences." The Jewish-dominated Communist International ordered the Minneapolis and Toledo strikes, "with their brutal murders." And in regard to Jewish control of education, Riley repeated his point that Jews were particularly prone to "advocate the evolutionary hypothesis."[13] By teaching young people that they were animals, Jews contributed to the "rotting morals" evoked by Hitler.

It is only with this plot in mind that one can understand Riley's continuing indifference to the fate of Jews in Germany even in late 1938 after *Kristallnacht*. On November 9 and 10, Nazi gangs killed ninety-one Jews, beat hundreds, ransacked thousands of Jewish-owned shops and homes, destroyed more than one thousand synagogues, and arrested thirty thousand.[14] The details of the attacks were widely reported in the United States. A typical headline from the *Dallas Morning News* read, "Hysterical Nazis Wreck Thousands of Jewish Shops, Burn Synagogues in Wild Orgy of Looting and Terror; Policemen Refuse to Halt Organized Riots in Germany."

But on November 20, ten days after this news reached Americans, Riley gave yet another sermon minimizing the Nazi threat and blaming Jews. The main enemy, he reminded his flock, were the Communists ruling Russia. Their leaders included "Jews from New York City." The Nazis, to their credit, had stopped them. And the Communists were far more sinister than the Nazis—a "hundredfold" more. In another formulation, Riley compared the threat of "one" dictator in Germany with 541 in Russia.[15] In the course of four years, the differential threat level had confusingly gone down by a power of ten—from a thousand to one hundred—and then back up to over five hundred. But Riley's followers got the point. They were to shed no tears for the Jews of Germany.

There were other leading fundamentalists in America—J. Frank Norris among them—who did not share Riley's Jew hatred or who did not express it publicly.[16] But from his base in Wichita, Kansas, Gerald Winrod was marching in step with Riley on the subject of the *Protocols*, Jews, evolution, and the Nazi regime. Around the time that Riley began writing about the Elders of Zion, Winrod discovered the *Protocols*. In 1933,

Winrod claimed that a group of some three hundred men had gained control "of the gold of the nations," were working secretly "in the shadows," and aimed to create "chaos" so that they could overthrow the existing order. Winrod reported that the group of men "are wealthy Jews." Much as Riley had done, Winrod reprinted pro-Nazi news from Germany without comment. In April 1933, he drew from a press report indicating the Jewish shops were closed in Essen, Germany, with Nazi flags hoisted above their doors. Dr. Wilhelm Frick, Hitler's minister of the interior, explained that the Nazis would clear Jews out of universities, the production of film and literature, and the press. The "Semitic sensualists," Frick said, would be stopped from spreading their "nefarious international poison." And so Winrod's *Defender* story ended.[17]

A month later, Winrod reiterated his claims about the *Protocols* and his support for the *Führer*. Once again, Hitler's actions—his "alleged Jewish persecutions"—toward German Jews were justified. After all, the *Protocols* tell us, the "Semitic men of finance" were seeking to create "chaos" and also to "overthrow the morals of the Gentiles." In contrast to the incessant negative characterization of the Soviet leaders as animals, Winrod uses the same language to praise the Nazi leader. Hitler, Winrod wrote, was "like a wild beast that refuses to be controlled."[18] Along the same lines, Winrod justified violent, Nazi-like tactics in the United States. Several months after the Minneapolis Teamsters strikes, Winrod reported that vandals had attacked a Minneapolis bookstore that sold "sinister Red literature." They broke the plate glass, stole books, burned them, and ransacked the property. On a highway on the outskirts of the city, where only ashes remained from the burned books, signs read "FIRST WARNING TO COMMUNISTS" and "BURNED COMMUNIST LITERATURE." Winrod issued no disapproval of this "bestial" behavior in the service of a righteous cause.[19]

At the same time that Winrod was praising Hitler, he was also sizing up President Franklin Roosevelt. Like many other premillennialists, Winrod came to see Roosevelt as the Antichrist.[20] Winrod joined them in suspecting that the blue eagle of the National Recovery Administration was the "mark of the beast" of the book of Revelation. Unlike most of his counterparts, however, Winrod focused on the specifically Jewish elements of the threat. For Winrod, the essential background was the nefarious activity of the Illuminati, a secret, freethinking organization founded in 1776 by Adam Weishaupt, who was trained as a Jesuit priest but was born Jewish. As Winrod explained in *Adam Weishaupt: Human Devil* (1936), the Illuminati's Jewish origins explained its ability to cause revolutions, wars, and economic depressions, including the 1929 stock market crash. Both the French and Russian revolutions, according to Winrod, were organized by Jewish plotters as an "onslaught against Christianity and the moral and social systems." The "Jew-ocracy of Moscow" was implicated in both evolutionary immoralism and the "socialistic" New Deal.[21]

Winrod's historical investigation led him to an irresistible conclusion: Franklin Delano Roosevelt was secretly Jewish. William Dudley Pelley's pro-Nazi Silver Shirts had already been circulating the allegation of Roosevelt's Jewish ancestry for a number of years.[22] But Winrod did his part to make it stick. A few weeks before the 1936 elections, he announced the news in his tabloid, the *Revealer*. Featuring a graphic of the Roosevelt family tree, the front-page story led readers to believe that the president was descended from Dutch Jews named Rosenvelt. Roosevelt's Jewish "racial" origins explained his "natural bent toward radicalism" and made it clear why he had appointed left-wingers, many of them Jews, to positions in the New Deal administration. As Winrod summed it up, the New Deal was not just a political challenge for fundamentalists but a "biological" one.[23] In a broader sense, that "biological" challenge, from Winrod's standpoint, encompassed not only Jewish racial characteristics but also their communistic ability to spread the immoralism of evolutionary science. In fingering Roosevelt as a closet Jew, Winrod outdid Hitler. The Nazi regime made much of Roosevelt as "a tool of the Jewish world conspiracy," tying him to Freemasons, bankers, and other evil forces. But they apparently saw no need to identify him as Jewish.[24]

Where Winrod and Riley felt no compunction about openly praising Hitler and damning Jews, their allies were more circumspect. Most notable was Dan W. Gilbert (1911–1962), the West Coast journalist who made a major contribution to the communist–evolution–free love nexus during the Depression decade.[25] Though scarcely remembered today, he spoke with an unusually clear and persuasive voice to a large audience of conservative Christians. Born in 1911 in Oakdale, California, where his father, Amos Lawrence Gilbert, ran a successful farm equipment company,

Dan Gilbert attended the University of Nevada and then launched his career as a newspaper reporter and columnist.[26] In the mid-1930s, the young Gilbert joined forces with Riley, teaching at Northwestern, preaching at First Baptist, and writing for the *Pilot*. He became a contributing editor in 1939, served as the general secretary for the relatively quiescent World's Christian Fundamentals Association, and during World War II relocated to Washington, where he edited the *National Republic* and ran the Christian News Bureau. After the war, he was a prominent voice in the National Association of Evangelicals and worked as a radio evangelist until his death in 1962.[27]

While Gilbert wrote during the 1930s and '40s on a wide range of topics, his main contribution to fundamentalist literature focused on the connection between communism, evolution, and new views of sexuality. His main targets were V. F. Calverton (1900–1940) and Samuel Schmalhausen (1890–1964), two communist-minded intellectuals who wrote widely on sex. Born George Goetz into a working-class family in Baltimore, Calverton founded the *Modern Quarterly* in 1923, a highly influential left-wing non-party journal of literary, artistic, and political commentary.[28] A Jewish socialist ten years Calverton's senior, Samuel Daniel Schmalhausen had begun his career teaching English at DeWitt Clinton High School in New York City. The school's progressive-minded biology teachers included George Hunter, whose *Civic Biology* stood at the center of the 1925 Scopes trial.[29] Thanks to socialist writer Upton Sinclair, Schmalhausen gained national attention when he was fired by the city's school board for encouraging sedition in his students.[30] Following his dismissal, he pursued a strong interest in Freudian psychoanalysis, set up his own psychology institute, and began to write on the subject of sex and society. His irreverent and often comical writing style led commentators to describe him as the "Groucho Marx of the Left."[31]

Dan Gilbert's first foray into this political minefield came with *Crucifying Christ in Our Colleges* (1933). Drawing on his experience at the University of Nevada as well as stories told to him by a variety of pseudonymous student informants, Gilbert constructed the book as a series of cautionary tales about the moral and political dangers of the modern, secular, pro-evolutionary university. Each chapter focused on a different student who had been led astray by the "ubiquitous anti-Christ, anti-God, anti-Bible, anti-moral professor." The epigraph from Dr. Frederick P. Woellner, an associate professor of education at UCLA, identified the source of various evil "fads" and "isms" as the excessive presence of "Communist teachers" in secondary and postsecondary schools. "We should kick them out without argument or delay," said Woellner. Gilbert evidently agreed.

The victims of these modern Pied Pipers of moral ruin were various. "Agnes" studied evolutionary biology and the cosmic evolutionism of Herbert Spencer. She lost her moral compass, life lost all meaning, and she killed herself. "Evelyn" had been a good Christian girl, but at college she learned about evolution and began living a hedonistic life. She had once sung in her church choir, but now her "sweet soprano" was repurposed to the delight of "jazz maniacs." She dropped out of college and was never heard from again. "Lester" read Joseph LeConte on evolutionary geology and soon joined the socialists. He figured that that since there was no eternal reward, he would fight for justice in the here and now. But when he realized that the solar system was a product of evolution and that the sun would eventually run cold, his outlook became grim, and he sank deep into alcoholism. "Jean" was put off by the sexual immorality she encountered as a freshman. But once she read Engels, Bebel, and the sociologist Lester Ward, she renounced Christianity, "gloried in night dancing," went camping with her boyfriend, became pregnant, had a botched abortion, and now lingered on the brink of death.[32]

Gilbert's message was unmistakable: the teachings of evolution and Marxism were inevitably and dangerously intertwined. They were sweeping the nation's campuses. They undermined any basis for a meaningful moral code. And they led to disastrous consequences for the nation's young people. To illustrate the content of such teachings, Gilbert quoted from Marxist authorities, biological evolutionists, and social scientists with an evolutionary perspective. He also quoted unnamed professors whose classes he had suffered through at Nevada. One Marxist economist recited the words to the "Internationale" in class. A sociologist told students that America would eventually recognize that abortions were "morally and socially right" and would repeal laws against them just as the "enlightened regime" in Russia had done.[33] To support his claim that evolution-inspired, atheistic communism led to crime, Gilbert quoted George Barry O'Toole, who compared workers to "starving swine" who push aside their bloated "coupon-holding capitalist" brethren at the troughs they had once monopolized.

Crucifying Christ put Dan Gilbert on the map. The book was favorably reviewed by evangelical publications. Gilbert was invited to speak by a variety of Christian groups, who welcomed his explanation for the apparent moral decline of the nation. At a convention of Gideons, one attendee reported, Gilbert told his tales of woe in "frightening detail." After his talk ended with a hymn and prayer, he did a "brisk sale" of the book, which sold for one dollar.[34] Christian educators took note. One prominent reviewer thought Gilbert had shown "conclusively" how students' Christian faith was "often destroyed" by their college experience.[35]

In 1935, Gilbert followed up *Crucifying* with *Evolution: The Root of All Isms*. What his second book lacked in extended, detailed personal portraits of evolution's victims it made up for with a heavy emphasis on sex. The "upas-tree" of "atheist-communism," according to Gilbert, had grown from five interrelated, evolutionary roots: Nietzscheanism, Marxism, Determinism, and Free-Love-Isms. It was the latter that formed the figurative and literal centerpiece of the book. In the one-page table of contents, the chapter on "The Root of Free-Love-Isms" took up about 40 percent of the vertical space, due to the detail provided on its subsections. They included "Schmalhausen's Philosophy of Sex Promiscuity," "Freud's Advocacy of Free Sex Indulgence," "Free-Love Sociology," "[Bertrand] Russell's Doctrine of Sex Freedom for the Young," "Sexual Animalism (as indoctrinated by some Professors of Anthropology)," and "Briffault's Immoralism." Just as Gilbert introduced *Crucifying* with an epigraph about the danger of communist-minded professors, his second book began by quoting Giuseppe Tuccimei (no doubt, thanks again to O'Toole) on how the evolutionary philosophy had produced the consequences of "socialism and anarchy."[36]

Before elaborating on evolution's sexual sins, Gilbert placed the "beast" doctrine in the context of both Nietzsche's and Marx's support for Darwin's evolutionary concept along with the notion that "might makes right." To support his claim that Marxists embrace any means to the communist end, Gilbert quoted Lenin as follows: "The dictatorship of the proletariat is nothing else than power based upon force and limited by nothing—by no kind of law and by absolutely no rule." He also has Lenin saying that "all children should be present at the executions and should rejoice in the death of the enemies of the proletariat—Marx took savage delight in contemplating the bloody extinction of the proletariat's class enemies." Summarizing the Marxists on this question, Gilbert concludes that anything that helps to "hasten socialism" is right, no matter how reprehensible.[37]

Gilbert's desire to paint Marxists in the darkest colors led to loose standards of accuracy. In the first case, Gilbert misquoted Lenin and took him out of context. Lenin was quoting himself, addressing the Constitutional Democrats (Cadets) after the failed 1905 revolution. He actually said that any form of "dictatorship means unlimited power, based on force, not on law." The distorted version of the quote was circulating in the national press while Gilbert was writing his book.[38] In the second case, it is impossible to find any credible source for the quotation about the bloodthirsty Marx.[39] Other comments by Gilbert about Marx are simply inaccurate, such as "Marx's theory was that human nature is all bad." One original contribution, however, was Gilbert's ability to weave into his antievolutionary scheme Marx's concept of profit as surplus value extracted from the worker. Such an idea, wrote Gilbert, "appears to rest largely on the presumption that man is a soulless beast . . . who knows not the meaning of charity, kindness, and justice."[40]

In Gilbert's eyes, Marx's "wolf pack" ethics were detestable. But it was the "barnyard ethics" of "sex anarchy" that truly condemned the theory of man's beast ancestry. Gilbert began his examination with the writings of Schmalhausen, whom the author identified, with some exaggeration, as the most "popular and persistent champion" of the "New Morality." This Groucho Marx of the Left celebrated all that Gilbert found repulsive and immoral: sexual promiscuity, homosexuality, masturbation, and various unnamed forms of "perversion." As Gilbert correctly noted, Schmalhausen's ideas about sex were based on the evolutionary idea of humanity's "ancient animal history" and on the observation of our primate "monkey" cousins in their natural habitats. The resulting quotations from Schmalhausen were meant to shock Gilbert's readers: Nature "objects neither to incest nor homosexuality nor playful perversions"; "Promiscuity is in the nature of things"; "animals enjoy the practices of masturbation and homosexuality"; "we are born perverts." In contrast to traditional moralists who pointed to the behavior of animals in the wild as an example of how human beings should *not* behave, writes Gilbert, Schmalhausen turned things upside down and used the same observations to argue that such behavior is "NATURAL and PROPER" for human beings.[41]

From here Gilbert turned to the man who, he claimed, had done more to promote sexual immorality in the nation's youth than any other: Sigmund Freud. Quoting mainly from textbooks that drew on Freudian ideas, Gilbert claimed that Freud had caused an "incalculable" degree of "moral ruin." Freud's viewpoint was based on evolution, as expressed in the master's formulation in *A General Introduction to Psychoanalysis* (1917) that "man is a pleasure-seeking animal." Since sexual desire was akin to hunger that needed to be satisfied, the frustration of such desire led to pathological results. Giving free rein to sexual impulses was healthy. Support came from psychologist Daniel Bell Leary, who explained that human behavior was based on two drives, hunger for food and "the race-preservative sex hunger"; from psychologist W. B. Pillsbury, who wrote that sex was "one of the strong impulses of every *normal* individual" (Gilbert's emphasis added); and from Barbara Low, who described Freud's concept of the unconscious mind as consisting of "primitive inherited impulses and desires," which "remain indestructible." As Gilbert summed up Freudian thinking for his readers, "Fundamentally and NORMALLY man is a sex-mad, pleasure-seeking animal."[42]

It was not only psychologists who were making this dangerous argument but sociologists and anthropologists as well. Columbia's F. H. Giddings, Gilbert reminded readers, had publicly defended Russian novelist Maxim Gorky when he arrived in the US without his wife and in company with his lover.[43] It made sense to Gilbert that an evolutionary sociologist like Giddings would write, "The new sexual relations of the future will be promiscuity on a higher plane." Joining Giddings in Gilbert's pantheon was Friedrich Engels, whom Gilbert dubbed as "one of the greatest" proponents of evolutionary immorality. In *Origin of the Family*, Gilbert charged, Engels affirmed the "brute ancestry" of humanity seven times and frankly admitted that it was the basis of his "case for free-love." Though sociologist Lester Ward was no Marxist, Gilbert also lumped him with Engels as someone who "substantially endorses the Marxist doctrine that 'marriage is legalized prostitution.'"[44] And he took aim at anthropologist Robert Briffault, whose review of studies of apes in the wild led him to argue that chastity was unnatural. Gilbert was particularly alarmed by Briffault's application of his conclusions to women, based on the following observation of apes: "The sexual activity of females is as pronounced and as promiscuous as that of the males."[45]

While not every author cited by Gilbert was a Marxist, it would have been difficult for a casual reader to miss the evil thread running between evolutionism, free love, and communist doctrine. It is surprising, then, that in his discussion of Schmalhausen, Gilbert failed to mention the man's Marxist sympathies. But he corrected this omission in a series of articles he wrote for Riley's *Pilot* in the late 1930s. Gilbert told readers that the "Soviet system" aimed to eradicate morality by teaching "animalism" as both an intellectual philosophy and as a "way of life"—that is, as evolutionary science and sexual promiscuity. Gilbert featured a 1930 Schmalhausen essay on the family, which explained modern neuroses as the logical result of the repression of sexuality embodied in "home, sweet home" and promised that such problems would disappear under "a communistic form of society."[46] In a transformed environment, according to Schmalhausen, the sex urge would be expressed in "acts of social compassion and humanistic love." Based on this damning evidence, Gilbert warned readers that the professors who were crucifying Christ and undermining the family on the nation's campuses were "advancing the communistic cause."[47] In this way, Gilbert's *Pilot* articles reinforced the red connection to evolution that he had written about in 1935. Unlike *Crucifying Christ*, Gilbert's *Evolution: The Root of All Isms* appeared in only one edition, but it was still offered for sale at Northwestern Bible conferences as late as 1942, seven years after publication.[48] Schooled by William Bell Riley on evolutionary evils, Gilbert, in his relentless and detailed focus on evolution, communism, and sexual immorality, worked to magnify and extend the reach of his mentor's ideas.

Given the anxiety Dan Gilbert sought to generate by the stories of "Evelyn," "Agnes," and "Jean"—young, educated, openly sexual, independent-minded women—one might expect to find but few women in the leading ranks of American fundamentalism in the 1930s. Indeed, their presence had declined, as the result of what historian Margaret Bendroth calls the "growing masculine dominance of fundamentalism."[49] And yet a number of women evangelists and writers did make some of the same connections men made between evolution, communism, the Jewish conspiracy, and Satan. Among them was one of the few women who exercised leadership in the WCFA: premillennialist writer and lecturer Elizabeth Knauss (1885–?).[50] Hailing from Davenport, Iowa, Knauss was active in the local WCFA branch in the early twenties, heading up its Young People's Gospel team, which traveled around the region to provide music for fundamentalist revival preaching.[51] By 1928, Knauss was the general secretary of the Iowa WCFA, and she spoke widely around the country about the dangers of Bolshevism.[52]

In the spring of 1931, Knauss visited the anthracite coalfields of Scranton, Pennsylvania, to give a series of talks at local churches. Communism was on the minds of local residents as the party had considerable success recruiting local coal miners into an Unemployed League, under the effective leadership of recently arrived party leader Steve Nelson.[53] That same year, Knauss published her views in *The Menace of Bolshevism in America and throughout the World*, a twenty-four-page pamphlet that went through at least six editions.[54] Over the next two years, a series of articles by Knauss, drawing heavily from the pamphlet, appeared in Riley's *Pilot*.

Knauss began her study of Bolshevism some six years earlier when she first encountered the *Protocols*. Quoting Henry Ford, Knauss proclaimed her judgment that the document was "authentic." A group of "apostate Jews," wrote Knauss, were now carrying out their "diabolical" plan. The fact that a large majority of Bolshevik leaders were of Jewish origin but concealed this fact with pen names was circumstantial proof.[55] To any Christian who might question why the spread of Jewish-inspired Bolshevism was truly dangerous, Knauss sought to clarify that their main objectives were to destroy home, church, government, and schools. Her evidence included stories about the dire results of the Bolshevik Revolution for Russian women and children, various quotes from Bolshevik leaders purporting to show their monstrous ambitions, a description of moral decline in the United States, and an account of the effects of the spread of Bolshevism in China.

Relying on published newspaper and magazine reports from Russia, and quoting generously from them, Knauss shared a nightmare vision with her readers that would have been familiar to *Defender* subscribers. She retold the story brought to light by George McCready Price in *Socialism in the Test-Tube* about the new public schools organized by Soviet commissar of education Lunacharsky. Adding a detail Price had omitted, Knauss noted that boys and girls occupied the same dormitories at the state boarding school. Knauss also made it clear that the horror of Bolshevism was about both morality and power. These young Bolsheviks, complained an author quoted by Knauss, were "learning to shoot straight at the mark of full equality of the sexes," as part of their future leadership training.[56] In this 1930s culture war, gender relations and class struggle were intimately intertwined.

Implying that "full equality" was not a desirable goal, Knauss said nothing about the real measures that the young Soviet Union had taken to equalize conditions for the sexes. She did, however, give full rein to accounts of the "nationalization" of Soviet women. According to a letter written by a former Russian aristocrat, from which Knauss quoted, Red Army commanders in one town had carried out a "drive" to capture women for the troops. Soldiers raped them and then killed them and threw their bodies in the river. Official Soviet documents, claimed Knauss, corroborated these accounts and spoke of "permits" issued to commissars to communize the women. She also referred to "hair raising" stories that were too terrible to relate to a "mixed audience," and to another instance of a particularly "bestial type" of behavior.[57]

To reinforce the familiar animalistic idea that the Bolshevists were beasts without any sense of human values, Knauss also shared with readers quotations attributed to Lenin and Lunacharsky that emphasized the value of hating family and community members. From Lenin: "Children must be taught to hate their parents if they are not communists. If they are, then the child need not respect them, need no longer worry about them." And from Lunacharsky: "Christian love is an obstacle to the development of the Revolution. Down with the love of one's neighbour. What we need is hatred."[58] Once again, these quotations were almost certainly fabricated.[59] But they circulated widely. In making his case in Congress for nonrecognition of Soviet Russia in the early 1930s, Senator Arthur Raymond Robinson (R–IN) repeatedly quoted Lenin and Lunacharsky's fondness for hatred. In a pamphlet published in 1930 and timed to coincide with Pope Pius X's prayer for Russian Christians, Father Edmund A. Walsh, the vice president of Georgetown University, included the same Soviet "utterances."[60] Like the spurious Lenin quotation employed by Dan Gilbert, this "factual" evidence effectively portrayed communists as amoral devils.

Not only were Knauss's lurid description of Russian conditions akin to articles appearing in Winrod's publication, but she shared as well his alarm at events in China. Knauss relied on the reporting of Rev. Edgar E. Strother, a Moody Bible Institute–trained missionary based in Shanghai.[61]

According to Strother, Moscow had fooled Americans into thinking that the Chinese revolution was an indigenous movement, whereas the Russian Communists were actually pulling the strings. As an example of the bestial behavior practiced by communists, Knauss shared Strother's story of a "group of Bolshevized Chinese farmers who recently stoned to death a Chinese pastor at Yochow," claiming that he was in league with the imperialists. This incident showed that the "mad dog" of Bolshevism had to be stopped.[62]

Strother was useful for Knauss's case in one other respect: he was a devoted believer in the *Protocols*. He arranged for the publication of the first China edition of the book, in English, printed in Shanghai. As quoted by Knauss, Strother showed how not only communism but modernist religion was also a creation of the Jews. Just as William Bell Riley had invoked the culpability of "Jewry," quoting Hitler, for the Crucifixion of Christ, so did Strother specify that the Jewish *Protocol* plotters were "JEWS OF WHOM JUDAS ISCARIOT AND THE HYPOCRITICAL SCRIBES AND PHARISEES WERE A PROTOTYPE."

Describing Bolshevism as a form of rampaging "Satanism" loose in China and Russia, Knauss also called attention to its inroads into the life of Christians in the United States. In capital letters, she warned readers that "BOLSHEVISM IS HERE." Spread by "multitudes of agents," its effects could be seen in various signs of moral decline: the support of college students for "companionate marriage," the "near nudity" on America's beaches, the appearance of atheist youth movements, and the teaching of evolution, which ushers young people into years of "pernicious, soul-destroying instruction." Summing up the significance of the multifaceted Bolshevist menace, Knauss concluded that it reflected the "spirit of ANTI-CHRIST."[63]

The strategies that Elizabeth Knauss pursued in underlining the dangers of communism to Christians around the nation in the early 1930s were fundamentally similar to those employed by Riley and Winrod. All three accepted the claims of the *Protocols*; they identified Bolshevism with Satan and immoralism; and they connected evolution and communism. One notable difference in Knauss was the disproportionate amount of her text that consisted of quotations from male missionaries and ministers. As she was one of the few women leaders of the WCFA, her emphasis on women and children as victims of communism (and at least obliquely, evolution) may have stemmed in part from her need to establish a politically acceptable arena of operation. And yet, Knauss had expanded the audience for the anticommunist, antievolutionist message. She was determined to carry on her work to those "still in ignorance who need enlightenment."[64]

If Elizabeth Knauss stayed within a relatively traditional mold in delivering her message, Aimee Semple McPherson (1890–1944), her diehard anticommunist and antievolutionist counterpart on the West Coast, broke that mold.[65] Born Aimee Kennedy on a farm in Ontario, she met Irish-born Pentecostal minister Robert Semple at the age of seventeen and was swept up in enthusiasm for a Pentecostal variant of fundamentalism. Eventually landing in Los Angeles, she raised the money to erect the giant fifty-three-hundred-seat Angelus Temple in 1923, headquarters for McPherson's International Church of the Foursquare Gospel. Drawing on Pentecostal and holiness traditions that identified four aspects of Jesus— savior, healer, baptizer, and coming king—Sister Aimee, as she was known to millions, took an ecumenical approach, welcoming those from a wide variety of denominations into her church.[66] Dressed in white, McPherson became a celebrity preacher known for her "illustrated sermons," elaborately staged pageants with actors in costume, props, and her own dramatic preaching at its center.

Aimee Semple McPherson was an atypical fundamentalist. Not only was her church nondenominational and ecumenical—unlike those of Riley and Norris—she boasted of her faith-healing ability and spoke in tongues, at a time when Baptist fundamentalists like Norris were going overboard to pin the charge of Pentecostalism on their rivals. McPherson was also more open to the value of Social Gospel–inspired political action (and supported Franklin Roosevelt until her death in 1944). Most strikingly, she was a proto-feminist who pushed the boundaries of female "respectability." The Bible college she established in the 1920s ordained women ministers. She was a living example of female church leadership. In 1921, she filed to divorce her second husband, Harold McPherson, claiming cruelty and desertion. Then, in 1926, she disappeared for months, later claiming she had been kidnapped and taken to Mexico. The crime story may well have been a ruse to meet up with a lover.[67] Finally, in 1936, McPherson clarified her stance on gender roles in the church, calling for an end to discrimination against women and stating that "sex has nothing to do with the pulpit and pants don't make preachers."[68]

In other ways, however, she fit well into the fundamentalist fold. McPherson never dropped her militant opposition to evolution, she welcomed William Jennings Bryan into the pulpit at Angelus, and she shared his sense that Darwinism had borne evil social fruits. As the *New Yorker* described her views in 1927, evolution is "poisoning the minds of the children of the nation. It is responsible for jazz, bootleg booze, the crime wave, student suicides, Loeb and Leopold, and the peculiar behavior of the younger generation."[69] While McPherson made efforts to attract a multiracial congregation at Angelus, she was friendly with the California Ku Klux Klan, which helped pay for the construction of the gleaming edifice. And in 1938, shortly after Gerald Winrod ran in the Kansas Republican primaries for US Senate—being denounced as the "Jayhawk Nazi" by his political opponents—McPherson invited him to Los Angeles for a series of sermons. They had evangelized together, and *Defender* articles routinely appeared in Foursquare publications. While she bent to the outcry his first few appearances caused, and canceled the rest, his views by 1938 were well known. McPherson was ambivalent about the place of Jews in American political and spiritual life. She had given voice to common racist stereotypes of Jews as subversive and money hungry, even if at other times she warned against their persecution.[70]

The key to her willingness to invite Winrod, however, was most likely their shared antievolutionism and anticommunism. In early 1934, once again flouting the gender stereotype, Sister Aimee staged a traveling debate with Charles Lee Smith, the notorious atheist who had funded Ivanov's ape-human breeding research. With a variety of props, including a large cardboard cutout of a gorilla, McPherson attacked the weak points of evolutionary science and contended that Darwinistic ideas encouraged immoral behavior.[71] That this behavior could lead in communistic directions was evident in a cartoon on the front page of the *Foursquare Crusader*. In "Communism in Operation," a giant octopus stretched out on a map of Europe, with a stereotypically bearded and hatted Russian Communist head perched awkwardly on the center of its body. Its three main tentacles, each of which grasps a lighted torch, are labeled "Atheism," "Evolution," and "Red Propaganda," the latter stretching across the Atlantic Ocean and slithering its way through the national capitol building,

Figure 9. Aimee Semple McPherson, 1934. In a contrast to her gorilla-fighting image onstage, Sister Aimee strikes a more traditionally feminine motherly pose with a baby monkey at the Seattle, Washington, zoo, before holding a debate on evolution with atheist activist Charles Smith as part of a nationwide tour, January 1, 1934. Courtesy of Getty Images / Bettmann.

a schoolhouse, a home, and a church. "Nothing is sacred to this devouring monster," proclaims the caption. "Communism aims at the destruction of our entire Christian civilization."[72]

McPherson's message was familiar to legions of Christians as a result of the national tour that she had conducted in 1934, featuring her "illustrated sermon" titled "America, Awake!"[73] The tour culminated in Los Angeles with a rousing election-eve rally to mobilize California voters against the gubernatorial candidacy of socialist Upton Sinclair, whose End Poverty in California campaign had gained him the Democratic nomination.[74] Her address began with a providential vision of American history in which our God-inspired forebears, from the Puritans to the Founding Fathers to Abraham Lincoln, built the nation on the rock of religion. This foundation, however, was now threatened by atheistic communists, whom McPherson described as "subtle, powerful, relentless, [and] diabolic." Evoking the language of official Soviet atheism by labeling her enemies the "militantly Godless," she pinned responsibility for the spread of atheism among America's youth, as did Dan Gilbert, on the schools. Young people had been seduced into joining unbelieving groups such as "The Circle of the Godless" at the University of North Dakota and "The Legion of the Damned" at the University of Wisconsin.

Although she said little about evolution, she did make it clear that animalistic behavior followed logically from the erosion of religious faith. Take it away, she told her audiences, "and man becomes a beast!" Destroy the "moral code," the family and church, and young people become animals. "Denizens of the forest will soon be greeting new playmates!" she cried in alarm, though not without a sense of humor. No wonder that the rate of "juvenile delinquencies" was on the rise. "Darwin's theorizing on evolution," along with the writings of Thomas Paine, Voltaire, and American freethinker Robert G. Ingersoll, "have usurped the place of the Bible on the intellectual throne."

Then came the climax, in which the villains appeared onstage, while an unsuspecting Columbia sleeps on the US Capitol steps. "I see Satan, entering from the right, smirking and gloating over the recumbent form," declared McPherson. Satan then removed the cornerstone of those steps labeled "Faith." Next came a "grinning Bolshevik stealthily approaching, with his cap pulled low over his eyes," she announced. He removed the other cornerstone, labeled "home." In place of it, the Bolshevik replaced the "bomb of atheism." Red dynamite, indeed! After lighting the "fuse of subversion" and replacing the American flag with the red flag of revolution, the Bolshevik proclaims, "We are against God, we are against Capital, we are for a Socialist revolution." But then Miss Columbia awakens. She "turns into a veritable fury," writes McPherson. In the stage version, Sister Aimee herself walked onto the stage at that critical moment, removed the red flag, and replaced it with Old Glory. The audience erupted with applause. The republic was restored.[75]

Given McPherson's near-celebrity status, her ecumenical spirit, and the drama of her illustrated sermons, she may have reached more Americans, in a face-to-face setting, with the message that evolution and communism were allied evils than did Riley, Norris, or Winrod. Unlike Elizabeth Knauss, Sister Aimee did not pen long, detailed analyses of events in China or the Soviet Union. She took a more simple, direct, and popular approach that resonated with the large audiences she attracted around the country. Precisely because she was less conservative in certain respects than were her prominent fundamentalist counterparts, she may have reached a wider swath of the American public with her Red Dynamite message.

Just as McPherson was making her name out West, Texas-based J. Frank Norris was expanding his fundamentalist base northward. In 1934, having pastored Fort Worth First Baptist for twenty-five years, Norris made a bid for national fundamentalist leadership by moving to Detroit to take over leadership of Temple Baptist Church. Located in a working-class area of the city inhabited by thousands of workers at the city's giant automobile plants, Temple Baptist had a relatively small congregation of 800 when Norris arrived. By 1939, over 6,000 Detroiters had joined the church. By the mid-1940s, Norris's combined congregation numbered some 25,000.[76] Some 35,700 subscribed to his *Fundamentalist*, the successor to the *Searchlight*.[77] All in all, Norris was one of the most powerful ministers in America.

He also would become one of the most effective popularizers of anticommunism during the 1930s and '40s. In late May 1936, a letter arrived in Norris's office from a young woman living in Philadelphia, Mississippi. Ila Fleming, whose father subscribed to Norris's *Fundamentalist*, was heading off to boarding school and wanted her own weekly copy of the paper. Not only did Fleming enclose payment for a six-month subscription, but she took advantage of a free book offer and asked Norris to send her *Sovietizing America through Churches, College, and Consumer Co-operatives.* Other books offered for sale on the form she returned included *The Gospel of Dynamite* (1935), which contained twelve sermons. One of these was "World-Wide Sweep of Russian Bolshevism, and Its Relation to the Second Coming of Christ."[78]

Just as T. T. Shields had spoken of the need to blow up certain individuals and institutions with spiritual "dynamite," so did Norris employ this widely used metaphor. As if to justify its use in a religious context, Norris offered a quotation from the book of Romans on the book's title page, suggesting that the word itself appeared in the Bible: "The Gospel— is the Power (Dynamite) of God. (*Rom. 1:16*)." While "dynamite" does not appear in the King James Bible, the Greek root of the word—δύναμις (*dunamis*)—is the basis for a variety of terms that do, ranging from "power" to "mighty works" to "wonderful works" to "miracles."[79] Given the explosive growth in attendance at both his churches, members of his congregations evidently responded to the "dynamite" of his style and the sense that he was filled, and filling them, with the power of God.

Not only did Norris continue preaching in the style he had perfected in Fort Worth, but he also continued to tie together the threats of communism, evolutionism, and racial integration, a theme to which he gave increasing prominence. In "The World-Wide Sweep," delivered in 1931, Norris drew upon chapter 5 of the Epistle of James, in which James warns the rich exploiters of their laborers of the coming return and judgment of Jesus. Reviewing recent world events, Norris interpreted them in light of prophecy. Revolutions; lack of faith (what he called "the black night of atheism"); conflict between the rich and the poor ("a fight to the finish"); and the rise of an "iron dictator"—all of these were predicted by the Bible. Citing Gog and Magog from the book of Revelation, Norris told his congregation that these predictions were coming true in Russia, where a "casteless society" was developing. Soviet leaders were campaigning under the red flag of communism and the black flag of atheism. They were "sweeping" through Asia, where China was the most "fruitful field." Given the extremes of wealth and poverty, it was the most likely place where Sovietism could enable the "bottom rail" to replace the top, an analogy that Norris identified with his Texas roots. Not only were the Russians out to level the social classes, but they aimed to erase "all national lines, all racial lines, that there is to be no color lines, the negro has equality with the whites." As if this were not evil enough, the Bolsheviks were also spreading atheism, and one of the "points of attack" was America, as evidenced by the spread of the "philosophy of evolution." "There is no such thing as Theistic Evolution," he stated.[80] In a sense, Norris agreed with his communist opponents that there was no dilly-dallying middle ground—one must take sides. If they chose God, Norris assured his flock, then they would be ready for Christ's return.

Since Norris was telling his followers that their duty was to accept Christ as the only way to save themselves from the mess that humanity had made of the world, one might imagine that he chose to abstain from taking sides in the labor battles raging among working-class Americans. But Norris stood at the center of the war for the hearts and minds of Detroit autoworkers in the 1930s. And he was hardly the first nationally known cleric to do so. In the 1920s, a young Reinhold Niebuhr, fresh out of seminary, denounced Henry Ford's treatment of workers and publicly supported their right to organize unions.[81] During the Depression decade, the famous and infamous "radio priest" Father Charles Coughlin gained a wide following among those who toiled in the auto plants.[82] Like Coughlin, Norris claimed that he was interested in the spiritual and physical welfare of working people. He supported their right to engage in collective bargaining but nevertheless urged them to reject the Congress of Industrial Organizations (CIO) and its most prominent leader, John L. Lewis, as their mortal enemy. Norris not only preached this message at Temple Baptist, but he also delivered revival-style sermons at the Baptist Tabernacle, on downtown Detroit's Woodward Avenue. Each week, those sermons were then rebroadcast on radio station WJR, and then reprinted in the pages of the *Fundamentalist*.

Not for nothing did Zygmund Dobrzynski, the national director of organizing at Ford for the CIO-affiliated United Auto Workers (UAW), devote an entire article in the UAW newspaper to Norris, whom he accused of betraying the cause of the labor unionism in the guise of friendship for workers. Using a New Testament analogy, Dobrzynski likened Norris to Judas who "sold out the ONE who had led the oppressed peoples of those days in protest against human bondage." Contrary to his claims in favor of democracy, Norris was a "raving minister" using the "pulpit as a mask to promote dictatorship."[83]

In his sermons delivered in Detroit, Norris did not directly address the connection between evolution, communism, and the CIO. He did, however, direct his fire at modernist clergy, whom he identified in one sermon as the "hidden hand" behind the landmark Flint sit-down strike of 1937 that succeeded in unionizing General Motors. The guilty parties included Rev. Harry F. Ward, who had been identified back in 1925 by Chattanooga's T. W. Callaway as a player in the communist context of the Scopes trial. According to Norris, Ward had sent one of his representatives to Detroit to aid John L. Lewis and the CIO. Norris identified Ward as "chairman and prominent ruling spirit of the ultra-radical, revolutionary, and I.W.W.-defending American Civil Liberties Union." He joined Callaway in describing the ACLU as a "supporter of all subversive movements."[84]

If Norris did not connect the dots between evolution, communism, and industrial unionism, Dan Gilbert was pleased to do so. In "The Rise of Beastism," published in 1938 in *Moody Monthly*, the young evangelist redeployed his argument about Marxist "beast" doctrine and applied it to the burgeoning movement for industrial unionism. In Gilbert's eyes, Darwin's ideas boiled down to the proposition that progress is achieved by the strong crushing the weak. Gilbert then argued that communist rule meant the "reign of brute force." As for the conduct of the labor movement, Gilbert claimed that until recently, working "men" had rejected this communist idea and relied instead on the tactic of peaceful picketing, aiming to win over public opinion. This was "the American way, the Christian way." But in the past year, the "amazing vogue" of the European-based sit-down strike heralded a rise in "beastism." Instead of appealing to public opinion, workers, led by "radicals and communists," were seizing private property and relying on "organized lawlessness and terrorism."[85] Not only was the sit-down strike a product of communist evolutionism, but it was a fulfillment of prophecy. The rise of beastism as reflected in the CIO was "a sign that the spirit of Antichrist is abroad in our land." The prophets had foretold that the Antichrist would rule by "brute force" and that his dictatorship would be based on terrorism and lawlessness. In the face of this "swelling tide of beastism," it was the duty of Christians to "exercise every effort" to resist that demonic force.

Dan Gilbert's application of the concept of "beastism" to the phenomenon of the sit-down strike was a remarkable expression of creationist politics. Gilbert managed to explicitly connect Darwinian evolution, Marxist communism, labor movement politics, and biblical prophecy. A seasoned journalist writing in one of the most popular evangelical publications in America with forty thousand subscribers, Gilbert knew how to communicate with the masses.[86] Though J. Frank Norris and Gilbert's close associate William Bell Riley were not collaborating closely at this point, Gilbert's piece complemented the fire-and-brimstone preaching that Norris did in Detroit to combat the CIO. If any of Norris's listeners read Gilbert's piece in *Moody*'s, one imagines that they would be nodding their heads in easy agreement.

From the onset of the Great Depression to the end World War II, conservative-minded Americans had numerous opportunities to learn about the intertwined evils of communism and evolution. Depending on who taught this lesson, those "evils" might or might not have been further interwoven with the world Jewish conspiracy, the CIO, sit-down strikes, sex education, dancing and drinking college students, Freud, or the Chinese revolution. Remaining consistent was the idea that what George McCready Price labeled "Red Dynamite" posed a mortal threat to American and world civilization. Over the next two decades, as the Cold War profoundly shaped culture and politics, and as evolutionary biologists consolidated the "modern synthesis" of Darwinism and population genetics, that threat would be reinterpreted by a new generation in distinctive ways.

5

BEAST ANCESTRY, DANGEROUS TRIPLETS,
AND DAMNABLE HERESIES

In April 1949, an aging J. Frank Norris stood once again to address the members of the Texas state legislature. When Norris had last occupied that speakers' platform in 1923, his main subject was evolution. Now, more than a quarter-century later, Norris was speaking as an expert on communism, which loomed larger than the beast doctrine in the public mind. A week earlier, the state House of Representatives had passed a resolution honoring Norris for his work in rooting out "subversives." Presenting the seventy-one-year-old Norris with a large printed copy of the resolution, the House clerk introduced him as an authority on "Communists," "fellow-travelers," and "all those who seek to destroy the things that are good in this land."

Norris did get around to the topic of evolution, but he began with the Red Menace. The most insidious threat facing Americans was not the out-and-out Communists but rather their allies who were "boring from within." As Aimee Semple McPherson had warned in 1934, Americans were "asleep" and needed to be awakened to the real danger. These

"Benedict Arnolds" were radical labor union leaders, professors, and clergy whose nefarious activities paved the way for Soviet control. Even if such people had not broken any law or openly revolted, they were "traitors," Norris charged.

They betrayed their country by following the immoral "religion" of communism. They did the devil's work by denying the word of God. They promoted materialism, or "blind force," ruling God "out of the Universe." Without initially invoking evolution by name, Norris explained the communist belief that "man is just a beast and dies like a dog." Worst of all, the betrayers sink "into the deepest depths of immorality." As evidence, Norris recalled that the University of Texas had fired ten "homosexualist" employees, and that in Houston, a man was advocating "some things I wouldn't even mention to a mixed audience." Fortunately, Norris related, the "real men" of Houston rode him out of town.

The sexual immorality of communism encompassed what Norris called, in an ironic nod to Communist political terminology, "the negro question." Claiming that he was the "best friend" of "the Negro," Norris stated that he believed in "social equality" and that Blacks should not be denied equal schooling. Norris nonetheless raised an alarm. The "pith of the whole business," Norris claimed, was that communists had prodded Blacks into desiring that the races merge into a "mongrel race." There was no need for Norris to specify that this radical kind of "social equality" was interracial sex and marriage.

Finally, Norris arrived at the topic of evolution. He attributed to subversive professors the idea that humans were mere "beasts." These educators of our children were not just purveyors of Darwinistic ideas; they were "evolutionists *and* communists, two sides of the same question." As before, Norris joked about evolution. He drew his audience's attention to its "moral effect." To the men, if your wife is jumping on you "with hatchet and tongs," that's the "hyena in her." And to the women, if you discover that your man has been lying about his whereabouts and running around town, that's just the "bearcat" in him. Redeploying his 1923 text, Norris gave his mocking summary of how some unknown thing had evolved into "so many bald-headed men" who ended up teaching evolutionary science. This time, however, he added an ugly detail about where they got their suits—"from a second-hand Jew joint down here at Austin." During World War II, Norris had proudly declared his solidarity with European Jews and supported American Zionist leaders.[1] But Christian Zionism and anti-Semitism were not mutually exclusive.

The themes evoked in Norris's speech—including the sexually charged connections between communism and evolution—were familiar. Yet his appearance before Texas lawmakers in 1949 marked the transition to a new era. Norris's influence was fading within his own church. Loyal lieutenant G. Beauchamp Vick had labored for decades for Norris and served as pastor at the Norris-founded Temple Baptist Church in Detroit. No longer able to tolerate Norris's authoritarian ways, Vick broke with him in 1950 and founded the Bible Baptist Fellowship (BBF) and the Baptist Bible College (BBC) in Springfield, Missouri.[2] With William Bell Riley's death in 1947 and Norris's in 1952, leadership of the fundamentalist cause fell to a younger generation of evangelists, including *Sword of the Lord* editor and Norris protégé John R. Rice (1895–1981). The man whom Jerry Falwell would decades later call "the most trusted man in fundamentalism," Rice played an underappreciated role in linking together generations of conservative Christian activists.[3] He also made his own distinctive contribution to the Red Dynamite tradition in a 1954 sermon (later published as a pamphlet) titled *Dangerous Triplets.* In that same banner year, the US Supreme Court decision in *Brown v. Board of Education* sparked renewed organizing both to break down the walls of Jim Crow and to keep them intact. The defense of racial segregation under the banner of "racial purity" drew from the book of Genesis in ways that offered surprising connections between creationism, "massive resistance" to civil rights, and anticommunism.

A generation younger than J. Frank Norris, John R. Rice was born in Cooke County, Texas, in 1895. His father, Will Rice, attended Southern Baptist Theological Seminary, joined the revived Ku Klux Klan, and was elected to the Texas State Senate in 1921 along with a raft of other Klan-supported candidates. Young John started his own career as an elementary school teacher near Dundee, Texas, while working on his father's ranch. Attending Decatur Baptist College in 1916, he met Lloys McClure Cooke, who would become his wife and close collaborator in all things spiritual. Rice was drafted in 1918 but did not see service overseas, thanks to a mumps epidemic in his Seventh Division unit. After the war, he and Lloys both attended Baylor, and upon graduation, John taught briefly at a Baptist college in the fall of 1920.[4]

In light of his subsequent fundamentalist leadership, Rice's next move was surprising: he headed north to attend divinity school at the liberal Baptist bastion of the University of Chicago. Of this move, Rice's grandson and biographer Andrew Himes speculates that "he was feeling a strong pull away from the rigidity of the small churches in Texas where he had grown up."[5] For the short time that Rice was in Chicago, he entertained a range of modernist ideas, including theistic evolution. But then, in May of 1921, Rice heard William Jennings Bryan deliver "The Bible and Its Enemies." Linking Darwin's ideas to their evil social and political consequences, Bryan clarified for Rice the stakes in the debate over evolutionary science.

Primed by Bryan's sermon to save the nation's children from evolutionary teachings, Rice began to question his career plans to become an academic, or, as he put it, a "great educator." The turning point was an encounter with a "drunken bum" at the Pacific Garden Mission in downtown Chicago, where none other than evangelist Billy Sunday had been converted to Christ. Kneeling next to this sinner, Rice showed him the way to God's grace. All of a sudden, Rice recalled, "I saw the transformation in his face, the evidence of wonderful peace in his heart." That evidence proved to Rice that his calling was in the ministry. That summer, John and Lloys left Chicago for Texas. They were soon married and enrolled at Southwestern Baptist Theological Seminary in Fort Worth, just miles from J. Frank Norris's First Baptist Church. John became an ordained Baptist minister. Ministerial assignments took the couple from Decatur to Plainview to Shamrock and, in 1932, back to Norris territory in nearby Dallas, where Rice pastored the Fundamentalist Baptist Church.[6]

Rice's criticism of the insufficiently fundamentalist Southern Baptist establishment had drawn Norris's notice by the early 1920s. In 1926, Norris offered Rice a weekly radio program on his Forth Worth station and called the younger man "a great preacher of the gospel of Christ." In 1928, Norris invited Rice to preach at First Baptist, and that same year, Rice followed Norris by taking his congregation out of the Southern Baptist Convention. The two evangelists joined forces in campaigning for Republican Herbert Hoover and against the Democratic candidate Al Smith, warning of Smith's beliefs in "social equality" for Blacks. In 1932, Norris called Rice "the greatest Bible teacher among us."[7] Rice had left academia behind, but he would become a "great educator" for thousands of conservative Christians for decades to come.

In 1934, still on good terms with Norris—who would break with him over the following year—Rice launched the *Sword of the Lord* as the newspaper of his church. Rice drew its name from the biblical story of Gideon, a Hebrew "judge" or leader of the Tribe of Manasseh. According to the book of Judges, God chose Gideon as his instrument to bring the Hebrew people back from their worship of foreign idols. With God's help, Gideon's small army prevailed over the much more numerous Midianites. The Hebrew people were restored to God. As Rice's masthead indicated, "And they cried, The Sword of the Lord, and of Gideon."[8] In 1934, Rice viewed himself as a modern-day Gideon, both a warrior and a weapon in the hand of God. In these early years, the full title of his newspaper was "The Sword of the Lord and of JOHN R. RICE." A short mission statement followed: "An independent religious weekly to preach the gospel, expose sin, spread premillennial Bible teaching, and foster the work of the Fundamentalist Baptist Church."[9]

Rice and Norris were marching side by side into battle, but Norris's patronage of Rice did not last. As readership of *Sword of the Lord* expanded, Norris grew jealous of Rice. As even Norris's admirers admitted, he was not one to share the limelight. In 1936, when Rice organized a citywide revival in Binghamton, New York, the first such attempt in the North for many years, Norris attempted to sabotage it by spreading false rumors that Rice had Pentecostal "holy-roller" tendencies. Rice succeeded despite Norris. To signal his independence from the "shooting parson," he not only answered Norris's charges in print but renamed his church Galilean Baptist, since "Fundamentalist Baptist" had Norrisite overtones. Rice also rewrote his newspaper's mission statement. Now more pointedly fundamentalist, it stood for the "Verbal Inspiration of the Bible, the Deity of Christ, His Blood Atonement, Salvation by Faith, [and] New Testament Soul Winning." It stood against "Sin, Modernism," and in a clear swipe at Norris, "Denominational Overlordship."[10]

Unlike Dan Gilbert, John R. Rice would take longer to make explicit connections between evolution, communism, and "sexual sin." But his articles, pamphlets, and sermons on moral decline shared with thousands of *Sword* readers formed a rich soil that would allow those connections to flourish. In a sermon Rice described as "dynamite," he denounced dancing as the "road to hell."[11] Divorce, unless granted for proven adultery, was a sign of "sex sin."[12] Rice also spoke out about "criminal abortion," which he viewed as plain evidence of "sinful and illicit" sexual relationships between married men and young unmarried women.[13]

During the 1940s, Rice devoted substantial attention to promoting proper gender roles for men and women in marriage. Without explicitly addressing "sex radicals" like Schmalhausen and Calverton, Rice was implicitly holding the line against a social evolutionary view of gender. In *Rebellious Wives and Slacker Husbands* (1941), and *Bobbed Hair, Bossy Wives, and Women Preachers* (1943), Rice spoke out against modern "feminist" notions of women's equality. For their part, men were ordained by God to lead their households. A man was to be "like a god in his home." The fact that many men were shirking this responsibility, Rice told his readers, explained the "train of evils" that was afflicting American society, from broken homes to misbehaved children to women who dressed "immodestly." Men who behaved in this way were "degenerate, weakling men, slackers and shrinkers and quitters, not willing to take the place of manhood."

For their part, women were to submit to the leadership of their husbands. They were to cover their bodies and to wear their hair long, so that they were clearly distinguished from men. A woman was obligated to obey her husband regardless of his character, his treatment of her, or even whether or not he had accepted Jesus Christ as savior. The woman whom others viewed as "rebellious" or "bossy" was not only acting with bad manners. She was falling victim to the evil impulses that emanated from Satan. As Rice wrote, "the heart of all sin is rebellion against authority," just as Satan rebelled against God in heaven. Moral rebellion was not just a personal matter but formed part of the whole social and political picture. The satanic spirit of rebellion, Rice averred, explained the "crime and lawlessness which plagues America and other governments." It is the desire of God, explained Rice, that children obey their parents, servants obey their masters, citizens follow their government, "even if administered by wicked and corrupt men," and women obey their husbands.[14]

By the time Rice wrote these words, he was gaining a national audience. Not only did he reach thousands through the pages of his newspaper, but he preached widely and could be heard in the late 1930s on the daily noonday radio broadcasts in Chicago. To pursue new opportunities to spread God's word and to build a national fundamentalist network, Rice, Lloys, and their six daughters moved north to Wheaton, Illinois, in the spring of 1940. As Rice explained the move, Chicago was "the center for fundamentally sound Christian work." It was strategically situated for Rice to stage revivals all over the Midwest; it sported the Moody Bible Institute and the Northern Baptist Seminary; and his new hometown was the seat of Wheaton College, the "strongest and largest of the independent Christian colleges in America." That same year, a young evangelical student arrived at Wheaton, attended Rice's citywide revivals, and subscribed to *Sword of the Lord*, where his own sermons would later appear in print for the first time. His name was Billy Graham. He was not the last prominent young evangelical leader to be mentored by John R. Rice.[15]

Figure 10. John R. Rice at a *Sword of the Lord* preaching conference, c. 1965. A fierce Baptist fundamentalist who battled evolution and communism, Rice was a living link between the old Christian Right of J. Frank Norris and William Bell Riley and the New Christian Right of Jerry Falwell. Courtesy of John R. Rice Papers, B.H. Carroll Center for Baptist Heritage and Mission, Southwestern Baptist Theological Seminary, Fort Worth, Texas.

When Rice addressed evolution during these years, it became clear that his grim assessment of humanity's future prospects informed his perspective. Evolutionary ideas, argued Rice, were the product of unfounded human pride, of the "crack-pot" evolutionist who believed that he could use the Bible to support modern scientific ideas and take a "personal, creating, supervising God" out of the equation. Such "proud and disdainful men," wrote Rice, failed to acknowledge the reality of human sin. They were oblivious to the fact that far from progressing, humanity was headed on a downward slide. Monkeys are not becoming men, he wrote, men are becoming monkeys. Rather than evolution, Rice concluded, we have "devolution" and "degeneracy." In Rice's pessimistic account, social and biological regression seemed to merge together.[16]

While Rice's antievolutionism was typical of his fellow fundamentalists, his views on racial and national oppression were not. Where Winrod and Riley had applauded Hitler's measures against Jews, Rice called out the *Führer* for "Jew-hate" and "concentration camps." By this time, Riley had also come out against Hitler as a Darwin-inspired exemplar of the "survival of the fittest." In a telling omission, however, Riley's pamphlet *Hitlerism: Or, The Philosophy of Evolution in Action* (1941) said not a single word about Hitler's treatment of Jews.[17] Rice denounced Hitler and the Ku Klux Klan and the Nazis as comparable in their brutality. "What difference is in high-riding Ku Klux Klansmen in America who beat offending Jews or negroes or foreigners," Rice asked, "than a German machine-gunner, strafing refugees in Poland with a machine-gun?"

Despite his seemingly evenhanded denunciation of evil human behavior as rooted in original sin, Rice still reserved a special critical edge for communism, both in Russia and in the United States. Bolshevik Russia, according to Rice, was a "land of slaughter, of poverty, of famine, or oppression, of violence and atheism." It was far from the "working-man's paradise" that "millions" of Russians had imagined it might become. It had become the opposite, a hell on earth. And Rice had an explanation for this that went beyond original sin. Citing Revelation 16, Rice claimed that Stalin, like Hitler, was surely aided by Satan; Russia and Germany both were "Godless and atheistic bandit nations."[18]

The road to John R. Rice's 1954 *Dangerous Triplets* sermon at Highland Park Baptist Church in Chattanooga, Tennessee, ran through nearly a decade of post–World War II preaching, editorializing, and organizing

Sword of the Lord conferences. Rice viewed himself as a "soul winner," one who more than anything else wished to help America's sinners find their way to Jesus Christ. For Rice, that task meant orienting his readers and listeners toward a Bible-believing faith and steering Christians away from a false modernist one. The pages of *Sword* in that postwar decade were filled with both seemingly nonpolitical appeals to individual sinners and explicitly political pieces that warned readers of the dangerous lure of false prophets. In this respect, Rice continued to follow in the footsteps of his erstwhile mentor, J. Frank Norris.[19] Navigating through these turbulent political waters, Rice allied himself, for a time, with fundamentalist separatists like Carl McIntire as well as their soon-to-be foe, the increasingly ecumenical, "neo-evangelical," and longtime *Sword of the Lord* reader Billy Graham.

John R. Rice's ally Carl McIntire (1906–2002) was one of the fiercest and most influential anticommunists of the twentieth century. A Presbyterian fundamentalist minister, McIntire trained at Princeton in the late 1920s under the conservative New Testament scholar J. Gresham Machen. When Machen left Princeton to found Westminster Seminary, rejecting both a liberalizing trend at Princeton and a populist fundamentalism based on premillennialism, McIntire joined him and then helped form a succession of new, independent Presbyterian sects, the last being the Bible Presbyterians in 1937. Pastoring a Bible Presbyterian church in Collingswood, New Jersey, for decades after, McIntire took the lead in creating a fundamentalist rival to the liberal Federal (later National) Council of Churches (FCC/NCC) called the American Council of Christian Churches (ACCC). What distinguished McIntire from other midcentury fundamentalists, including John R. Rice, was McIntire's refusal to join the relatively conservative National Association of Evangelicals (NAE). An exponent of "secondary separation," McIntire demanded not only that the member churches of his group be fundamentalist, but that their parent organizations be as well. Since many NAE member ministers pastored local churches that remained in the Southern and Northern Baptist Conventions, that put them beyond the pale. McIntire started the weekly *Christian Beacon* newspaper in 1936, began to publish books in the mid-1940s, and quickly gained a national following as a highly articulate and uncompromising foe of modernism and communism.[20]

McIntire's collaboration with John R. Rice supports biographer Markku Ruotsila's contention that McIntire was a "pivotal transitional and transformative figure" in the long history of the New Christian Right.[21] McIntire's second book, *The Rise of the Tyrant* (1945), first captured John Rice's attention. Over the next four years, Rice reprinted excerpts from this book as well as *Christian Beacon* articles in *Sword of the Lord*. As many *Sword* readers (and Rice himself) were NAE members, and others sympathetic to various degrees with the FCC/NCC, McIntire's writings spurred both angry and supportive letters to editor Rice. In a commentary on the articles as a whole, Rice acknowledged that he had been, on occasion, "irritated" by McIntire's extremism. And yet Rice agreed with the essence of McIntire's message and happily shared it with readers.[22]

That message was summarized by the headline of the very first article Rice reprinted: "The Modernist-Communist Threat to American Liberties." Just one week after the Japanese surrender, McIntire noted that America had no longer anything to fear from foreign foes but now faced an internal enemy—what Norris had labeled "Benedict Arnolds." Singling out Methodist bishop G. Bromley Oxnam, the FCC president, McIntire charged him with promoting economic collectivism. Like the Social Gospelers and Christian Socialists decades earlier, mainline Protestant leaders were relatively friendly to reforms that would curb the power of big business. But such Christians, in McIntire's judgment, were misusing the Bible and their ecclesiastical authority. If Christians wanted to protect the "true church," McIntire wrote, they needed to defend the "profit motive, competition, private enterprise, and the individual," all of which rested on biblical foundations. McIntire's identification of greater control of the economy by the state with tyranny was the same message contained in works by secular authors, most notably Friedrich Hayek's *The Road to Serfdom* (1944) and John Flynn's *The Road Ahead* (1949). What distinguished McIntire's approach was his biblical defense of capitalist economic institutions.

In a subsequent *Sword* article, "Private Enterprise in the Scriptures," McIntire demonstrated how the Bible endorsed capitalism. God's command to the Hebrew people, "Thou shalt not steal," in McIntire's view, was about "private enterprise." It underlined the divine origin of private property and economic individualism. When the FCC pushed for greater collective economic responsibility, it was "attacking the eternal truth of

God." The Bible endorsed not only private property, but profit. Why else would Solomon, the wisest of all men, have said, "In all labour, there is profit"? Confirmation came from the New Testament, where Luke tells of a nobleman who sends his servants to trade, and on their return, wants to know "how much every man had gained by trading." "Here is private enterprise," wrote McIntire, "the profit motive." In case any reader might miss the meaning of these details, editor Rice added a long subheadline in all capital letters that began: "A DISTINGUISHED CHRISTIAN LEADER SHOWS THAT THE SCRIPTURES ESTABLISH CAPITALISM AND FREE ENTERPRISE, AND ARE AGAINST COMMUNISM AND SOCIALISM."[23] If John R. Rice had ever harbored any suspicion that the Bible endorsed socialism—as had a young William Bell Riley—McIntire helped to bury those suspicions for good.

While Rice did not yet in the late 1940s explicitly join together the dangers of communism and evolutionism, he made it clear that modernist Christians who promoted evolution were dishonest, ungodly, and downright evil. In an article on modernism that gave extended coverage of evolution, Rice invoked Matthew 7:15 on evil fruits and Second Peter 2:1 to describe modernist evolution-believing Christians as dangerous, hypocritical "false prophets" and as sneaky "fifth-columnists" who were stealthily introducing "damnable heresies" within Christian churches. Riley had quoted these same passages two decades before in his sermon on "Bestial Bolshevism." The false prophet was a metaphor well suited to the battle between fundamentalists and modernists in Christian congregations—for spotting the enemy within.[24]

For any Christians who were considering partaking of the evil fruit of theistic evolutionism, Rice made it clear that they needed to choose: evolution or the Bible. The two were "irreconcilable." At the same time, Rice did make a gesture, if only a confused one, toward a scientific consideration of evolutionary science. Evolution, he maintained, remained far from a "proven fact." It was, rather, "a theory, a hypothesis, a guess." No transitional forms—"missing links"—had been found, charged Rice. The discoveries of the bones of human ancestors were either unconvincing or outright fakes. Rice mistakenly attributed to Darwin (rather than Lamarck) the claim that giraffes had acquired long necks in the course of their lives by reaching for leaves high up in the trees and described this idea as "natural selection." He mocked this Darwinist idea, in Weismann-like fashion, noting that after a "thousand generations" of sheep with tails cut off or "thirty-five centuries" of circumcising Jewish boys, sheep still had tails and Jewish boys were born with foreskins.[25]

To his growing pro-capitalist and openly antievolutionary views Rice added an increasingly vehement anticommunism. Rice condemned the "socialistic" New Deal and its legacy in the Truman administration. On the eve of the 1948 presidential election, he answered a letter from a *Sword* reader asking about his stance on Christian Nationalist Crusade leader Gerald L.K. Smith. Rice restated his previous stance on Smith's virulent "antijewish propaganda" as "wicked" and "unchristian." But he also took the occasion to urge readers to vote the Republican ticket of Dewey and Warren. Roosevelt and his New Dealers had encouraged labor strikes and radicalism in the labor movement, given Stalin "far too great a hold on the world," stifled "free enterprise," and promoted "class and race hate in America."[26]

Rice also published philosophical pieces by others who addressed the one big question facing Christians in the Cold War world: Christianity or Marxism? In "Karl Marx or Jesus Christ?" the address given by Wheaton College president (and fellow Wheaton resident) Raymond Edman to graduates in 1959, Edman cautioned students that Marxism was not a mere academic theory but rather a "religion" to "millions" around the world. Comparing the two men as personalities, Edman found that Marx was "inconsiderate," "hostile," and cold, while Jesus was devoted, loving, compassionate, and "unspeakably sweet in spirit." Their philosophies were similarly contrasting. Marx's "religion" preached materialism, godlessness, "brute force," and destruction, while Jesus offered spiritual values, a change in heart, and regeneration "by the grace of God."[27]

For all of the space that John R. Rice provided to fellow anticommunist Carl McIntire, Rice had a greater affinity for his Baptist forebears, not only his erstwhile mentor J. Frank Norris, but also William Bell Riley. In an April 1947 piece on the occasion of Riley's eighty-sixth birthday, just months before the aging fundamentalist's death, Rice paid tribute to the man's "rich achievements for God." Rice admired Riley's defense of the fundamentals of the faith, his opposition to evolution—"science falsely so-called"—his warnings about communism and "un-Americanism," and his building up of an empire for God in the Northwestern complex of schools.[28] Rice followed up this endorsement of Riley with reprints of his articles. "Atheism, the Enemy of Civilization" identified atheists with "men of reprobate morals" and pointed to the immoral "bitter harvest" of evolution, including the "debauch of infidelity" in Bolshevik Russia.[29]

One aspect of Riley's ministerial message that Rice could not publicly praise was the Minneapolis preacher's fanatical devotion to the *Protocols*. In "The False 'Protocols' and Wicked Anti-Semitism," Rice rejected this position, noting that Henry Ford had publicly apologized for endorsing the czarist forgery. Communism, Rice clarified for his readers, was not a plot of the Jewish "race," but rather was the product of humanity's sinful nature and our temptation by Satan. But neither did Rice denounce Riley for his position. Nowhere were *Sword* readers reminded that the man most responsible for popularizing a set of ideas that Rice had identified as "wicked" was the same man whom Rice had thanked for living a life in service of God. Rice's silence on Riley's pro-Nazi politics helped obscure that inconvenient fact for future generations of evangelicals.

Another aspect of Riley's legacy initially garnered great enthusiasm from Rice and only later ambivalence and then outright rejection: the appointment as his successor at Northwestern of a young preacher named William Franklin "Billy" Graham. The evolving relationship between Rice and Graham sheds additional light on the history and politics of creationist anticommunism. Born in the waning days of World War I in Charlotte, North Carolina, to a Methodist father and Presbyterian mother, Graham accepted Christ as his Lord and Savior in 1934 during a weeks-long tent revival held on his family farm by the fiery Baptist moralist, antievolutionist, anticommunist, and *Protocols*-based conspiracy theorist Mordecai Ham. As Graham has related, he was aware, at the tender age of sixteen, that Ham had been accused of being "anti-Semitic." But Graham claimed that "I had no way of knowing if that was true; I did not even know what that term meant."[30] Perhaps so, but Graham may have absorbed Ham's ideas nonetheless.

That possibility emerged in a spectacular way decades later, when secretly recorded tapes of Graham's 1972 meeting with then-president Richard Nixon and his chief of staff H. R. Haldeman came to light. All three men expressed typical anti-Semitic ideas about Jews dominating the media. Graham spoke of Jewish culpability for pornography. In a subsequent telephone conversation with Nixon, Graham reiterated Jewish responsibility for pornographic literature and "obscene" films. Graham attributed this activity to a particular subset of Jews that the Bible refers to as "the synagogue of Satan."[31] These claims were common currency for Christian fundamentalists in the 1930s. But they also evoked the sermonizing material that Mordecai Ham, channeling Henry Ford's *International Jew*, was spouting in Elizabeth City a decade earlier.

In the sermons that persuaded Graham to accept Christ into his heart, Mordecai Ham dealt with the standard fare of "money, infidelity, the Sabbath, and drinking."[32] After selling Fuller paint brushes in California for a summer, Billy Graham spent one semester at Bob Jones College. In a sign of his future departure from fundamentalism, Graham chafed under the tight restrictions on student conduct and soon left. He obtained a college degree from the Tampa-based and unaccredited Florida Bible Institute, affiliated with the Christian and Missionary Alliance (CMA). He began preaching at a small CMA church near Tampa and was ordained as a Southern Baptist minister. Soon, some visitors with connections to Wheaton College heard Graham preach in Florida and offered him a scholarship to Wheaton, where he arrived in the summer of 1940.[33] There he met Ruth Bell, his future wife, and daughter of L. Nelson and Virginia Bell. Stationed in China as Presbyterian missionaries, the Bells returned the following year to the US, where L. Nelson Bell would help lead a movement to resist modernism in the Presbyterian Church in the U.S.A.[34] Newly married, Graham used his Wheaton degree and connections to start a career in radio preaching and then public preaching with Youth for Christ in Chicago and throughout Europe.

While Youth for Christ made Graham into a national figure and attracted the attention of William Bell Riley, another crucial influence on the young evangelist was John R. Rice. Once ensconced in Wheaton, Graham became close friends with Rice. Graham read *Sword of the Lord* and was impressed by Rice's ability to organize citywide "union" revival campaigns, soon a hallmark of Graham's preaching. Some of Graham's first published sermons appeared in *Sword*. In the late 1940s, Graham appeared as a speaker at several *Sword* conferences, alongside the Bob Joneses, Rice, and other prominent evangelists.[35] Soon after Riley tapped Graham to lead Northwestern Bible College, Graham returned Rice's favors by placing John R. Rice on the school's board of trustees.[36]

The sermon Graham preached at the 1948 *Sword* conference on the various "crises" afflicting America is a good reference point for the fundamentalist standard that he would soon abandon. It also places him in the pantheon of those alarmed by the twin evils of evolution and communism. First and foremost, Graham told his audience, there is a "philosophic" crisis. Based on his work at Northwestern, Graham said he had "found out" that we were living in an age of "materialism . . . evolution and naturalism." "If you study geology or any other of the sciences," Graham explained, "you will find that the basis for all the false teaching today is evolution, which denies the existence of God Almighty." Closely related was the "moral" crisis. It found expression in an alarmingly high divorce rate, a crime wave, the growing incidence of prostitution, the end of Prohibition, and billions spent on gambling. In addition to these "obvious" evils, which he illustrated with examples, Graham cited, without any specifics, "social unrest and industrial strife on every hand in America." There was also a political crisis, exemplified by the near-triumph of the Chinese Communists and the probability of renewed war with the Soviet Union. This raised the prospect of the end of humanity, whether it be through atomic weapons, a "death ray," or "germ bombs." Finally, there was a religious crisis. In North Africa, the "Arab race" was uniting under the banner of Islam; Roman Catholics were repressing Protestants in Franco's Spain. Most threatening was the "fanatical religion" of communism, growing by "leaps and bounds." In the face of these multiple crises, the only chance America had was to "repent" its evil ways. Citywide revivals organized by modern-day Jonahs raised up by God were the key.[37]

Billy Graham's revival campaign preached in Los Angeles the following year made him into an evangelistic superstar. For the next several years, John R. Rice and other fundamentalist stalwarts were proud to have nurtured and encouraged the young Graham. Even as Graham no longer spoke at *Sword* conferences, Rice continued to publish his sermons and publicly praise him.[38] Though he was unalterably alienated from Rice, J. Frank Norris was also initially enthusiastic about "Billie" Graham. In February 1950, Norris traveled to hear Graham preach a revival in Columbia, South Carolina, and invited the evangelist to preach at Norris's churches in Fort Worth and Detroit. He described Graham to a fellow Baptist as "the greatest soul winner of this hour" and contributed money to the Billy Graham Evangelistic Association. In a March 1950 letter to

Graham, Norris gushed with praise for the young evangelist, telling him
that in comparison with the famed Billy Sunday (whom, Norris boasted,
he had met "three different times"), you are "different and I think far
ahead."[39]

Norris's love for Billy Graham did not last long. Even as Norris
was effusing about "Billie" in 1950, Graham was wary of the "shooting parson." In a 1950 letter to Norris, Graham confessed that he had
not answered earlier letters from Norris because he feared that Norris
might publish his reply in the *Fundamentalist*. Graham allowed as well
that he had differed with Norris's tactics and that he remained faithful to
the Southern Baptist Convention.[40] Norris had made headlines in 1947
when he appeared at the Southern Baptist Convention annual meeting in
St. Louis and confronted its new modernist president Louie Newton.
Pastor of the liberal Druid Hills Baptist Church in Atlanta, Newton had
drawn fire after his recent visit to Soviet Russia. In *An American Church-
man in the Soviet Union* (1948), Newton wrote that "Baptists stand for
the same thing as the Russian Government—renouncement of, and resistance to, coercion in matters of belief." After Norris was physically
removed from the meeting hall, he fired off a telegram to his Fort Worth
congregation, boasting that he had achieved the "greatest victory in [the]
history [of] fundamentalism."[41] Over the next few years, as it became
clear that Billy Graham was willing to work with all denominations, as
well as with open modernists, to build the biggest possible revival meetings, Norris joined with separatist extraordinaire Carl McIntire in denouncing Graham's compromises.[42]

It was during this transitional period, in which Norris and McIntire
were pulling away from Graham, but Rice was not, that Rice was at
the height of his influence and made his distinctive contribution to the
Red Dynamite tradition. The circulation of *Sword* reached ninety thousand in 1953. *Sword* evangelism conferences, held in Winona Lake, Indiana, the home of Grace Seminary, and Toccoa, Georgia, continued
to draw top evangelical preachers and big crowds. Now living in Murfreesboro, Tennessee, Rice had joined and helped lead Highland Park
Baptist Church in Chattanooga, a prominent independent (non-SBC)
fundamentalist "mega" church located forty miles from Dayton, the site
of the Scopes trial. Pastored by Lee Roberson, a longtime friend and

associate of Rice's, who spoke often at *Sword* conferences, Highland Park was influential in fundamentalist circles. Rice called it "the greatest soul-winning church in the world." When members of Highland Park were recruiting the up-and-coming Roberson to pastor their church in 1942, the local welcoming committee included T. W. Callaway, the man who had "outed" John T. Scopes as a red-diaper baby in 1925. Roberson published the *Evangelist*, launched a weekly radio program called *Gospel Dynamite*, and preached to thousands in his church each Sunday. As fond of the explosive metaphor as J. Frank Norris, Roberson published a book of his sermons, titled *It's Dynamite*, with *Sword of the Lord* publishers in 1953. Roberson's Highland Park "complex" included Temple Seminary (later Tennessee Temple University), of which Rice was appointed vice president, and a Department of Evangelism, which Rice headed, starting in 1954.[43]

At Highland Park in the summer of 1954, Rice offered his own Red Dynamite sermon. His July 27 sermon was titled "Dangerous Triplets: 1. Russian Communism, 2. New-Deal Socialism, 3. Bible-Denying Modernism." For anyone in the congregation that morning who had been reading *Sword* over the past half decade, many of Rice's references would have been familiar. He opened with Matthew 7:15 on false prophets, followed up with 2 Peter on the same, and cited Raymond Edman on the immorality of Marx. He presented McIntire-esque arguments to prove that the Bible endorsed capitalism, with one section titled "Investment and Interest in Invested Capital Approved by Jesus."[44] Rice provided "damning" quotations from socialist-leaning religious modernists like G. Bromley Oxnam. And he lambasted various New Deal projects, including the Tennessee Valley Authority, as handouts for "deadbeats."

The task that Rice set himself in this sermon was not only to highlight the evils of the separate phenomena of socialism, communism, and modernism, but to show that they were intimately, genealogically related. Rice knew that some Highland Park members did not see these connections. They resisted the idea that "modernists" were in the same category as the obviously immoral and evil communists. It was one thing to lambaste Soviet leaders or even Franklin Roosevelt, but quite another to denounce other Baptist preachers. Right after the opening words of his sermon, quoting from Matthew, Rice said, "You say, 'why all this shouting about modernism?'" Then a bit later, Rice added, "Someone says, 'I don't like your calling these men names and saying things about them.'" Near the end of the sermon, Rice acknowledged the uncomfortable predicament of false prophets close to home: "You say, 'But my friends.'"

To help Chattanooga's fundamentalist faithful face the painful truth and accept the need to break friendships, Rice offered four reasons that modernists, communists, and socialists were "alike." Their kinship revolved around the question of whether humanity was the creation of God or the evolved product of nature. First, all three "triplets" hated the Bible. Second, they believed in evolution, that man came by slowly evolving from lower forms of life." That claim, Rice told his audience, is "part of the particular doctrine of Karl Marx." As proof of that connection, Rice asked if anyone knew to whom Marx dedicated the *Communist Manifesto*. "To Charles Darwin!" he shouted. Then he gave a synopsis of Marxist evolutionary thought: "He says that the human race is evolving upward and that therefore, little by little, society is growing into a better state and eventually we will not have any private property, and everybody will own everything together. And he says that evolution is absolutely certain to bring communism over the whole world. So Stalin taught, so all communists believe." Rice's Marx sounded remarkably un-Marxist. He was more like the early Socialist Party's version of Spencer, envisioning a future society as the gradual and inevitable product of organic growth. Revolutionary change was nowhere in evidence. Rice made an uncharacteristic error in substituting the *Manifesto* for *Capital*. But the basic point was accurate. Marx was a Darwin fan.[45]

Rice's third reason amplified the second by noting that the triplets all opposed the notion of God's creation of humanity and the related concept that we are fallen creatures whose hearts were "desperately wicked." Instead, the triplets believed that "man" was "essentially good." The "system" was wrong. And we could improve the world through a "collective state," as the Bolsheviks had done in Russia and as liberals had tried to do through the New Deal. This was yet another reason why the triplets "run together," as Rice put it.

In laying out the fourth and final reason tying the "dangerous triplets" together—"a certain wicked immorality"—Rice reached the emotional climax of his sermon, doing his best to make sure that Highland Park's members understood what was at stake. The case was simple for the communists. Since they rejected the Bible, they had no moral standards:

"If lying will win, if murder will win, if stealing will win, if rape will win, that is alright. Communists have no morality." Rice had prepared the punch behind this point earlier in the sermon by quoting selectively from the *Communist Manifesto* to "prove" that Marx and Engels were in favor of establishing a legalized "community of women."[46] This charge drew strength from the long-standing story of the Bolshevik "Bureau of Free Love." In the published version of Rice's sermon this section is titled "Communism Would Abolish the Family As We Know It."

The modernists were trickier. They pretended to be men and women of God. And then these false prophets used their clerical authority to break down the faith of Christian believers. Between the modernists and communists, concluded Rice, there was a "kinship" of immorality. What is the difference, Rice asked, between spies who steal America's secrets and betray their country—no doubt referring to the Rosenbergs, executed the previous year—and Baptists trained at the SBC seminary who then betray the true Christian faith? They were both guilty of crimes akin to "murder." They were both "wicked, rebellious sinners, against God and against Christ, against the Bible and against Bible morality, and Bible doctrine and Bible truth." These, concluded Rice, are the "dangerous triplets" you are facing in the world today. Just as Rice's mentor J. Frank Norris, in his 1923 address to Texas legislators, had held evolutionists responsible for war, crime, and immorality, John Rice all but said that evolutionists deserved the death penalty.

If Billy Graham had been present that day in Chattanooga, he might well have agreed with his "old friend." But he and Rice would soon part ways. Their break was precipitated by Graham's changing stand on evolutionary science and its relation to the Bible. In March 1955, *Sword* published an article in which Graham listed "Books Which Have Most Influenced My Life." Coming in at number three was *The Christian View of Science and Scripture* (1954) by Bernard Ramm (1916–1992). A Baptist theologian with a serious interest in science, Ramm did graduate work at the University of Southern California, where he wrote a master's thesis on James Jeans and Arthur Eddington, the cofounders of British cosmology, and a dissertation on the philosophical implications of Einstein's "new" physics.[47] Though he began his teaching career at the fundamentalist Bible Institute of Los Angeles, by the 1950s he had taken a position at Baylor

University, Rice's alma mater, and was looking for a middle way between a discredited fundamentalism and outright modernism.

Ramm's *Christian View* was one facet of the broader liberalizing neo-evangelical trend. Leading neo-evangelicals Carl F. H. Henry, L. Nelson Bell, and Harold Ockenga, among others, formed the National Association of Evangelicals in 1942 and soon filled the pages of *Christianity Today*, established in 1956.[48] Like them, Ramm held on to biblical "inerrancy" but refused to defend what he called the "pedantic hyperorthodox" interpretation of scripture.[49] Fearing that "the masses at large" thought that one needed to choose between the Bible and modern science, Ramm offered a way out of this dilemma by harmonizing science and scripture in a unique way. Ramm's work won over Billy Graham, prompting the break with Rice, but by the end of the decade his work also provoked a counterreaction. It led to the publication of Whitcomb and Morris's *The Genesis Flood* and a revival of young-earth creationism that has continued to this day.

Ramm began by reiterating his evangelical bona fides. Scripture was inerrant, inspired by God, and the data of the natural world were imbued with a "purpose and teleological ordering." But he quickly followed with his key argument that the language of scripture was filtered through the cultural idiom of its time and place. That idiom was, in his words, "*popular, prescientific,*" and, using terminology drawn from the philosophy of science, "*non-postulational.*" The Bible's words, that is, were inspired by the Spirit of God, but they do not "*theorize as to the actual nature of things.*"[50] This formulation allowed Ramm to construct an interpretation of scripture that took into account the findings of modern science in the twentieth century.

God could perform miracles, in Ramm's view, but in other cases, science had a better explanation. Did God actually make the sun stand still, as narrated in Joshua 10? Clarence Darrow asked this very question to William Jennings Bryan at the Scopes trial.[51] Rather than repeat Bryan's simple affirmation of the words of Joshua, Ramm preferred to cite the English astronomer E. W. Maunder. His study of this passage had convinced him that what Joshua had requested was to keep the sun from shining. In response, God sent a hailstorm, which obscured the sun and refreshed Joshua's soldiers, thus giving them the illusion that they had marched far less than they actually had. For details of Maunder's argument, Ramm referred readers to his "astronomical, geographical, exegetical, and historical data." In the final analysis, God *had* intervened—miracles occurred—but there was also a *natural* explanation underlying the apparently *supernatural* account.[52]

In sections on geology and biology, Ramm presented his perspective on evolution, which he called "Progressive Creationism." Crucially, Ramm accepted an old age for the earth. He rejected the notion that God created the world in 4004 BC, based on Bishop Ussher's calculations. Rather, Ramm offered, we should accept the claim, based on modern geological research, that the earth is "at least four billion years old." This claim flew in the face of the arguments advanced by George McCready Price, whose influence Ramm called "staggering." To Price's contention that dating rocks by fossils and fossils by rocks amounted to circular reasoning, Ramm affirmed the validity of radiocarbon dating. Ramm also concluded that the Noachian flood was not universal but local.

At the same time, Ramm did not wholly accept the day/age thesis, nor was he enamored of the gap theory, popularized by the Scofield Reference Bible and then by antievolutionist Harry Rimmer. Instead, Ramm argued that God created the heavens and earth, and then returned at various times to perform "great creative acts, *de novo*." Having rejected Price's young-earth view and yet seeking to maintain fidelity to Genesis, Ramm argued for what he called "Pictorial Day" creation. As he put it, "creation was *revealed* in six days, not *performed* in six days."[53] Whatever exactly that meant, it allowed Ramm to both affirm the inerrancy of the biblical account and to accept the findings of modern geological science about the antiquity of the earth.

When it came to biology, Ramm attempted to carry out a similar balancing act. He allowed for the possibility that "*root-species*" created by God could have given rise to other species by "the unraveling of gene potentialities or recombination." But he asserted that this sort of change was only "horizontal radiation," development within the "root-species," or, as Price might have said, within the created "kind." There could be no "vertical" progress without divine intervention. Ramm was emphatic about the impossibility of human evolution: we humans have a "*mental or spiritual nature which must come from above and not from below*." Ramm distinguished his view from theistic evolutionism. On the other hand, Ramm knew many Christians who did believe in God-directed evolution. He was not prepared to conclude unequivocally that evolution was "metaphysically incompatible with Christianity."[54]

And yet, having opened the door to the possibility of theistic evolution, Ramm made a clear point about the moral and political dangers of naturalistic evolution that George McCready Price, William Bell Riley, J. Frank Norris, and Dan Gilbert would have appreciated. Surveying the varieties of evolution, Ramm identified an "antichristian" type that "atheists and naturalists and materialists" used to "club" Bible-believing Christians. He related this kind of evolution to communism: "Dialectical materialism, the official philosophy of Russia, glories in evolution as the scientific doctrine of creation which frees man from faith in God." So that there were no doubts among his readers about where Ramm stood politically, he added that "evangelical Christianity will always be at war" with this "use of the theory of evolution."

In defending theistic evolution from its fundamentalist detractors, Ramm suggested that they ought to be paying less attention to evolution and more attention to "atheistic philosophies, atheistic psychologies, and atheistic sociologies." According to Ramm, the "hyperorthodox" had not uttered a "squeak" against them.[55] That was untrue. Riley, Norris, Gilbert, and others had been pillorying atheistic social sciences for decades. At the same time that Ramm sought to provide a middle way between fundamentalists and modernists, he affirmed his own anticommunist credentials and urged conservatives to step up their attacks on communist- and evolutionary-tinged social science.

John R. Rice was not persuaded. He temporarily continued to ally himself with Billy Graham, accompanying him on a whirlwind evangelistic tour to Scotland in the spring of 1955. But that summer, Rice laid down the gauntlet with a review of Ramm's book. Titled "Shall We Appease Unbelieving Scholars?" Rice's response took issue with Ramm's view of the Bible as mediated through Hebrew culture. Rice thought that Ramm was influenced, "perhaps unconsciously," by the modernists and "neo-orthodox" types. More pointedly, Rice told his readers that "we cannot recommend and we will not sell this book."[56] It was only a matter of time before Rice split with Graham. The penultimate act was a simmering conflict from 1954 to 1956 between the Southern Baptist Convention and Roberson's Highland Park Baptist Church. A week after Roberson pulled Highland Park out of the SBC in March 1956, Graham sided with the SBC. In October, *Christianity Today* began publication, a clear alternative to not only the modernist *Christian Century* but to Rice's *Sword of the Lord*. The next month, Rice publicly broke with Billy Graham. As Graham came to embody the neo-evangelical worldview, including Bernard Ramm's proto-theistic evolutionism, Graham's willingness to view evolution and communism as dangerous twins receded. He remained an anticommunist, but antievolutionism faded from his sermons.[57] Conflict over the politics of evolution helped to precipitate a signal break in the fortress of American fundamentalism.

In the same year that Ramm published the book that divided Graham and Rice, the US Supreme Court announced a nationally polarizing decision in *Brown v. Board of Education of Topeka, Kansas*. "Separate but equal" was now "inherently unequal." The 1954 *Brown* decision capped more than a decade of renewed civil rights organizing starting with the March on Washington movement that pressured President Franklin Roosevelt to issue an executive order in 1941 banning racial discrimination in government war contracting. Under pressure from African American activists and their allies, the Democratic Party added a civil rights plank to its 1948 party platform. President Truman ordered the desegregation of the Jim Crow army that same year. Like the initial opposition to Roosevelt's order, the rise of "massive resistance" to *Brown* was justified on the grounds of states' rights. But that sentiment drew its popular power from the association of civil rights activists with communism, as well as fears of interracial sex and marriage—"race mixing" and "mongrelization."[58] These themes infused a theological defense of segregation, which provided a popular and powerful vehicle for mobilizing white southerners.[59] In this light, it should not be surprising that resistance to civil rights had an antievolutionary subtext as well.

Public discussion of racial desegregation, with its communistic and sexual overtones, was well under way in the decade leading to *Brown*. Two years before J. Frank Norris spoke to Texas legislators in 1949 about racial desegregation by alluding to the dangers of a "mongrel race," he received a supportive letter from Texas preacher Ranald McDonald. Norris had just made headlines for confronting Louie Newton at the SBC convention in St. Louis, and McDonald had read an article attacking Norris. A widely reprinted piece from the *Atlanta Constitution* by antisegregationist editor Ralph McGill called Norris "a Ku Klux yelper and a loud-mouthed shouter in many demagogic political and hate rallies." McGill's article, McDonald told Norris, proved that the writer was a Communist. Such subversives always label their opponents as Klan members. Elaborating on the point, McDonald explained Communist logic as follows: "If you refuse to aid Negro rapist [*sic*] in raping your neighbor's [*sic*] wives and daughters you are a Ku-Kluxer." McDonald ended his brief letter assuring that he was wholeheartedly on Norris's side—"ten billion percent."[60] McDonald's equation of racial desegregation with rape and his desire to go exponentially beyond 100 percent for Norris conveys how intensely the issue was felt seven years before *Brown*.

While McDonald mentioned God and Christianity as he lashed out at integrationists, he did not explicitly ground his comments in biblical verse. But many did.[61] On the high end of education and "respectability" was Judge Horace C. Wilkinson (1887–1957), whose 1948 piece in the *Alabama Baptist* on the dangers of desegregation found its way into Norris's files. Educated at the University of Alabama law school, the "distinguished" Wilkinson was described by the *Alabama Lawyer* as "one of the State's most active and successful trial lawyers." After serving as assistant attorney general in the early 1920s, with a populist reputation, he helped lead the state Ku Klux Klan, launching vicious tirades against Jews, Blacks, and Catholics. Dispensing patronage as a local politico in Birmingham during the 1930s, Wilkinson was appointed to the bench and emerged as a power broker in the Alabama Democratic Party. Speaking to a group of Birmingham businessmen in 1942, he raised alarms about the gains that Blacks had made during the war, and emphasized the threat posed to both working-class whites and their employers. He also pointed the finger at "activist" judges who had incited Blacks to "murder and ravish and rob." He ended his talk by calling for a League to Maintain White Supremacy.[62]

Writing in the *Alabama Baptist* in 1948, Wilkinson took up the challenge from a growing number of white southern Baptists who wondered if racial segregation might be "unchristian." The proper way to examine this question, according to Wilkinson, was to ask another one: is "racial purity" unchristian? As his fellow Baptists knew, the answer could not possibly be yes, since racial purity was a gift of God. Wilkinson cited Bible verse in support of this claim. He began, appropriately, with Genesis and set the key text in bold type for emphasis: "And God said, Let the earth bring forth the living creature **after his kind,** cattle and creeping thing, and beast of the earth **after his kind:** and it was so." Segregation had arisen by "divine command." It was embedded in the act of creation. God had commanded that every separately created "kind" must "reproduce 'after his kind.'"

Wilkinson did not draw out the issue of evolution, but his insistence that each "kind" remained constant over time was significant. The word "kind" held no status in modern biological classification systems, the closest term being "family." Yet since it appeared in the King James Bible, creationists felt compelled to employ it. In 1944, a creation scientist coined the term "baramin"—after the Hebrew *bara* (created) and *min* (kind)—to evade this problem.[63] Though Wilkinson may have been oblivious to this fact, he was employing, for different purposes, the precise argument that antievolutionists had been making and would be making in the years ahead.

While Wilkinson considered Blacks to be fellow human beings with a common "Adamic" origin, he also observed that God had separated "the different races" at the Tower of Babel. God kept them separate so that each race could "maintain its racial integrity." So that "racial purity" remained intact, the races needed to be kept separate in all spheres. "We know beyond a reasonable doubt," wrote Wilkinson, "that social association and political intimacies between people of different races inevitably brings about and leads to intermarriage." Intermarriage meant interracial sex and pollution of racial purity. Surely, argued Wilkinson, God did not intend for this to happen. Any movement in the direction of breaking down these barriers must be opposed as ungodly. Segregation was sacred.[64]

Reverend Carey Daniel Jr. (1915–1987) agreed and amplified the prosegregation message for millions, with a nod to the politics of evolutionary science. Pastor of the First Baptist Church in West Dallas, Texas, Daniel served as vice-chairman of the Dallas chapter of the Texas Citizens' Council, which organized resistance to the *Brown* decision.[65] By 1954, Carey Daniel was not only the pastor of a prominent church. He was well connected. His cousin, Price Daniel, was the junior US senator from Texas. A protégé of the senior senator Lyndon B. Johnson, Price was a tireless proponent of "states' rights," whether they concerned tidelands claimed by the federal government or the issue of racial segregation. In what became a landmark case, Daniel defended the state of Texas in its refusal to admit African American student Heman Marion Sweatt to the University of Texas law school. Sweatt lost at the state level but won in the US Supreme Court, which ruled in 1950 that the "separate but equal" facilities for Black law students at the Texas State University for Negroes were distinctly unequal. When a group of nineteen US senators and eighty-two US representatives signed the "Declaration of Constitutional Principles" (better known as the "Southern Manifesto") in 1956, in defiance of *Brown*, Daniel was among the signatories.[66]

Despite Price Daniel's record as a supporter of segregation, pressure generated by his brother Carey probably helped to ensure that he would sign. Not only did Carey write to Price, warning him not be "soft" on integration, but Carey knew how to get the people of Texas fired up about the issue.[67] On May 23, 1954, the Sunday following the announcement of *Brown*, Carey Daniel preached a sermon at First Baptist in West Dallas. It was published in the form of a twelve-page pamphlet the following year as *God the Original Segregationist*. Daniel ranged over the Old and New Testaments to demonstrate that God, Moses, Jesus, the apostle Paul, and even Mother Nature, "God's second book," were advocates of racial purity. In "other objections answered," Daniel showed how racial segregation did not violate the golden rule, constitute disobedience to civil authority, illustrate the increasingly discredited "curse of Ham," contradict the Declaration of Independence, or amount to racial hatred. Over the following decade, Daniel's pamphlet sold more than a million copies.[68]

Daniel began with the book of Genesis. Unlike Horace Wilkinson's account of God's segregationist intentions, which began in the Garden of Eden, Daniel's began at Babel. In this respect, his story was less explicitly "creationist." There was no talk of "kinds." And yet Daniel's focus on Nimrod, who commanded the Tower of Babel to be built, was an anticipation of Babel-based explanations for the origins of evolutionary thinking that emerged over the next several decades in the works of Henry Morris. Nimrod, Daniel tells us, was Ham's grandson, and his name means "Rebel" or "Let Us Rebel." He was a "two-fold rebel, a double-dyed anarchist," who literally elevated "man" as if he were a god and resisted God's plan for scattering the races.[69] This political language fit well with

Daniel's contention, later in the piece, that there was "no Negro problem" until "Communist-inspired pressure groups" created one. Daniel misread history, but communists and socialists active in the labor movement had played an outsize role in the early decades of the Black freedom struggle, demonstrating the ineluctable connection between "cultural" and class battles.[70]

Like many biblical commentators, Daniel also described Nimrod as an agent of Satan, "a mouthpiece of the Devil." Indeed, Daniel counterposed Nimrod to God by labeling this section of the pamphlet "Nimrod the Original Desegregationist." He half-jokingly added that "I might have done better if I had entitled this section 'Satan the Original Desegregationist.'"[71] In similar terms, Henry Morris would describe Nimrod, in league with the devil, as the original evolutionist. Daniel's words might sound bizarre to secular ears, but Daniel knew his audience. By linking civil rights groups with the devil, and through guilt by association tying the Communists to the Evil One as well, Daniel had raised the religious stakes as high as possible. If Satan was behind the *Brown* decision, he needed to be stopped at all costs.

As *Time* magazine reported the following year, on the eve of the 1956 presidential election, Daniel led by example when he offered to turn over the buildings of the West Dallas First Baptist Church to establish an all-white school in the event that desegregation came to Texas. Explaining his plan to the press, Daniel noted that in setting up such a school, he would not only combat segregation but "correct several other evils" as well. His curriculum would condemn the United Nations and its "oneworld ideology" (often referred to as a modern Tower of Babel). Daniel's school would also present "evolution as a damnable heresy and not as scientific fact."[72] As members of Daniel's flock knew well, he was not just taking a random swipe at evolution, but was referring to 2 Peter and its warning about false prophets spreading "damnable heresies," just as John R. Rice had warned at Highland Park Baptist in Chattanooga two years earlier.

In 1956, Carey Daniel's offer to turn his church into a segregated private school with a distinctly right-wing Christian curriculum was the beginning of the development of "segregation academies" that sprouted across America in the coming decades. Many of these would serve to inculcate in young people a suspicion toward if not outright hostility to evolutionary science. Starting in the 1960s, such institutions, and like-minded teachers in public schools, would draw on a newly aggressive and influential "creation science" movement, led by Henry Morris and John Whitcomb Jr., culminating in the founding of the Institute for Creation Research in 1970. In the 1950s, evolution was still a subterranean political issue. But through public battles over religious modernism and racial segregation, Americans learned about the connections between communism, animalistic and interracial sex, atheism, and Satan. With such a combustible set of raw materials, the "anti-Darwin" fuse would not take long to light.

6

FLOOD, FRUIT, AND SATAN

In 1961, young-earth creationism exploded onto the American scene with the publication of *The Genesis Flood* by John Whitcomb Jr. and Henry M. Morris. By any standard the book was a blockbuster—more than two hundred thousand copies sold through twenty-nine printings by the mid-1980s.[1] Over the course of the 1960s, Morris took the lead in reviving an organized antievolution movement. By 1970, he succeeded in creating the infrastructure for the Institute for Creation Research (ICR). It served as the standard-bearer in the public battle against evolutionary science for the next thirty years. The success of Morris's efforts was an ironic testament to how much ground evolutionary science had gained, and not only in the universities and modernist churches. The Soviet Union's successful launching of *Sputnik* in 1957 sparked a massive effort by the US government, through the National Science Foundation (NSF), to retool high school biology education. One result was the NSF-funded University of Colorado–based Biological Sciences Curriculum Study. By 1963 the initiative had produced a series of attractive biology textbooks, infused with the modern evolutionary synthesis, that were adopted in more than half the nation's public schools.[2]

As Morris and his counterparts struck back, they were careful to declare that their interest in spreading creationist ideas was educational and not political. They recruited creationist scholars, most with PhDs, to staff the Creation Research Society and write articles for its journal that reflected their scientific training. And yet, the process that led to the ICR at the end of the tumultuous decade of the 1960s was thoroughly imbued with political concerns. The text of Morris's early works—including *Genesis Flood*—includes explicit references to communism as a fruit of evolution. The political context in which Morris built his organizational base was critical to its success. A rich network of anticommunist activists nourished Morris's Red Dynamite ideas. Among them were Fred Schwarz, David Noebel, James Bales, John N. Moore, R. J. Rushdoony, and Tim LaHaye. The establishment of the ICR reminds us that the "radical" 1960s generated social movements on both sides of the political spectrum. While young-earth creationists were just getting started on an organizational level, they would soon help to redefine the political contours of the American religious landscape.

Whitcomb and Morris, the duo who produced *The Genesis Flood*, were in some ways unlikely candidates for the role. The son of a US Army colonel, John Whitcomb Jr. was born in 1924 in Washington, DC, into a nominally Christian Episcopalian family. As a boy, he lived in China, where his father's regiment was stationed to protect American interests in the chaos of the Chinese Civil War. Back in the US he attended the McCallie School, an exclusive Christian prep academy in Chattanooga. After being ruled ineligible for West Point because of poor eyesight, Whitcomb attended Princeton starting in the fall of 1942 and loved his science courses, especially one in evolutionary geology. As he later recalled, "I was a total evolutionist."[3]

Whitcomb's road to creationism began during his freshman year, when Donald Fullerton, the leader of the Princeton Evangelical Fellowship, "confronted me with the Gospel of Jesus Christ in my dormitory room." Soon, as Whitcomb remembers, "I surrendered to the claims and the authority of the Lord Jesus Christ." Two months later, however, the war interrupted his education. He arrived in Europe in the fall of 1944, serving in an artillery unit in the battle of the Bulge. Discharged in 1946, Whitcomb returned to Princeton, became a history major, and continued his spiritual studies with Fullerton, who taught Whitcomb gap creationism based on the Scofield Reference Bible, which allowed for an unaccountably long period between Genesis 1 and Genesis 2.[4] When Whitcomb graduated in 1948, he was ready to serve God, and he set off for Grace Theological Seminary. After receiving his bachelor of divinity degree from Grace in 1951, Whitcomb was hired to teach the Old Testament. He was still teaching the gap theory. But he already had doubts about it after reading a new book written in defense of young-earth flood geology. That volume was titled *That You Might Believe* (1946); the author was twenty-eight-year-old Henry Morris.[5]

Born in Dallas in 1918 into a Southern Baptist family that moved around the state before finally settling in Houston, Henry Morris was baptized at the age of eight.[6] Morris hoped to study journalism at the University of Texas in Austin, but as the Depression undermined his father's real estate business, Morris had to settle for Rice Institute, which he could attend for free. Morris started at Rice in 1935, barely a decade after Fort Worth–based J. Frank Norris had blasted that school for harboring evolutionists. If student newspaper reporting is any indication of prevailing views in the 1920s, Norris was not popular there.[7] But Rice was congenial enough to the young Morris, who recalled that at this point he was, like John Whitcomb, "a theistic evolutionist and Sunday-morning Christian." Pursuing the more practical field of civil engineering, Morris graduated with honors in 1939.

It was only after he graduated and moved south to El Paso, where he worked for three years as an engineer for the International Boundary and Water Commission, that Morris's thinking began to change. Living away from his fiancée Mary Louise that first year, he had plenty of time to think and read the Bible. Soon Morris came to reject his theistic evolutionism and instead became convinced that God had created the earth in six twenty-four-hour days. After he and Mary Louise married, they joined a fundamentalist Baptist church in El Paso and started teaching Sunday school. Morris joined the Gideons in 1942. When he was invited to return to Rice as an instructor of civil engineering, Morris joined the Inter-Varsity Christian Fellowship, began advising the student Rice Christian Fellowship, and held regular Bible classes at his home.[8]

Morris, like Whitcomb, was still attracted to gap creationism, allowing as it did for both a literal six-day creation and an old earth. The best popularizer of this view was Dr. Harry Rimmer, a prolific author and spellbinding speaker. Rimmer's recently published work, *The Theory of Evolution and the Facts of Science* (1941), helped erode Morris's confidence in evolutionary explanations of the earth's history.[9] But Rimmer's writings also introduced Morris to the work of George McCready Price and put Morris on the path to young-earth creationism.

As Morris was trying to sort out the age of the earth and the length of the days of creation, he read Price's *New Geology*. Morris was thrilled. Here was "scientific" confirmation of both a six-day creation and a young earth. "It was a life-changing experience for me," Morris wrote. Surely unaware of Price's frustrations in his own quest to become a noted writer, Morris, the would-be journalist, was taken with Price's knowledge of science and scripture, his "careful logic," and his "beautiful writing style." Price's work, recalled Morris, "made a profound impression on me." After reading *New Geology*, Morris joined the young-earth Deluge Geology Society. Over the next few years, Morris hunted down Price's other books and eventually "read most of" them as well.[10]

Whether those books included Price's politicized *Poisoning Democracy* (1920), *Socialism in a Test Tube* (1921), or *The Predicament of Evolution* (1925) is unclear.[11] What is certain is that by the time Morris published his first work of Christian apologetics in 1946, *That You Might Believe*, he had absorbed the concept that evolution and communism were closely allied evils. Later republished as *The Bible and Modern Science* by the Moody Bible Institute, Morris's book was aimed at college students. They were suffering, in the author's view, from the "anti-Christian nature of the teaching" in secular institutions and even many "so-called Christian schools." Morris's inaugural work contained a chapter on evolution and another on the biblical flood. But in his very first chapter, Morris placed these topics in a contemporary political context. Given Morris's later insistence on separating his antievolutionism from politics, the attention he gives in this first work to the broader world is notable.

Morris described a chaotic world in moral decline: "The almost hopelessly confused political and international picture has its social and economic counterpart in a prodigious number of 'isms' and 'ocracies,' in the apparently irreconcilable conflict between capital and labor, in the fields of world trade and finance, in racial conflicts, in seemingly every phase of man's economic activity. . . . Moral barriers are falling on every hand, and few people in the world any longer seem to know or care about the differences between wrong and right."[12]

Even in the realm of physical science, old certainties were crumbling under the impact of Einsteinian relativity and Planck's quantum physics. So, too, was Christianity afflicted, with a profusion of "denominations, sects, and cults."[13] To confront this massive confusion, Morris offered the absolute certainty of his Christian fundamentalism.

In his chapter on the bankruptcy of evolution, Morris began by comparing the claims and logic of evolutionary science with the "fact of a creation already completed by GOD, as revealed especially in Genesis I."[14] The bulk of his attack on evolution focused mainly on the lack of factual support for that theory. He reviewed the "so-called proofs" for evolution, from Mendelian genetics to comparative anatomy to embryology to the fossil record. In later editions, Morris would add a section on creationist R.E.D. Clarke's "law of morpholysis," based on the second law of thermodynamics, which argued that nature tends to move from order to disorder. This argument against evolution from "entropy" would become a staple of creationist thought. In every single case, Morris concluded, the evidence was inadequate to support evolutionary claims.[15]

Before advancing this negative argument, however, Morris did provide positive "evidences" for what he called the "atheistic and satanic character" of evolutionary thought. These were the "evil social doctrines" that evolutionary thinking had "spawned." Morris pointed to the "deadly philosophies" of Nietzsche and Marx, who were both "profoundly influenced" by Darwin. The former brought the world German militarism, Nazism, and Mussolini's Fascism. From Marx, Morris wrote, "the world has inherited socialism, communism, and anarchism." There were also the "immoral doctrines" taught by Freud and Bertrand Russell. Summing up the cumulative impact of evolution, Morris considered it "unthinkable that a theory of any kind could have such far-reaching and such deadly effects as has the theory of evolution."[16]

In attributing social and political evils to evolution, Morris could draw on a long tradition including Price, Riley, Norris, O'Toole, Gilbert, and others. In the list of sixteen recommended books at the end of his chapter on evolution, Morris cited recent works by Price, O'Toole's 1925 book, and perhaps most significantly, *Evolution: The Root of All Isms* by Dan

Gilbert, published in a second edition only four years earlier in 1942. Gilbert stood out in the degree and explicitness with which he attributed both communism and sexual immorality to evolutionary teachings. In making the Red Dynamite argument in the 1930s and '40s, Gilbert spoke with an unusually clear voice. Morris was listening.

When Moody republished *That You Might Believe* in 1956, the list of recommended readings no longer included Dan Gilbert's work. Perhaps the publishers were clairvoyant. For in 1962, Gilbert, the supreme critic of the immorality of modern sex education, hit the headlines in the worst possible way. In a hotel outside San Bernardino, California, Gilbert was shot to death by Robert Marrs. Marrs's wife Martha had been having an affair with Gilbert and was pregnant with their unborn child. Robert Marrs was later acquitted of first-degree murder.[17] Gilbert was not the first nor would he be the last prominent evangelical to go down in a flaming inferno of hypocrisy.

When Henry Morris published *That You Might Believe* in 1946, Gilbert still provided inspiration, and Morris was on track to follow in his footsteps. When he arrived later that year in Minneapolis to begin graduate study at the University of Minnesota, none other than Gilbert's mentor, the aging antievolutionist warrior William Bell Riley, summoned the young Morris to his office. Riley had read Morris's book, and he was considering Morris for the job of leading Northwestern—a job that he would soon give to Billy Graham. It could not have hurt Morris's chances that he had promoted Riley's young protégé Dan Gilbert. While Morris politely declined the job offer, Riley was correct to see in Morris's writings a continuation of what the older man had started. Not only did these two men share a fundamentalist Baptist faith, but they agreed that evolution posed great dangers for American society, morality, and politics.[18]

Morris may have been in step with Riley, but in the ranks of the American Scientific Affiliation (ASA), in which evangelical Christians attempted to honor both God and the latest scientific advances, Morris's young-earth perspective was losing out. Signs of this included *Creation and Evolution* (1950) by ASA member Russell Mixter, a professor of biology at Wheaton College. Mixter rejected a universal Noachian flood and accepted an old earth. Along with ASA colleague J. Frank Cassel, Mixter urged their ASA colleagues to adopt theistic evolutionism. "Evolution is a fact," wrote

Cassel in 1951.[19] Then came Ramm's 1954 work that drove a wedge between Billy Graham and John R. Rice.

Now working on his graduate degrees at Grace Theological Seminary, John Whitcomb resented Ramm's book and looked for a way to strike back. Whitcomb had read *That You Might Believe* at Princeton. When Henry Morris appeared at Grace to speak in 1953 in defense of flood geology, Whitcomb was inspired to focus his doctoral dissertation on that subject.[20] In framing his approach to geological history, Whitcomb followed the lead of Cornelius Van Til, who taught apologetics at the fundamentalist Presbyterian Westminster Theological seminary. Whereas Christian fundamentalists had generally taken an inductive, fact-based "evidentialist" approach, as reflected in the writings of both George McCready Price and Henry Morris, Van Til instead argued for "presuppositionalism." Drawing on the writings of Dutch Reformed neo-Calvinists, Van Til contended that human beings had no autonomous reasoning ability apart from the mind of God. Our starting assumptions—or "presuppositions"—predetermine our conclusions. The only starting point that can grant humans true knowledge and understanding, argued Van Til, was total acceptance of the divine revelation of scripture.

Whitcomb began the "Genesis Flood" dissertation with a bold statement of his own philosophical "starting point"—"the infallible Word of God." Since Whitcomb believed that "revealed truth" was the only way to construct an accurate history of the earth, he made "no apology" for building his "entire case" on interpretation of biblical text. If he failed to make a convincing case, Whitcomb wrote, he would lose on the "battlegrounds of the Hebrew and Greek text of Scripture" and "not on the steep and slippery heights of mountains." The word of God, not evidence from the natural world, would settle the question.[21] Whitcomb proceeded to show how scriptural evidence could be aligned with the young-earth conclusions that Price had reached fifty years earlier and that Morris was reviving.

While Whitcomb claimed he did not need to employ geological facts from the natural world to make a positive biblical argument, he could not ignore the "steep and slippery heights." Like nearly all creationists before and since, he felt impelled to attack facts deployed by evolutionists. Not only did Whitcomb cite and quote Price on thrust faults, but he relied on George Barry O'Toole, "a scientist of note," whose critique of fossil dating methods and evidence of horse evolution Whitcomb found compelling. Whitcomb described O'Toole's 1925 book as "the most devastating attack ever made against the theory of organic evolution," and quoted long passages from it.[22]

In 1957, Whitcomb completed work on "The Genesis Flood" and sought to get it into print. He boldly stood on biblical authority. But he had produced a book manuscript on geological history without even a pretense of scientific method. The presuppositionalist needed an evidentialist. Whitcomb embarked on a search that eventually led him back to Morris. By then, Morris was teaching at Virginia Polytechnic Institute (VPI), where he headed up the Civil Engineering Department. What began as a plan for Morris to contribute a modest quantum of additional material ended with a book that he primarily authored.[23]

Among the creationist experts that Whitcomb had contacted for help was none other than George McCready Price. Then eighty-six, Price politely declined to offer his editorial services, but he did express his pleasure at seeing a new generation of creationists "coming forward to carry the torch of truth." Since Whitcomb had essentially updated Pricean flood geology, contacting Price made sense. And yet Price proved to be a public relations problem. As Morris informed Whitcomb, most scientists considered the self-taught geologist a "crackpot." Beyond Price himself, Adventism was suspect in the eyes of many evangelicals. In the end, Whitcomb and Morris carefully omitted all but a few incidental references to Price in the revised manuscript and removed all mention of his Adventism. In a letter to Price's biographer in 1964, some eighteen months after Price's death, Morris rewrote history. While "the direct influence of his writings were not significant in the preparation of our book," Morris explained, "the indirect influence was quite substantial."[24] The marginalization of Price has endured in creationist collective memory. Decades later, John Morris, himself a leading creationist, claimed that his father Henry Morris "wasn't following [George] McCready Price . . . [who was] a lay scientist and not a hard scientist. . . . When he was writing *The Genesis Flood* . . . he wasn't reading theologians. . . . McCready Price was a preacher, I think."[25]

With the benefit of hindsight, we know that Pricean flood geology was silently present in *The Genesis Flood*. It also may be that his Red Dynamite thinking was there as well. Even late in his life, Price was still known as someone who had made this connection. In the 1950s, Price kept up a correspondence with James Bales, a professor of Bible and theology at conservative Church of Christ–affiliated Harding College in Searcy, Arkansas. Harding president George Benson had persuaded a group of fundamentalist businessmen to fund the Harding American Studies Institute, which transformed the nearly bankrupt school into a major center for anticommunist organizing. It attracted a wide range of conservative figures, including the young Ronald Reagan, who lent his voice-over talents to a series of anticommunist films.[26]

With a doctorate in the history and philosophy of education, Bales had become active on the evolution issue in the late 1940s, linking up with a British creationist organization called the Evolution Protest Movement (EPM). Initial publicity for EPM, in 1932, had sought to awaken the British public to the dangers of evolution in ways that resonated with Price's early writings: "Christianity sanctifies the individual and the home; Evolution glorifies the herd and is the parent of Socialism and Communism; In Russia the theory of Evolution has supplanted Christianity. Darwin is the new Messiah."[27] Bales was increasingly outspoken about communism, publishing *Atheism's Faith and Fruits* in 1951 and later working with Billy James Hargis's anticommunist Christian Crusade.[28] In 1954, Bales contacted Price about his anticommunist activism, and Price suggested he read *Poisoning Democracy*. Bales thanked him for the suggestion and stated that Price was "exactly right" about the connection between evolution and communism. In a 1959 letter, Bales again complimented Price for his foresight.[29] Decades after Price had invented the "Red Dynamite" label, his views seemed more relevant than ever.

Even though Henry Morris shared Price's anticommunist perspective, Morris's job in *The Genesis Flood* was to showcase science, not politics. Compared to Morris's discussion of communism in *That You Might Believe*, the politics is easy to miss in *The Genesis Flood*. It comes at the very end of the book, after some four hundred pages devoted to detailed analyses of the size and capacity of the ark, Lyell's uniformitarianism, volcanism, the antediluvian vapor blanket, radiocarbon dating methods, and dinosaurs. Not only did Whitcomb lean heavily on O'Toole's critique of evolutionary science, but Morris tipped his hat to O'Toole's organizational strategy, as the author saves for the afterword his castigation of evolution for its baleful social, moral, and political consequences. And yet last is not least.

In the book's seventh and final chapter, which bears the distinctly apolitical-sounding title "Problems in Biblical Geology," readers encounter a summary section titled "Modern Significance of the Genesis Flood." It is here that Whitcomb and Morris explain the "Importance of the Question." Was the earth created, or did it evolve? Was the flood local or universal? These are "*not* mere academic questions," the authors insist. To the contrary, argue Whitcomb and Morris, "one's conviction about them may have deep influence upon his whole philosophy of life, and therefore, perhaps even upon his ultimate destiny."[30] Moreover, it is not just that evolutionary thinking and behavior may consign individuals to eternal damnation. Beyond this, as Whitcomb and Morris proceed to demonstrate, evolution creates a moral, social, and political hell on earth.

In order to join together evolution and communism, they begin by affirming that there are only two basic philosophies: God-centered and man-centered. In the second, human beings are capable of providing for themselves. Underlying that philosophy, the authors charge, is "the concept of evolution!" Any notion that humans can improve their earthly conditions—by their own efforts—is powered by evolutionary thinking and is an insult to God. This blasphemous idea, Whitcomb and Morris argue, is found in all non-Christian religions and philosophies, "be these ancient idolatries or primitive animism or modern existentialism or atheistic communism!"[31]

Noting that theistic evolutionism was more popular in America than the atheistic variety, Whitcomb and Morris observed that the same was not true in the Communist world. Atheistic evolutionism formed "the backbone of the whole scientific structure of Communistic philosophy." The authors then quoted Engels's funeral oration for Marx about his collaborator's affinity with Darwin and repeated the still-accepted story that Marx offered to dedicate *Das Kapital* to Darwin. Linking communism to other "man-centered" philosophies, Whitcomb and Morris assigned it a special place: "Communism is the most dangerous and widespread philosophy opposing Christianity today."

Even if communism was not sweeping the United States in the early 1960s, humanistic social sciences were thoroughly "permeated" by evolutionary thought. Since these fields of study had a "more immediate impact on actual human relationships and conduct" than did biological science, they posed a greater threat to society "than most people realize." Morris and Whitcomb followed the example of Riley, Norris, Gilbert, and others who had shone a spotlight on social science. They concluded the discussion of the "Importance of the Question" by making the ultimate source of that danger unmistakable: the "pride and deception of the great adversary, Satan himself."[32]

The Genesis Flood made Morris both famous and infamous. At VPI, the book quickly isolated Morris among his colleagues. The liberal pastor of the local Blacksburg Baptist Church relieved Morris of his Sunday school duties. "He was pretty much run out of town by the open-minded liberal faculty there at Virginia Tech," recalled son John Morris. But before leaving VPI, Henry Morris joined with other fundamentalist-minded colleagues to found the College Baptist Church (now the Harvest Baptist Church), affiliated with the Independent Fundamental Churches of America.[33]

Morris was suddenly in demand as a creationist speaker. While the bulk of the attention garnered by the book focused on the scientific-sounding support the authors provided for a six-day creation, their political message continued to be heard. On September 10, 1962, Morris spoke to some five hundred members of the Houston Geological Society. The bulk of his talk concerned the false "presuppositions" of uniformitarian geology. But the way Morris ended the talk was strikingly un-geological. Morris stated that "there is much more at stake here than simply a matter of geologic interpretation." Evolutionary science, he argued, had invaded "nearly every aspect of human life" and was the basis of Dewey's progressive education, Nietzschean ethics, Fascism, and Nazism. Significantly, he added, "*even more seriously* . . . modern Communism today is grounded squarely on the theory of evolution." He then cited Jesus's warning about "evil fruit" and urged his listeners to seriously investigate "the nature of the tree itself."[34] The chairman of the group ended the meeting abruptly without leaving any time for discussion, later apologizing to Morris after some of the Houston geologists lodged a complaint. Still, some of the attendees spoke with Morris afterward, and a few of them expressed "full agreement."[35]

With the Cold War in full swing, most Americans took the threat of communism seriously in early September 1962, only weeks before the

Cuban Missile Crisis. That morning's front-page headlines reported on anti-Castro Cubans in the US appealing to the Kennedy administration to help overthrow the Castro regime; the possibility of war with Cuba; and the downing of US-made U2 aircraft sent by Taiwan into Communist Chinese airspace.[36] In January 1963, Morris sent Price a copy of the talk, writing that he thought Price "might be particularly interested in the enclosed paper" and letting him know that the Houston crowd was "surprisingly friendly." This was the last letter Morris wrote to Price, who died about two weeks later, at the age of ninety-two.[37]

Morris continued to connect evolution and communism. In *The Twilight of Evolution* (1963), reprinted twenty-six times through 1990, Morris further developed his political "fruit" argument. The book was based on a series of sermons that the newly famous Morris delivered to the annual meeting of the Reformed Fellowship in Grand Rapids, Michigan. *Twilight* had one big advantage over Whitcomb and Morris's blockbuster: it was only ninety-seven pages long. Unlike *The Genesis Flood*, which began with the arcane details of the biblical account and the accompanying creationist science, *Twilight* began with the "fruits" of evolution. In the book's first chapter, "The Influence of Evolution," Morris made a revealing and strikingly pragmatic argument. After establishing that evolution reigned supreme in the nation's halls of science, Morris clarified why Christians should care. If evolution were "merely a scientific theory affecting the interpretation of the data of biology, geology, and astronomy," Morris explained, "we would not be too concerned." If evolutionary science could be reconciled with the Bible (although Morris insisted this could not be done), Christians "would be quite content to leave the subject to these scientists to work out on their own."

But evolution had massive consequences. The pro-evolutionary social science disciplines of sociology and psychology had a profound effect on "social orders." Political philosophies and movements animated by evolution had similar effects. As examples, Morris offered up communism, socialism, militarism, "and even the anti-Christian aspects of modern capitalism and colonialism." The attempt to distinguish and salvage a godly, moral, Christian capitalism from the exceptional instances of exploitation, attributed to anti-Christian evolutionary ideas, would continue to feature in Morris's work. But here, Morris devoted more detail to socialists than to unnamed evolutionist robber barons. Again citing Marx's admiration for Darwin, Morris pointed to socialist evolutionists of the

Second International and "modern Soviet writers," all of whom grounded their doctrines of "class struggle" and "atheistic materialism" in Darwinism. Hitler and Mussolini—identified with "racism and militarism"—also appeared in Morris's evolutionary pantheon. In Morris's eyes, evolution also undergirded Freudianism, behaviorism, and "Kinseyism." Morris summed up the "tremendous" and "deleterious" impact of evolution by citing the Gospel of Matthew and its "bitter fruit."[38]

Finally, evolution was evil. Morris expanded his argument, presented briefly in *The Genesis Flood*, that Darwinism was based on Satan's rebellion against God. Here Morris posed the question of how evolutionary thinking could persist if it contradicted both science and the Bible. His reply: "The answer is *Satan!*" The deceiver had "fathered this monstrous lie of evolution, for he is the father of lies." Citing John, Ephesians, Luke, and Corinthians, Morris provided textual evidence for the deceptive character of the devil. Just as George McCready Price had warned about a coming despotism, Morris warned that the communist-inspired United Nations—and "its multitudinous tentacles"—foreshadowed a Satan-inspired, human-centered world government that would culminate in the Antichrist. Meanwhile, the "evil progeny" of evolution—which included socialism and communism—was spreading "in terrifying profusion" around the world.[39] Morris could not have made it clearer why Bible-believing Christians needed to stop the evolutionists in their tracks.

Even before the publication of *The Genesis Flood*, Morris had been thinking about setting up a new creationist organization. In 1963, the Creation Research Society (CRS) was born. Walter Lammerts (1904–1996), a Missouri Synod Lutheran and PhD in genetics who taught horticulture at UCLA, served as the group's first president and editor of the *Creation Research Society Quarterly* (*CRSQ*). Lammerts represented a new breed of young-earth creationists who did not lack for academic bona fides.[40] Members of the CRS board of directors held numerous advanced degrees in science. These included Bolton Davidheiser, a Johns Hopkins–trained zoologist, and Duane Gish, who had earned a PhD in biochemistry from the University of California at Berkeley.[41] "Voting members" of the CRS, who elected the board, were required to have an advanced scientific degree. Without that credential, one could still be a "sustaining member." By 1967, the CRS had roughly two hundred voting members and six

Figure 11. Henry M. Morris, founder of the Creation Research Society and the Institute for Creation Research. The coauthor of *The Genesis Flood* (1961), which claimed that science supported young-earth creationism, Morris argued that evolution mattered because its "evil fruits" were so harmful to humanity. Institute for Creation Research. Used by permission.

hundred sustainers.[42] The organizational base for the ICR (and the future breakaway group, Answers in Genesis) was under way.

Articles in the early *CRSQ* centered on evolutionary topics such as fossil bacteria, the origins of the tetrapod limb, and microflora of the Grand Canyon. The prevalence of "real" science might suggest that Henry Morris's own interest in the social and political consequences of evolutionary thought was neglected. It was not, however, owing to the exertions of several key contributors. Prominent among them was John N. Moore, who served for nearly twenty years as the managing editor of *CRSQ*. As Morris relates, Moore had earned an MA in biology and an EdD in science education and taught natural science at Michigan State University (MSU).[43] What Morris does not mention is Moore's anticommunist politics.

During the early 1960s, when CRS was getting off the ground, Moore was anything but quiet about his extra-scientific views. Cosponsor of the student Conservative Club at MSU, he opposed allowing Communist speakers on campus, and his actions made local headlines in early 1961. Moore stepped up the pressure later that year when he sent a letter to the Van Buren County Farm Bureau, charging that MSU was dominated by communist ideology and was muzzling conservatives. The Farm Bureau reprinted his list of charges in its next newsletter, and the MSU faculty senate agreed to investigate Moore's claims. Moore was a regular speaker at meetings of the Daughters of the American Revolution, holding forth on the value of the American flag and the dangers of "ultra-liberalism" on campus. Joining with members of the Michigan Republican Conservatives in 1963, where the crowd listened to a taped address by presidential hopeful Senator Barry Goldwater, Moore chaired a forum about education. And at Hillsdale College in 1964, Moore spoke about the "consequences of scientific methodology" when used in the social sciences, arts, and humanities.[44]

Most likely that methodology was "evolutionary," since Moore's contributions to the *CRSQ* harped precisely on the noxious effects of evolutionary thinking on academia. In "Neo-Darwinism and Society," Moore began with a quotation from Richard Weaver's *Ideas Have Consequences*, which called attention to a false sense that society has been progressing. To the contrary, as Moore summarized Weaver, ever since the rise of empiricism, in which experience rather than transcendent values and ideals rule, society has been on the downgrade. Unfounded faith in human reason and a "philosophy of materialism" thus led to Darwinism in biology, Marxism in economics, and both behaviorism and psychoanalysis, the latter two based on "naturalistic 'drives' or 'urges' of sexual behavior."

While Moore tended to write in a humorless, formal manner, he emphasized the affinity of Darwinism and Marxism in earthier terms: "Darwinism was clutched to the bosom of Karl Marx and Frederick Engels." This broad change in thinking—moving away from the authority of God—had caused what Moore called an "indoctrination of the intelligentsia." Having established the general downward trend, Moore examined three academic fields reflecting it—history, economics, and literature. Each one, Moore concluded, was built on a false, evolutionary, materialist foundation. The negative impact on public policy, Moore told *CRSQ* readers, has been "colossal."[45]

Considering that he addressed a subject that often excited passionate debate, perhaps Moore's relentless "scientific" logic and formalism offered a refreshing change of tone for his readers. He could not be accused of being unclear, as he included diagrams demonstrating the causal links between Darwinian ideas and their social and political consequences.[46] He could be persuasive, as we know from the published account of one young evangelical by the name of Zola Levitt. Born into an Orthodox Jewish family, Levitt became a born-again Christian in 1971. By the mid-1970s, Levitt was a well-known premillennial dispensationalist who specialized in the subject of Israel's role in the end-time.[47] While he was still sorting through the implications of his newfound faith, Levitt met Moore, who helped convince him that he must reject evolution.[48] Levitt's resulting book, *Creation: A Scientist's Choice*, is an extended answer to a question that Levitt posed to Moore—what does it matter whether or not one believes in evolution? Moore's response did not focus primarily on the evidence for the findings of evolutionary biologists. Rather, it reproduced his claim that ideas have consequences. As Moore told Levitt, "World conditions are a good example of what occurs when people live by evolutionary thinking." Communist ideology, Moore told Levitt, originated with evolutionary thinking. Stalin's work camps were a form of the principle of the "survival of the fittest." As Levitt summarized this view, "Somehow Darwin's *The Origin of Species* led to Solzhenitsyn's *The Gulag Archipelago*." More broadly, the way we behave toward other human beings is a product of how we think of their essential nature. "If we believe we descended from animals," Levitt learned from Moore, "then we will tend to behave like animals, in accordance with such lineage."[49]

As one might expect, Levitt's book received high marks from a *CRSQ* reviewer. CRS supporter G. Richard Culp fully endorsed Moore's argument that evolution held responsibility for "crumbling moral standards" and that evolution-inspired "agnosticism" had led to the "monstrous systems of fascism and bolshevism." Culp hoped that Levitt's book would be widely read and would help young people find Jesus. Some readers might have been surprised to see a second review in the same issue, this one by CRS president Walter Lammerts. He joined Culp in endorsing the book's basic antievolution message. He parted company, however, with Moore's attribution of communism to evolution. Communism arose, in Lammerts's view, as a natural reaction to czarist oppression. "Capitalism" at the turn of the century had been very "harsh." But fortunately, Lammerts explained, conditions improved owing to the efforts of progressive politicians such as Robert LaFollette and "progressive capitalists" such as Henry Ford.

Lammerts did not fit the stereotype of creationist conservatism: a civil rights–supporting "Kennedy Democrat" in the fold of young-earth creation science who had little patience for the John Birch Society.[50] And yet he fit the description of an anticommunist Cold War liberal. Lammerts attributed improvements in workplace safety and higher wages not to the organized efforts of the labor movement but rather to an anticommunist and Jew-baiting auto magnate. Lammerts remained on relatively good terms with Henry Morris, who had made Moore's "ideas have consequences" argument in his own writings since 1946. It may have been a matter of style—Morris subtly wove his anticommunism into a broader tapestry of "creation science," while Moore was making headlines and wildly waving his political flag.

Somewhere in between the two was Bolton Davidheiser, the evolutionary zoologist turned antievolutionist who briefly served on the CRS board in the late 1960s.[51] In 1969, Davidheiser's *Evolution and Christian Faith* was published by the Presbyterian and Reformed Publishing Company, which had also brought out *The Genesis Flood*. In a favorable appraisal of the book in *CRSQ*, a reviewer placed Davidheiser's work alongside Morris and Whitcomb's creationist landmark: "What the *Genesis Flood* did with respect to geology, Dr. Bolton Davidheiser has done with respect to biology." The book ranged widely, providing a long history of evolutionary thought going back to the ancient Greek materialist Empedocles, whose role as the "father of the evolution idea" was recognized by many evolutionists. Just as Morris and Whitcomb had supposedly demonstrated the validity of a young earth, Davidheiser had reviewed the scientific evidence for biological evolution and convincingly showed it to come up short.[52] For all his interest in comparing Davidheiser's book to *The Genesis Flood*, the *CRSQ* reviewer chose not to comment on one intriguing similarity—its politics came at the end. After devoting some 350 pages to biology, Davidheiser turned to the subject of "Social Darwinism." Here, the author linked Darwin's ideas with American robber barons, with European imperialists and militarists, with Nietzsche, Mussolini, and Hitler.

Davidheiser's reviewer—Rousas J. Rushdoony (1916–2001)—was not unaware of the relationship between evolutionary biology and politics. Rushdoony later became infamous as the founder of Christian Reconstructionism. Unlike the premillennialists who populated Baptist and Methodist (and Seventh-day Adventist) congregations and foresaw a terrible apocalypse preceding the second coming of Christ, Rushdoony, like many Presbyterians and others in the Reformed tradition, was a postmillennialist. Jesus would return for his thousand-year reign only after human beings had *reconstructed* a modern-day version of what Adam and Eve had lost. Christians would first establish God's kingdom and reassert "dominion" over the earth, following God's command in the book of Genesis. Published in 1973, Rushdoony's Calvin-inspired *Institutes of Biblical Law* provided a blueprint for his distinctive vision. He called for the reestablishment of biblical law (including stoning as punishment) as the basis for a decentralized Christian theocracy (in his words, "a theonomy," for God's law) in America. Since Rushdoony's Reconstructionism (or Dominionism) occupied a place outside mainstream Christian evangelicalism, his early role in the rise of young-earth creationism has received little notice.[53] But it was significant and helps illuminate the Red Dynamite story.

Scion of an Ottoman Armenian immigrant family, Rushdoony was born in New York shortly after his parents narrowly escaped the genocide carried out by the Ottoman Turks against the Armenian people during World War I. He arrived in 1934 as a freshman at the University of

California with leftist sympathies. But as he was exposed to the variety of materialist and evolutionary "isms" that had alarmed the young Dan Gilbert—who had recently left the University of Nevada—Rushdoony began to challenge his professors.[54] Unlike Gilbert, Rushdoony seemed to relish academic disputation, as he remained at Berkeley for his MA in English literature. He attended a local, theologically modernist seminary, after which he was ordained as a minister in the mainline Presbyterian Church U.S.A. Moving rightward, Rushdoony worked closely with medieval historian Ernst Kantorowicz, a German-Jewish refugee who arrived in the US in 1940 and was fired from Berkeley in 1950 for refusing to take a McCarthyist loyalty oath. Unlike other such refugee scholars, however, Kantorowicz was nowhere near being a communist. He was a theologically minded historian and a staunch German nationalist. After fighting in the German army in World War I, he had volunteered for the ultra-right Munich *Freikorps*, which battled communists in the streets and included in its ranks a young Adolf Hitler.[55]

Rushdoony's intellectual turning point came after he left Berkeley, while he was serving as a missionary to Paiute and Shoshone people in northern Nevada. Along with his family's experience of persecution by Ottoman authorities, and his alienation from theological modernism, the plight of Native Americans helped deepen a Christian-infused hostility to the modern secular state. At the same time, Rushdoony had also absorbed a relatively optimistic Calvinist postmillennialism, which dictated that saved Christians must construct God's kingdom on earth *before* the second coming of Jesus. This eschatology also bespoke an appealing confidence that such a reconstructed America could be built. As he worked to create a little Christian commonwealth among the Native American people of northern Nevada, he continued to harbor academic ambitions. Rushdoony authored a book manuscript on Christian ritual and British politics and submitted it to the University of Chicago Press. They turned him down, and Rushdoony began to give up hope of making his mark either on the academy or on the reservation.

Thanks to a minister friend, he came upon a copy of the writings of Cornelius Van Til.[56] The uncompromising nature of Van Til's position was hardly new to Christian fundamentalism, but the way it was recast as a theory of knowledge appealed to those, like Rushdoony, who were seeking more secure intellectual foundation for their theological ideas. Starting in 1952, when Rushdoony and his family left Nevada for Santa Cruz, California, where he began pastoring a Presbyterian church, he gradually became more outspoken. In 1958, he left the Presbyterian Church U.S.A. and joined Van Til's Orthodox Presbyterian Church. That year, he published his first book, in which he offered the following Tilian formulation: "All facts being created facts, factuality can only be understood in subordination to God. But to understand factuality, man needs a norm and this Scripture provides."[57]

In 1962, Rushdoony left pastoring for good, and he began full-time work for a series of libertarian conservative think tanks, including the Center of American Studies.[58] After internal battles left Rushdoony on the sidelines, he moved in 1965 to start Chalcedon Inc. in Vallecito, California (later the Chalcedon Foundation), named for the AD 451 Ecumenical Council of Chalcedon. At this historic meeting, the church fathers clarified the nature of the human and divine character of Jesus Christ: the two were in perfect union—coexisting but unmixed. Since the human and divine were separate, this fact clarified that no man—even a king—could ever be divine. But despite this clear guidance, human beings had disregarded Chalcedon. They had set themselves up as gods, through their elevation of autonomous human reason and the power of the secular state.[59] For more than three decades, at the helm of the Chalcedon Foundation, Rushdoony sought to undo humanity's perverse handiwork.

Considering that the rejection of his first book manuscript changed the course of R. J. Rushdoony's life, it is appropriate that his contribution to the rise of the modern young-earth creationist movement also concerned a book seeking a publisher. In 1960, Rushdoony was called upon by the Moody Institute Press to review *The Genesis Flood* manuscript. Despite Rushdoony's positive review, Moody decided not to publish. Undeterred, Rushdoony suggested that the authors approach Presbyterian and Reformed Publishing, a small press owned by his friend Charles H. Craig, who had published Rushdoony's first book two years earlier. Craig admired George McCready Price and was thrilled to publish a book that seemed to confirm Price's "flood geology."[60]

After providing this crucial bit of young-earth creationist midwifery, Rushdoony continued to collaborate with Henry Morris and the CRS. In 1965, the year Rushdoony founded Chalcedon, he contributed a critique of evolutionary thought to the pages of the *CRSQ*. Rushdoony used the case study of "evolutionary scientist" Sigmund Freud to demonstrate, in Van Tilian fashion, that only a Christian faith could produce true science. Gilbert and other antievolutionists had held up Freud as an example of how evolutionary thinking justified sexual animalism. Rushdoony was more philosophical. Freud had used discredited Lamarckian arguments about the inheritance of acquired characteristics to explain features of human psychology—an explanation Rushdoony described as relying on the "miraculous." Freud did so, Rushdoony suggested, because there was no credible evolutionary alternative, only a mysterious and equally fantastical notion of some life-force built into the universe.[61] For Rushdoony, this meant that anyone attempting to argue from a non-Christian perspective inevitably and unconsciously "reintroduces His attributes" in a distorted form. We cannot think, that is, without God. If we want to reach the truth, we need to accept the true source of all knowledge—creation by the "totally self-conscious ontological Trinity."[62]

If Rushdoony's presuppositional philosophy might have been a bit obscure for *CRSQ* readers, he soon showed that he could pitch creationism to a broader audience. In 1966, Rushdoony wrote a pamphlet and helped produce an accompanying educational filmstrip (reviewed in *CRSQ*), both titled *The Necessity of Creationism*. The seven-page pamphlet was published and distributed to thousands of readers the next year by the Bible Science Association, which worked closely with the CRS and specialized in carrying "creationism to the masses."[63] As many antievolutionists had done before him, Rushdoony began by posing the question of why his readers should care about evolutionary science. Short of accepting Darwin wholesale, one could easily join with Bernard Ramm and "harmonize" science and the Bible. What, as he put it, is "at stake"? Rushdoony offered three reasons why readers should reject both Darwin and Ramm. First, if one allowed for any one part of the Bible to be questioned, the entire fabric fell apart. Second—and again invoking Tilian arguments—the anti-God "presuppositions" undermine a true, Christian science.

Third, and most expansively, Rushdoony explained to readers how evolution could "affect the mind and welfare of man." According to evolutionary thinking, man was no longer created in the image of God. Rather than the active subject called up by God to exercise "dominion" over the earth, he was the passive product of earthly evolution. Humanity then replaced God with the tyrannical, "total state," as in ancient Egypt,

whose rulers followed a proto-evolutionary philosophy. Evolution produced "slavery," and modern "totalitarian" thinkers, such as Karl Marx,
welcomed Darwin's ideas as "the necessary foundation for socialism." "If
men put their faith in evolution," Rushdoony wrote, bolding his text for
emphasis, "**they will then look to scientific socialist planners for salvation
rather than to Jesus Christ.**"

In such conditions, man became "the primary experimental animal."
Since he is no longer held to account by an eternal godly moral code, his
morals were reshaped by the social environment. Thus, evolution ushered
in a "Pavlovian world," in which humans would be conditioned to behave
in an immoral and "lawless" fashion. Without Christian morality and
law, evolutionarily inspired Americans would run amok. To illustrate the
consequent moral decline, Rushdoony quoted, in horror, from a handout
produced at UCLA by Bruins for Voluntary Parenthood and Sexual Liberty, a group supporting abortion rights, birth control, and freedom from
repressive sexual codes: "Where there is no victim, every act is morally
right." He also pointed to the "moral breakdown" in America's big cities,
where people lock their homes up tight to protect themselves from ni-
hilistic, anarchistic, evolutionary-thinking "monstrous new barbarians."
Like John R. Rice, J. Frank Norris, and others, Rushdoony saw only one
solution: a renewal of "Biblical Christianity." Unlike the Baptist premillennialists, including Henry Morris, Rushdoony did not invoke the signs
of the end times. He was girding for the grim struggle to retake America
that lay ahead.

In a review of the filmstrip version of the pamphlet, *CRSQ* editor
George Howe acknowledged the challenge Rushdoony faced in bringing
"the dry bones" of philosophical discourse to life, but judged that the
effort had succeeded. The audiotape accompanying the filmstrip closely
followed the text of the published pamphlet, including Rushdoony's
claim about "socialist planners" versus "Jesus Christ." The review indicates some of the graphic aids employed in conveying Rushdoony's ideas.
They included an "evolution tree" (presented twice during the filmstrip)
whose evil "fruit" was nourished by "anti-god and anti-biblical roots."
Howe helpfully summarized the rest: thirty-three photos, including six
portraits; fourteen Bible passages; eleven color diagrams; eighteen pictures of the Bible or other books or magazines; eight still lifes; four great
works of art; and five historical pictures, including scenes from ancient

Egypt and a portrayal of Noah's ark. Howe concluded that high school students would "find this filmstrip of great value."[64] In view of the dazzling array of high-technology educational tools available to teachers in the early twentieth-first century, it is easy to underestimate the impact of the lowly filmstrip and cassette tape. But they were at the cutting edge of "audiovisual" teaching technology in the mid-1960s. Rushdoony, who went on to become a major proponent of Christian homeschooling, may well have helped Henry Morris and his counterparts to see the potential in making creationism accessible.[65]

As Morris and Whitcomb began to gather a cadre of scientific creationists and like-minded thinkers in the mid-1960s, they gained allies from the broader world of Christian fundamentalism. Among the earliest was David Noebel (1937–), who would play a major role in illuminating the connections between evolution and communism and keeping those fires burning well into the twenty-first century. Born in Oshkosh, Wisconsin, in 1937, Noebel grew up in the nondenominational Lakeview Church, where he served as president of his youth group and committed his life to Jesus at the age of fifteen. After high school, he headed off to Milwaukee Bible College and then transferred to Hope College in Holland, Michigan, where he graduated cum laude with a degree in philosophy in 1959. Noebel was ordained as a minister in 1961, but the twenty-four-year-old was not done studying philosophy. He began pursuing his doctorate at the University of Wisconsin, under Arthur C. Garnett (1894–1970), a prominent philosopher of religion.[66]

By his own account, the liberal Madison campus tested Noebel's faith.[67] But Noebel passed the test and by the early 1960s had become a devout Christian conservative and a staunch anticommunist. In 1961, Noebel began pastoring the Fundamental Bible Church in Madison. At a neighboring church that year, Noebel spoke on "The Threat of Communism," showing the congregation the recently completed film *Communist-Led Riots against the House Committee on Un-American Activities in San Francisco, May 12–14, 1960*, better known by its short title, *Operation Abolition*. Produced by the US House Committee on Un-American Activities (HUAC), the film contained footage of recent student demonstrations in San Francisco organized against local HUAC hearings. According to the film's narrators, the students were dupes of "professional Communist agitators," including leaders of the International Longshoremen's Association subpoenaed to testify.[68]

Having gained some press, Noebel took a further step into politics in 1962 when he challenged Representative Robert Kastenmeier, the incumbent liberal Democrat, for the Second Congressional District seat. Running as an independent Republican, Noebel made his main issue "international communism." He vowed not to settle for "stalemate" but to fight for "victory."[69] Noebel lost the Republican primary election but garnered more than one-third of the party faithful vote and won suburban Waukesha County, the largest Republican stronghold in the state.[70] Noebel's electoral experience would serve him well as he developed into a skilled anticommunist activist.

As Noebel honed his political skills, a primary influence on the young activist's thinking was an Australian-born surgeon, marriage counselor, and psychiatrist who had an unlikely name for a Christian anticommunist powerhouse: Fred C. Schwarz. Dr. Schwarz helped Noebel and a whole generation of anticommunists think about the links between Marxist philosophy and evolutionary science. The son of a Viennese-born Jew who converted to fundamentalist Baptism, Schwarz first came to the US under the auspices of Presbyterian Carl McIntire and Baptist T. T. Shields, of the militantly separatist fundamentalist ACCC and its international branch, the ICCC. In 1953, with the encouragement of Billy Graham and others, Schwarz founded the Christian Anti-Communism Crusade (CACC).[71] Linking Christianity and anticommunism was hardly novel, but Schwarz had a talent for clear and provocative analysis.

Understanding how the communists thought, as Schwarz saw it, would enable anticommunists to predict their next moves and ultimately defeat them. A conservative who challenged communism without understanding the philosophy of dialectical materialism, Schwarz insisted, was akin to a dairy farmer who aimed to maximize milk production but had no interest whatsoever in cows.[72] Schwarz quickly developed a reputation as an authoritative speaker. His first two books—*The Heart, Mind, and Soul of Communism* (1952) and *Communism: Diagnosis and Treatment* (1956)—educated anticommunist Americans about the roots of communist thought in accessible fashion but also with serious attention to Marxist classics like Engels's *Origin of the Family, Private Property, and the State.*[73]

Schwarz's big break in came in 1957 when HUAC invited him to testify on "the communist mind." Consistent with Schwarz's attention to philosophical matters, he made the evolutionary foundations of communism clear. According to Schwarz, the communists were guided by "scientific laws" that rested on materialist foundations: atheism, the "material animal nature of man," and economic determinism. To underline the evolutionary point, Schwarz quoted the US Communist Party chief William Z. Foster: "Henceforth, the evolution of human species must be done artificially by the conscious action of man himself."

Schwarz indicted Communist evolutionism in two ways. First, Communists denied that humans were created in God's image, that we had a spiritual nature. Second, Communists' evolutionary view of humanity rendered us as little more than "expendable animals" to be used and killed for higher Communist purposes. As Schwarz told the committee, it was not just that Communists killed. The "tragedy" was that they committed murder and then transformed it into "a moral and righteous act." Communism was, thus, "ruthless and amoral." Summing up what he called "the great evil" of communism, Schwarz pointed to the "philosophic, basic concepts" of a movement that "materializes and bestializes man."[74] Thus did Schwarz invent a verb-form for the "beast" vocabulary that had been evolving over the previous century. But his political logic was old school, echoing the charge that John R. Rice had made in his 1954 "Dangerous Triplets" sermon—that communists were committed to whatever tactic would enable them to "win," whether it was murder, lying, stealing, or rape.

If twenty-year-old David Noebel had not yet heard of Schwarz in the spring of 1957, he most likely had heard of him by the end of that year. Thanks to the efforts of the Milwaukee-based Allen-Bradley Company, Schwarz's testimony was widely reprinted. For two decades, Allen-Bradley's owners had contended with workers organized into Local 1111 of the United Electrical Workers, a labor union with significant Communist Party influence. The company supported the John Birch Society and the American Enterprise Association (precursor of the American Enterprise Institute), a "non-partisan" research group that promoted the value of the free-market system and the dangers of labor unionism.[75] Allen-Bradley's owners viewed Schwarz's teaching on the nature of the communist "disease" as worth their investment.

In the fall of 1957, the company spent roughly $150,000 placing the entire text of Schwarz's testimony—as a full-page ad—in nearly thirty major newspapers and distributing hundreds of thousands of copies around the country. Released on December 22, the ad bore an alarmist headline: "WILL YOU BE FREE TO CELEBRATE CHRISTMAS IN THE FUTURE?" Subheadlines answered the question for readers. Among these: "NOT UNLESS You and other free Americans begin to understand and appreciate the benefits provided by God under the American free enterprise system." The ad rhetorically asked readers if "preservation of your life" was worth taking forty-five minutes to read Schwarz's words. If readers were dubious, Allen-Bradley's advertising managers helpfully selected sections of text and instructed readers to "pay attention to the material in bold type." Among the bolded sections was Schwarz's charge that communism "materializes and bestializes man." Schwarz reported to ACCC members that Allen-Bradley spurred others to follow suit. They included the American Legion and a host of companies, including the Southwestern Savings and Loan Association of Houston, which ordered ten thousand copies of Schwarz's testimony to send to all its account holders.[76]

This was only the beginning of Schwarz's rapid-fire ascent to anticommunist fame. In the early 1960s, he organized a series of anticommunist schools around the US. In Southern California, he mobilized an impressive coalition of politicians, businessmen, Hollywood actors, and celebrity entertainers to put on a series of giant spectacles. One high point was "Anti-Communism Week" in Los Angeles, endorsed by forty-one area mayors. It featured "Youth night" at the Los Angeles Sports Arena, attended by a crowd of sixteen thousand young people. Speakers included Roy Rogers, John Wayne, and Ronald Reagan, whose speech brought the crowd to their feet. Closing the show, and echoing antievolutionist Gerald Winrod's "funny" comment decades earlier about his own children, singer Pat Boone declared, "I don't want to live in a Communist United States. I would rather see my four girls shot and die as little girls who have faith in God than leave them to die some years later as godless, faithless, soulless Communists."[77] Fred Schwarz and his followers helped thousands of conservative Christians understand in their bones why anticommunism mattered.

Along with Dr. Schwarz, the other major influence on David Noebel's career was Billy James Hargis. He would catapult Noebel into national prominence and cement his reputation, for good or ill, as an expert on the fruits of evolution. Like J. Frank Norris, Hargis grew up in the American Southwest into a working-class family. Born in Texarkana, Texas, in 1925, and adopted by Jimmie and Laura Hargis, young Billy found Christ at the age of nine. The family attended the Rose Hill Christian Church, affiliated with the Disciples of Christ. At the age of ten, while his mother was undergoing life-threatening surgery, Hargis prayed to God and promised that if his mother lived through her ordeal, he would devote the remainder of his life to Jesus. In his own inimitable way, Hargis kept his promise.[78]

In 1943, after graduating from Texarkana High School and finishing one year of classes at the Ozark Bible College in Bentonville, Arkansas, the future home of the Walmart empire, Hargis was ordained a Disciples of Christ minister. In his early sermons, Hargis gained practice in the tradition of Oklahoma "bawl and jump" preaching but was not yet a conservative firebrand. Under the influence of the Disciples reform tradition, Hargis touted the virtues of pacifism and world government. But settling for several years in Sapulpa, Oklahoma, near Tulsa, Hargis became acquainted with A. B. McReynolds, an influential Disciples of Christ soul-winner and conservative activist. Feeding Hargis a steady stream of anticommunist writings, including John Flynn's *The Road Ahead*, McReynolds convinced Hargis that "the Federal Council [of Churches] was perverting the Gospel, undermining the Church, getting her involved in left-wing political activities, and I wanted no part of it." The young preacher decided to take a stand.

In 1948, Hargis began publishing *Christian Echoes*, which informed readers of the dangers posed by "deadly satanic-inspired Communism." Hargis attracted national attention, as he increasingly took to the radio waves. J. Frank Norris took note of Hargis in the pages of the *Fundamentalist*: "He is the Pastor of a great church. . . . He is a Pre-Millennialist and belongs to that great crowd of believers and evangelists of all denominations. He is a man of courage and deepest consecration." Two years later, Hargis left his Sapulpa pastorate, renamed the paper *Christian Crusade*, and launched the eponymous Tulsa-based organization that was the center of his work for the next two decades.[79]

The "Crusades" against communism operated by Hargis and Schwarz occupied some common ground. Both leaders had an internationalist

vision and collaborated with separatist fundamentalist Carl McIntire. In 1953, Hargis received publicity when, with McIntire's ICCC,
he launched from West Germany six thousand balloons attached to
Bible verses into Eastern Europe. With the triumph of the 1959 Cuban
Revolution, Hargis paid increasing attention to Fidel Castro and Soviet
designs on Latin America. He enlisted as a Christian Crusade lecturer
Fernando Penabaz, a prominent Cuban attorney and editor who had
turned against the revolutionary government, came to the US in 1960,
and joined Hargis on a nationwide anticommunist speaking tour called
Operation: America Awake.[80]

Hargis also helped publicize the case of Rev. Richard Wurmbrand, a
Jewish-born Romanian and former Communist who converted to Christianity. An ordained Anglican and Lutheran minister, Wurmbrand was
imprisoned twice for sedition by the Stalinist Romanian regime, until he
finally won amnesty and moved to the US. Testifying in 1966 before a
US Senate subcommittee investigating Communist persecution of religion, Wurmbrand stripped to his waist to reveal the scars he said were the
marks of his torture by his Romanian Communist captors. Wurmbrand's
account of the ordeal—*Tortured for Christ* (1967)—won him a wide hearing among American conservatives. The Christian Crusade distributed
Wurmbrand's book, sponsored joint speaking tours, and featured the pastor's story in the *Christian Crusade*, with a front-page cover photo of the
anticommunist Christian martyr shaking hands with Billy James Hargis.[81]
Among the American evangelicals who were captivated by Wurmbrand's
story was Henry Morris, who would later cite Wurmbrand on the satanic
character of Marxism.

Despite their common interest in global anticommunism, Billy James
Hargis and Fred Schwarz were very different people. Schwarz offered a
well-researched, carefully articulated summary of Marxist philosophy.
Hargis was a persuasive speaker. But he had little patience for study that
seemed irrelevant to his immediate needs. Looking back at his short stay
at Ozark Bible College, Hargis admitted that while the dean was a "genius" in teaching the Greek language, he himself had no interest in taking
his course "since I didn't intend to preach to any Greeks."[82] What Hargis
may have lacked in personal inclination, however, he more than compen-
sated for in recruiting a team of prominent and dedicated conservative
educators, led by the young Rev. David Noebel, to staff a summer training

school in the Rocky Mountains that would arm the nation's youth to do battle with socialism and Satan.

In June 1962, Hargis announced that the Christian Crusade had bought the Grand View Hotel in Manitou Springs, Colorado, nestled in the foothills of Pikes Peak, and renamed it the Summit Hotel. The following summer, David Noebel held the first session of the Crusade's Anti-Communist Youth University (ACYU). Also called the Summit School (and later rechristened Summit Ministries), the ACYU began classes in the summer of 1963. Thirty-six young people, age fourteen and up, attended the first summer session. Noebel, affectionately known by the students as "Doc," served as dean and one of the instructors. In the early days, he also did building maintenance, while his wife Alice did the cooking. Cost per student, which included room, board, tuition, and books, was $100 (approximately $865 today). That first summer, some 70 percent of the students were on full scholarship. Scholarship funds were contributed by *Christian Crusade* subscribers and other donors.[83]

A foundational Bible passage for the ACYU, which later provided the title for one of Noebel's best known books, came from 1 Chronicles 12, in which King David is gathering tens of thousands of "mighty men" to wage war. One of the smallest bands stands out in importance: "And of the children of Issachar, which were men that had *understanding of the times*, to know what Israel ought to do." Wisdom was a powerful weapon. As one of Noebel's supporters later commented, "Doc" aimed to equip young Christians with the knowledge so that they could resist the "attractive, but false worldviews penetrating and destroying Western Civilization."[84] Noebel emulated Fred Schwarz by insisting that a thorough examination of the enemy's philosophy was essential to doing Christian combat.

The multifaceted communist enemy was addressed in a wide range of talks to Summit students that first summer in 1963. Hargis, who would be a regular feature at Summit, spoke on "The National Council of Churches," "International Communism," and "Christ vs. Anti-Christ (God out of Schools)." Journalist and intelligence agent Edward Hunter, who accused the Chinese Communists of "brainwashing," spoke on this topic and delivered a two-part lecture on "Dialectical Materialism."[85] Noebel spoke on "The United Nations," "The Book of Revelation," "Internal Communism," and "The Seed of Woman (A Tale of Two Cities)."

Another member of the faculty was General Edwin Walker (1909–1993), who had become an ultraconservative cause célèbre when he resigned his commission after clashing with the Joint Chiefs for his promotion of John Birch Society views to his troops. In 1962, Walker was arrested (and then briefly institutionalized under orders from US Attorney General Robert F. Kennedy) after he helped to organize armed attacks on civil rights activists supporting the admission of James Meredith at the University of Mississippi. Walker—described by Noebel as a "heroic patriot-soldier"—spoke at Summit on "The Extremist," "A Definition of a Communist," and "Building Americanism."[86]

Finally, "Evolution" was the subject of teaching conducted by W. O. H. Garman (1899–1983). Born in Philadelphia, Garman embraced the fundamentalism of the Philadelphia School of the Bible. This training girded him for the more modernistic teachings at the Pittsburgh Theological Seminary, from which he emerged in 1925 as an ordained United Presbyterian minister. In an early sermon, Garman focused his fire on communists, labor unionists, members of "other devil created organizations" as the "onrushing forces of the Anti Christ."[87] In addition to his pastoring duties, Garman led several national fundamentalist organizations. The first, from 1942 to 1981, was the Associated Gospel Churches, a group formed for the explicit purpose of getting fundamentalist chaplains into the US armed forces. The second was the Independent Fundamental Churches of America, which Henry Morris and his supporters had joined in Blacksburg after publication of *The Genesis Flood*. The third was McIntire's ACCC, over which Garman presided from 1947 to 1950. Garman was fully on board with the central message of Noebel's venture, offering a lecture on "Communism and Our Churches." Based on Garman's writings, it is reasonable to conclude that he, like Morris himself, viewed evolution and communism as related evils. In a pamphlet published by the ACCC, Garman denounced the Federal Council of Churches and "their modernistic and communistic activities."[88]

Summit students would have been hard pressed to avoid the Red Dynamite message. The curriculum prominently included books like *Communism: Its Faith and Fallacies* (1962) by Harding College's James Bales. Bales explained that communists denied the existence of God and maintained instead that "matter created life and that life's manifold forms, including man, have evolved without the operation of any force or forces beyond those which we see working in matter today." Bales challenged the idea that, in dialectical fashion, matter (and then life) might have emerged from non-matter.[89] Students also read *The Naked Communist* (1958) by Mormon fundamentalist W. Cleon Skousen, who accurately informed students that "Dialectical Materialism is an evolutionary philosophy."[90] Finally, Dean David Noebel was adept at tracing the connections between the Marxism and evolutionism, as he would later detail in *Understanding the Times* (1996). Its chapters on Marxism, secular humanism, and biology, Noebel has claimed, reflect the accumulated experience of his Summit teaching.[91] America's anticommunist youth were well equipped in the Colorado mountains to do ideological battle.

By the middle of the 1960s, Henry Morris and his colleagues at the Creation Research Society had made gains, but they were missing a key ingredient: an organizer who could provide a strong administrative platform for Morris's young-earth creationist views and make them accessible to the Christian masses. Enter Tim LaHaye (1926–2016). The man who would decades later become the millionaire coauthor of the *Left Behind* novel series and a mover and shaker in the high politics of the Christian Right had a modest start in life. Born in 1926 in Detroit, LaHaye was the son of Frank and Margaret (Palmer) LaHaye. He was raised as a fundamentalist Baptist, his mother serving as a fellowship director at a local Baptist church.[92] His father, the son of French Canadian immigrants, worked at a Detroit Ford plant as a mechanic. When Tim was only a toddler, the family moved out to the small town of Farmington.[93]

Tim LaHaye might have grown to adulthood in Farmington but for the tragic event that took place on April 13, 1936, just two weeks shy of his tenth birthday: his father, age thirty-four, died of a heart attack. As Tim LaHaye later related, his despondency was mitigated only by the comment of his pastor, who said, "'This is not the end of Frank LaHaye; because he accepted Jesus, the day will come when the Lord will shout from heaven and descend, and the dead in Christ will rise first and then we'll be caught up together to meet him in the air.'" "All of a sudden," LaHaye recalled, "there was hope in my heart I'd see my father again."[94] That hope powered his dispensationalist conviction that no matter how badly things might seem, in the end, the righteous will be reunited in heaven.

We have only a glimpse of how Tim LaHaye's upbringing may have shaped his political views. With a father who worked at Ford while the battle over the UAW was raging, Tim might well have heard J. Frank Norris's sermons over the family radio. We do know, however, that already as a young man, LaHaye was critical of Roosevelt's New Deal. After Frank LaHaye's death, the family moved back to Detroit, where twelve-year-old Tim sold newspapers on the streets to help his mother make ends meet. She worked at a hospital and later in a war-industry plant, all the while studying at the Detroit Bible Institute. When the child-labor provisions of the federal 1938 Fair Labor Standards Act cost Tim this job, LaHaye recalls, "I remember coming home and telling my mother what I thought of politicians who mess around with our lives." If LaHaye was unhappy with Dr. New Deal, he was perhaps fonder of Dr. Win-the-War. At the age of eighteen, after finishing night school and briefly attending a Chicago Bible institute, he joined the Army Air Forces and served the last year of the war in uniform.[95]

The religio-political war that LaHaye would wage for decades to come began the next year when he enrolled in 1946 at fundamentalist Bob Jones University. If Tim LaHaye had any lingering thoughts about the validity of evolutionary ideas, they would have been crushed at Bob Jones, where instructors enforced a militant antievolutionism. On that Greenville, South Carolina, campus LaHaye developed a different kind of crush, on fellow student Beverly Jean Ratcliffe, a like-minded Detroit native who embraced a fundamentalist Baptist dispensationalism, and the two married in 1947. She would go on to lead the Christian conservative charge alongside her husband, as author, speaker, and organizer. After graduating in 1950, the LaHayes moved with their growing family from South Carolina to Minnesota, and then in 1956 to the San Diego area. That year, Tim LaHaye became pastor of Scott Memorial Baptist Church, in the San Diego suburb of El Cajon, and the LaHayes began to make connections in this hotbed of Christian conservatism. In the first of many joint activist ventures, Tim and Beverly began cohosting a half-hour weekly radio and then television program titled *The LaHayes on Family Life*.[96]

By the early 1960s, LaHaye had firmly committed himself to anticommunist activism. He appeared in an early John Birch Society film explaining why he joined. It was not until he read Birch material that he realized that the "liberal takeover" of church and country were "parallel." La-Haye hesitated to join the Birch Society since he preferred not to "mix religion and politics." But then he realized that if the communists were not stopped, Christians would not be free to preach the gospel. Having volunteered for service against the national enemy in World War II, LaHaye felt he could do no less, now that Americans confronted "the greatest enemy our country has ever faced."[97]

Among the evils perpetuated by the communists, in the minds California conservatives, was "the smut problem." According to El Cajon–based California assemblyman and Methodist minister E. Richard Barnes, a veteran "anti-obscenity" crusader who had headed up the San Diego chapter of Fred Schwarz's Christian Anti-Communism Crusade, the communists had mounted a "Satanic assault on the whole structure of American life." Around 1964, LaHaye teamed up with Barnes to launch California League Enlisting Action Now (CLEAN) to oppose pornography and sex education. In 1966, CLEAN campaigned for Proposition 16, which cracked down on the pornography industry and enlisted the support of the Republican candidate for governor Ronald Reagan.[98]

LaHaye had become sufficiently convinced of the urgency of the communist threat to American youth that he saw the need for Christian parents to band together and create their own educational institutions. Since the mid-1950s, when Carey Daniel offered his church in Dallas to house a private, fundamentalist, antievolutionary segregated school system, the movement for private Christian schools had grown. It was boosted by US Supreme Court rulings in *Engel v. Vitale* (1962), making school prayer unconstitutional, and *Abbington v. Schempp* (1963), doing the same for Bible reading in school. In 1965, LaHaye founded Christian High School as an adjunct to Scott Memorial Baptist. Only sixteen seniors graduated in 1967. But as LaHaye and others convinced conservative Christians that keeping their children in public schools was a mistake, enrollment boomed. By 1979, there were some five thousand such schools in the US, with more than one million students.[99] In San Diego County, by 1981, there were nine schools and some twenty-five hundred students in what became the Christian Unified School District, still under the aegis of Scott Memorial and the LaHayes.[100] Scott Memorial grew as well during these years. It tripled in size between 1956 and

1971, and later, under pastor David Jeremiah, it became one of the first "mega-churches" in the US.

In the 1960s, Scott Memorial's growth was due in large part to the increasing popularity of the work that Tim and Beverly LaHaye did to popularize a Christian approach to psychology and marriage counseling. Building on a decade of their broadcasts on family life, Tim LaHaye published the *Spirit-Controlled Temperament* in 1966, the first of a long series of Christian self-help books issuing from the LaHayes over subsequent decades. In his inaugural effort, LaHaye offered a fundamentalist Christian spin upon the ancient Greek temperament theory. All human beings, that is, were born with one of the four basic temperaments—sanguine, choleric, melancholic, and phlegmatic. By identifying their distinctive temperaments, readers could gain greater insight into themselves and achieve greater satisfaction in their relationship with others. LaHaye drew on the work of Norwegian theologian Ole Hallesby, whose work on temperament theory was published in English just four years earlier as *Temperament and the Christian Faith* (1962). Following Hallesby, LaHaye claimed that when someone was reborn in Jesus Christ, that person could acquire a new temperament or combination of old and new.[101]

In a supreme irony, LaHaye, who would come to be identified with modern creationism, had based his first book on ancient proto-evolutionary thinkers. Temperament theory rested on the claim by Empedocles that there were four elements: fire, air, earth, and water. As the creationist Bolton Davidheiser had noted, Empedocles could be considered the original evolutionist. LaHaye did encounter some resistance from fellow Christian counselors.[102] But this did not seem to matter to his readers. Given the cultural tumult of the 1960s, they were eager for practical tools to help them navigate the storm. *Spirit-Controlled Temperament* made the Religious Best Sellers List, based on a national sample of bookstores affiliated with the Christian Booksellers Association.[103] LaHaye was in high demand as an authority on Christian psychology. Addressing Christians about the intimate realm of their personal relationships, LaHaye always kept the big apocalyptic picture in mind. In 1971, when, on the invitation of the city's ministerial association, LaHaye gave a weeklong series of talks in the community building of Greeley, Colorado, his nightly topics came in pairs, such as "Keys to Wedded Bliss" and "When God Destroys Russia."[104] No dimension of human experience was too small or too large for Tim LaHaye to illuminate for his conservative Christian flock.

As LaHaye continued to raise his public profile, he crossed paths with Henry Morris. In early 1970, they both spoke at a conference named for the esteemed early fundamentalist Reuben Torrey, editor of the original *Fundamentals*, at Biola College (formerly the Bible Institute of Los Angeles), which Torrey had helped lead in its heady early days. Both men had been thinking about establishing an institution of higher learning that could teach creation science. For LaHaye, a San Diego–based college would complete the chain that he had begun with Christian high school five years earlier. He was familiar with Morris's writings and asked him to teach creationism at his planned college. In his written response to LaHaye, Morris expressed great enthusiasm for taking the fight against evolution to a higher level. "I do believe that the issue of Biblical creationism is the most urgent issue confronting Christianity today," Morris wrote. "The evolutionary system is at the root of most of the spiritual and moral problems that have arisen to hinder the gospel and its proclamation today." Whether or not LaHaye had read Morris's work that linked communism and evolution, he surely agreed that communism was an evil fruit of the satanic turn the country had taken.

Morris not only underlined the importance of creationism, but he also emphasized the kind of educational institution that was needed. It would not be enough to create another Christian "college." Rather, Morris envisioned a "Christian *university*," which could offer a graduate program in scientific creationism. Such a "nerve-center" of "Biblical truth," wrote Morris, would "prepare and send out solid Christians in various strategic professional fields as well as posts of full-time Christian service."[105] In the fall of 1970, Christian Heritage College opened its doors on the grounds of Scott Memorial Baptist Church. LaHaye was president, and Morris served as academic vice president. Morris wrote a statement of faith that reflected a premillennial, pre-tribulation, Baptist fundamentalism, but also specified tenets of Morris's young-earth creationism: creation of all things in "six natural days" and a worldwide flood. The initial entering class at CHC was only a "handful." The initial degree programs were meager in number: Bible, Education, Mission, and Liberal Arts.[106] But the convergence of LaHaye's and Morris's Christian ambitions would soon facilitate the transformation of Morris's CRS into the Institute for Creation Research and change the face of creationism for decades to come.

Trees, Knees, and Nurseries

When the Institute for Creation Research opened a new facility for its Museum of Creation and Earth History in Santee, California, in 1992, the exhibit that visitors saw before entering the gift shop featured two contrasting trees. The Creation Tree bore a variety of colorful "fruits." They represented "Genuine Christianity" and "Correct Practices," such as "true science," "true history," and "true government." The Evolutionary Tree offered more dangerous fare. From its branches hung "Harmful Philosophies" and "Evil Practices" that the ICR attributed to evolutionary thinking. At the top of the evil-practices list was promiscuity. Of the twelve items that followed, eight concerned sexuality, gender relations, or family life. They included pornography, abortion, homosexuality, and bestiality. "Communism" topped the list of harmful philosophies. Related items included humanism, atheism, amoralism, behaviorism, and materialism. Before arriving at the trees, visitors passed through the Hall of Scholars. On one wall of the corridor were proponents of evolution, who included familiar "social Darwinists" of the Right Andrew Carnegie and

John D. Rockefeller Jr. At their center and placed higher in elevation than the rest was a portrait of Karl Marx. Visitors learned that "although he was a professing Christian in his youth, he became an atheist and (according to some) a Satanist in college."[1]

For secular-minded Americans who were distressed by the public campaign against evolution in the 1970s and '80s, the ICR's linkage of evolution, communism, Marx, and Satan may have seemed bizarre. But to those who had been inspired by the writings of Henry Morris, Tim LaHaye, and other critics of evolution over the previous two decades, it would have made perfect sense. The attribution of a host of evils to evolutionary thought had become a staple of conservative Christianity. The prominence of sexual issues in this mix of "evils" was well established. In the period leading up to 1992, the language employed by antievolutionists focused less on "communism" or "socialism" and more on "secular humanism." This reflected the waning Cold War and the prominence of issues like abortion and homosexuality, on which the Stalinist Soviet Union was not in the forefront of progressive thought. By the end of 1991, the Soviet Union had disintegrated. But the focus on "secular humanism"

Karl Marx (1818–1883)

Karl Marx is considered to be the chief founder of Communism. Although he was a professing Christian in his youth, he became an atheist and (according to some) a Satanist in college. His philosophies of history and economics were squarely based on evolutionism. In fact, he wanted to dedicate his book *DAS KAPITAL* to Charles Darwin, who had given him what he thought was the scientific foundation for Communism.

Figure 12. Karl Marx (1818–1883), museum display panel, 2012. Based on Henry Morris's writings, this display panel tightly connects Marx, Darwin, and (possibly) Satan at the Creation and Earth History Museum, Santee, California, established by Morris's Institute for Creation Research. Photo by author. Courtesy of Light and Life Foundation.

represented a fundamental continuity, rather than a break, in Christian anticommunist thinking.

The evolution of evolutionary science in the 1970s also drew critical attention in ways that highlighted the red connection. As the public war over teaching evolution proceeded, one of the leading public defenders of evolution was Harvard paleontologist Stephen Jay Gould. Gould and his colleague Niles Eldredge diverged from the evolutionary mainstream with a set of ideas that they called "punctuated equilibrium," or "punk eek" for those in the know. As creationist critics observed and as Gould acknowledged, his idea of long periods of "stasis," interrupted by big sudden changes (on an evolutionary timescale), was roughly compatible with a Marxist view of history. Gould was also part of a group of left-wing evolutionary biologists who argued that the emerging sociobiological model of human behavior provided cover for racist and sexist ideas. Both these intramural scientific conflicts heightened the long-standing connection that creationists had made between Marxist politics and Darwinian science.

As Henry Morris and his cohorts in San Diego County built up their "creation science" infrastructure, they contributed to and were strengthened by a broader rise in Christian conservative activism, exemplified by the emergence of the Moral Majority in 1979. The election of Ronald Reagan in 1980 raised hopes that secular humanists could be defeated. As conservative evangelicals gained confidence that political efforts could make a difference, the prominence of "dominionist" ideas increased. Once associated with the relatively marginal R. J. Rushdoony, they now became more mainstream. D. James Kennedy, Jerry Falwell (a protégé of John R. Rice), Francis Schaeffer, Tim LaHaye, and others emphasized the importance of conservative Christians getting involved in electoral politics. The ICR collaborated with all of them in broadcasting its political message about the fruits of evolution.

Before taking independent organizational form, what became the ICR was the "creation science" division of the fledgling Christian Heritage College (CHC) in El Cajon, California. By 1978, CHC enrollment had grown to some five hundred students. Many came by way of LaHaye's Christian High School and through contact with the rapidly growing Scott Memorial Baptist. Others like Janet Laughton and Arthur Gutierrez came from the public schools. They had attended San Bernardino High School, met at CHC, and were married at San Bernardino First Baptist soon after graduation.[2] Early CHC instructors included Morris, Duane Gish, and a variety of others recruited based on their conservative theological credentials.[3]

Like all CHC students in their sophomore year, Laughton and Gutierrez would have taken six hours in "scientific creationism," taught in the early 1970s by Henry Morris himself. In the absence of textbooks written from Morris's young-earth perspective, Morris created handouts that developed into a series of texts, published by his in-house publishing arm, Creation-Life Publishers. These included Morris's own *Scientific Creationism* and *The Troubled Waters of Evolution*, as well as Duane Gish's *Evolution? The Fossils Say NO!*[4] In the first few years of CHC, Morris had only two antievolution teaching counterparts: Gish, the Berkeley PhD who had worked with Morris in the CRS, and Harold Slusher, a longtime creationist professor of physics at the University of Texas–El Paso, who specialized in critiquing the validity of radiocarbon dating.[5]

As Morris has explained, using a genetics analogy, the ICR had a "hybrid origin." In 1970, a group of activists in the Los Angeles–based Bible Science Association (BSA) merged with the CHC creationists to form the Creation-Science Research Center (CSRC). The BSA group, led by Nell Segraves and her son Kelly, developed a strong interest in getting creationist textbooks into California public schools. The previous year the state board of education had approved the inclusion of creationism in the state's "Science Framework." Morris was sympathetic to the BSA group's goals, but over time, a rift developed. As Morris put it, he and Gish favored "educational and scientific means," whereas the Segraves group preferred "political and promotional efforts." In 1972, the CSRC split in two; the Segraves duo took with them the CSRC name, and Morris renamed his CHC-affiliated faction the Institute for Creation Research.[6]

As Morris and the ICR forged ahead through that first decade, they attracted a bevy of credentialed creationists to staff the growing number of classrooms in the hills of El Cajon. Richard N. Bliss became the ICR's director of curriculum development in 1976. With a master's in biology from the University of Wisconsin and a doctorate in science education from the University of Sarasota, Bliss had taught biology in Racine, Wisconsin, and then served as director of science education for the city's public schools.[7] The inclusion of Bliss continued an upward trend.

It strengthened Morris's claim that the ICR staff were carrying out legitimate scientific research. And it underlined the split with the Segraves group. The ICR, Morris seemed to be saying, would not dirty its hands in the rough-and-tumble world of politics.

And yet the output of the ICR in its first two decades undermines Morris's claim that he and his colleagues were not engaging in "political and promotional efforts." One example was the organization's monthly newsletter, *Acts & Facts*, founded by Morris in 1972. Convinced that the publication's name was a gift from God, Morris recalled that it seemed to come to him from "out of the blue."[8] Regardless of its ultimate source, the name did seem to neatly and cleverly convey the essence of scientific creationism. It harked back to the commonsense empiricism that animated his creationist geologist hero George McCready Price. In harmony with this vision of science, Morris continued to refer, in the decades ahead, to the basis of scientific creationism as "true facts." At the same time, the ICR used a presuppositional framework to talk about evolution. For the only facts that could be true were those produced by a God-governed mind.

The ICR communicated with its flock through *Acts & Facts*. In comparison with the *Creation Research Society Quarterly*, which required subscribers to wade through obscure scientific terminology, *Acts* was friendlier to the average reader. More like a magazine than a journal, it was brief, on the order of eight pages; it was small, coming in at five by eight inches; it was in color; it featured photos, illustrations, and display ads. And it was free.[9] By 1984, some seventy-five thousand people received *Acts* each month. According to an ICR survey, 90 percent of *Acts* readers were high school graduates; 73 percent had graduated from college; 35 percent had master's degrees; and 8 percent held doctorates. Contrary to liberal perceptions of creationists as ignorant yokels, ICR supporters were above average in their level of formal education. As for their occupational profile, the survey revealed the following: business, 21 percent; pastor, 19 percent; teacher, 18 percent; industry, 17 percent; homemaker, 7 percent; and student, 6 percent.[10] Though the categories "business" and "industry" were vague, the results suggest that readers were above average in income.

Even as the ICR sought to reach a broader audience, the group could not ignore science. In presenting its presuppositional perspective on scientific topics, *Acts* sought to strike a balance between easily accessible material and more challenging fare. On the latter side of the spectrum was the June 1979 article by creationist astronomer Donald DeYoung. With a PhD in physics from Iowa State, DeYoung taught at Grace Seminary along with John C. Whitcomb Jr. and was serving as a visiting professor at CHC. His article appeared under the heading of the *Acts & Facts* "Impact" series, which featured "Vital Articles on Science/Creation" and formed the centerpiece of each issue. "Defects in the Jupiter Effect" critiqued predictions by some scientists that a rare alignment of the planets would cause massive earthquakes on earth in 1982. Young adorned his article with a diagram of the expected planetary configuration and featured a table with the mass, tidal effect, distance from sun, and angle in degrees between planet and earth. Based on his calculations, Young rejected the Jupiter effect predictions (he was correct) and then drew broader antievolutionary conclusions for his readers.[11]

Articles of this ilk may have prompted the ICR to include in its 1984 survey a question to readers about the technical level of the "Impact" articles. Nearly half the respondents described the articles as "sometimes too technical." Only 9 percent thought they were "sometimes too simple." Fully 41 percent checked "other." Their written comments, according to Henry Morris, suggested they believed the articles were just fine. Nonetheless, based on the full survey results, Morris reported, the ICR might want to modify "Impact" for "greater ease of understanding."[12] Morris faced a twofold challenge. The ICR needed to popularize its coverage. Creationists, in this regard, were not different from evolutionists, going all the way back to Ludwig Katterfeld and his *Evolution: A Journal of Nature* during the 1920s. In a second respect, however, Morris and his cohorts faced a more daunting task. To bolster the claim that creationists were true scientists, they had to speak the language of science. And yet, as Morris and Whitcomb acknowledged in *The Genesis Flood*, the "importance of the question" hinged on politics.

Acts & Facts made that clear in a number of ways. In 1977, "Impact" featured a series of pieces based on John Moore's articles from *CRSQ* demonstrating the impact of evolutionary thinking on the social sciences and humanities (and thus on impressionable young minds). In addition to informing readers about links between Darwin, Marx and Engels, and a variety of academics, Moore explained that Darwinism "sanctioned" a long line of twentieth-century dictators: Hitler, Mussolini, Lenin, Stalin, Khrushchev, "and current U.S.S.R. leaders as well." A chart in Moore's second article was titled "Relationships between Western Thought and Society and Marxist Communism and Totalitarianism." It revealed to readers the ineluctable connections between science, scientism, Western secularism, and Marxism.[13] At the same time, the ICR also began to use the language of secular humanism, warning of the harmful effects on society of teaching evolution to young people. In a 1977 article on evolutionary humanism, Morris charged that such teachings had "generated an amoralistic attitude in society" with "devastating results."[14]

Scientific creationists did not just publish articles; they took their show on the road. In January 1979, the ICR organized a creation conference in Coral Gables, Florida (their third held there), that stressed the "social impact" of evolution. Along the lines of Morris's repeated queries to readers seeking to establish the relevance of often seemingly obscure ideas, the official title for the conference was telling: "Evolution—What Difference Does It Make?" Speaking for the ICR, Richard Bliss delivered a keynote address to some seven hundred Miami students. Bliss's talk was titled "Have You Been Brainwashed?" echoing the language of the John Birch Society. He covered the ground laid by Moore's 1977 articles and emphasized the "harmful effects of evolutionary teaching on the minds of young people." More than two thousand people visited at least one conference session. Three local television stations and two radio stations covered the proceedings.[15]

The political nature of the Coral Gables conference stemmed not only from its ideological content but from the practical function it served— encouraging lobbying to change the public school curriculum. In 1980, Florida became just the latest state in which creationists sought to give "equal time" to creationism and evolution in the classroom. Until the US Supreme Court decision in *Edwards v. Aguillard* (1987) put a legal stop to this tactic, it seemed to hold promise. Reporting on Florida House Bill 107, *Acts & Facts* proudly noted that the "seeds sown" in the successive creation conferences "are bearing fruit today."[16] Even after *Aguillard*, the ICR continued to push along these lines and did not fail to mention the "social impact" of evolution. In 1988, when California adopted a set of evolutionary science standards, the ICR sent the state superintendent an "open letter," reprinted in *Acts & Facts*, describing the dangerous effects of the "religion" of evolution. According to the ICR, evolution had spawned a series of "secular religions," which included "Atheism, Humanism, Communism, Nazism and Social Darwinism."[17] Despite Morris's break with the Segraves group, the ICR was clearly engaging in "political and promotional efforts."

Back at home in El Cajon, the ICR curriculum at Christian Heritage College ensured that students would learn about the deadly social and political consequences of evolutionary thinking. In 1974, Creation-Life Publishers brought out *The Troubled Waters of Evolution*, which Morris used at CHC. With a three-color cover portraying Darwin's *Origin of Species* sinking in a whirlpool of creationist truth, the book was an updated and expanded version of *The Twilight of Evolution* aimed at a broader audience. A brief introduction came from the pen of Tim LaHaye. He was identified neither as a founder of CHC nor as an owner of Creation-Life Publishers, but rather as president of Family Life Seminars, for which he was a minor celebrity in conservative Christian circles.

LaHaye's introduction says virtually nothing about the "scientific" content of Morris's book. Having little interest in or knowledge of the evolutionary literature, LaHaye singled out the social and political impact of evolution:

> It is the platform from which socialism, communism, humanism, determinism, and one-worldism have been launched. All influential humanists are evolutionists, from Darwin to Huxley, Freud, and Pavlov to Rogers and Skinner. No one theory of man has ever influenced so many people. Accepting man as animal, its advocates endorse animalistic behavior such as free love, situation ethics, drugs, divorce, abortion. . . . It has devastated morals, destroyed man's hope for a better world, and contributed to the political enslavement of a billion or more people.[18]

LaHaye's précis of Morris's book accurately cataloged the range of evils attributed to evolution issuing from conservative Christians from William Bell Riley to Dan Gilbert to David Noebel. Summing up the value of Morris's new work, LaHaye boasted to readers that it "may well prove to be the most amazing book you have read in the last decade!"[19]

After presenting a thumbnail history of creationism and an entropy-based critique of evolutionary science, Morris got to the heart of the matter in a chapter called "Troubled Waters Everywhere." The trouble was that evolutionary concepts had infected all arenas of thought, from the physical and life sciences to the arts and humanities to Christianity itself. They also undergirded social and political practices such as racism (a topic of increasing attention from the ICR) and imperialism. A section on "Totalitarian Ideologies" examined Nazism and communism as products of Darwinian ideas of the "survival of the fittest." As anticommunists had done since the 1940s, Morris argued that rather than constituting opposing "right-wing" and "left-wing" political movements, they were "variants of the same species," namely, "evolutionistic, totalitarian collectivism."[20]

Morris wove references to Marx and communism into other chapters in ways that elucidated for CHC students and others just how dangerous was the combination of atheistic, materialistic communism and Darwinism. Retelling the tale of how evolutionary ideas acquired their grip on humanity, Morris wrote about the nineteenth century in terms of conspiracy. All kinds of "revolutionary movements" were undermining respect for God, among them "Illuminist conspirators." Along with the Illuminati, wrote Morris, Marx and Nietzsche were "acquiring disciples—perhaps also financial backers, as student of conspiracies have frequently suggested." All these evil types were "evolutionists of one breed or another," Morris informed his readers.[21]

Bolstering Morris's later attempt to link Marx with Satan in the ICR creation museum, *Troubled Waters* gave detailed attention to the satanic origins of Darwinism. Morris began by observing that all "pagan" origin myths from the ancient world had a naturalistic focus. He traced them back to biblical Babel. As Carey Daniel had earlier argued in pinning race mixing on the devil, the central agent of the prince of darkness at Babel was King Nimrod. Morris knew all about Nimrod. As a freshman at Rice Institute in 1935, the budding creationist had penned a poem, "The Tower," in which "Babel's fools entwine their souls in humanistic schemes / In proud, utopian Nimrod-dreams."[22] Perhaps Morris was inspired by book 12 of John Milton's epic poem *Paradise Lost*, in which an unnamed Nimrod appears as a "mightie Hunter" who "from Rebellion shall derive his name."[23]

Now writing as a mature creation scientist, Morris drew his evidence of satanic influence from the story of Babel in Genesis 10 and 11;

descriptions of Lucifer in Isaiah, Revelation, and Ephesians; and admittedly inconclusive myths and archaeological fragments. Treating the potential involvement of Satan as a legitimate research question, Morris acknowledged that he could not be sure of his conclusion. Like all good scientists, he presented it in tentative fashion: "It therefore is a reasonable deduction, even though hardly capable of proof, that the entire monstrous complex was revealed to Nimrod at Babel by demonic influences, perhaps by Satan himself."[24]

To bring home the ways in which evolution was "sociologically harmful," Morris contributed a final chapter with a Shakespeare-inspired title: "Boil and Bubble." Invoking the witches of *Macbeth* not only rhymed with the book's title. It conveyed Morris's allegation that evolutionists had turned the world from true Christianity to paganism. Morris's analysis of evolutionary approaches to population, ecology, racism, and, most of all, the "sexual revolution" showed that evolution was "false and deadly." It was a "corrupt tree" issuing "evil fruit." Those fruits were "materialism, collectivism, anarchism, atheism, and despair in death." Such trees, as Jesus had said, must be "hewn down and cast into the fire."[25] Thus did hundreds if not thousands of young Christian students learn the truth about evolution, ICR-style.

In between the publication of *Troubled Waters* and Morris's sequel—*Evolution in Turmoil*—the field of evolutionary science was rocked by conflict over a new set of ideas known as "punctuated equilibrium" (PE). Developed by paleontologists Niles Eldredge and Stephen Jay Gould, PE presented a challenge to the reigning "modern synthesis" of evolution that had been consolidated in the 1940s. By shining a spotlight on this intrascientific battle, Morris sought to use one side's arguments against the other and thereby undermine the credibility of both. He quoted out of context and distorted Gould's ideas. The fact that Gould had become a genuine scientific celebrity—appearing on the cover of *Newsweek* in 1982—meant that such a strategy could pay real political dividends.[26] Morris homed in on PE for a more specific political reason that was grounded in reality. In a chapter of *Turmoil* suggestively titled "Evolution and Revolution," he gave detailed attention to Gould's Marxist-oriented politics and how they informed his scientific ideas.[27] In doing so, Morris updated the Red Dynamite tradition for the waning years of the Cold War.

Given that Gould once wrote that his politics were a "private matter," one might think that his Marxism was a figment of Henry Morris's fevered creationist imagination.[28] In its original incarnation, PE came without political baggage. Gould and Eldredge first published their ideas in an academic volume called *Problems in Paleobiology* (1972). They were grappling with a conundrum that had plagued evolutionary paleontologists since Darwin's time: the apparent "imperfection" of the fossil record. Trained in the modern evolutionary synthesis, they were convinced of the reality of natural selection and the existence of transitional species. But they also knew that in any given location, one could never find a complete sequence of finely graduated intermediate forms. While paleontologists tended to explain a "gappy" fossil record as an incomplete record of a fuller story, Gould and Eldredge argued that those gaps were a kind of positive evidence for how evolution took place. Following Ernst Mayr's idea of "allopatric" or geographic speciation, Gould and Eldredge focused on the migration of small subpopulations of a species ("peripheral isolates") to a new area where they branched off from the original parent species. Since new species developed in a different location, intermediate forms did not appear in the "core" area. But if the new species then spread back into that core area, the newly evolved variant would show up in the fossil record as distinctly different from the original species. Thus, concluded Gould and Eldredge, "Many breaks in the fossil record are real." Rejecting "phyletic gradualism," in which whole populations of organisms evolved through a series of barely perceptible gradations, Gould and Eldredge argued for long periods of "stasis" or equilibrium, where little change occurred, "punctuated" by "rapid and episodic" speciation events.[29]

In their early attempts to reorient paleontology along the PE axis, Gould and Eldredge only hinted at a broader philosophical or political agenda. In their first article, the authors nodded their heads to philosophers of science Paul Feyerabend and Thomas Kuhn: "The expectations of theory color perception to such a degree that new notions seldom arise from facts collected under the influence of old pictures of the world."[30] In Kuhnian terms, Gould and Eldredge were challenging the dominant "paradigm" of gradualism. But in a subsequent 1977 article that garnered wider attention (and provided the basis for one of Gould's popular columns in *Natural History* magazine), Gould and Eldredge made no bones about the political implications of PE. Calling for a rethinking of the entire field of biology, Gould and Eldredge claimed that biologists' preference for gradualism was not based on "objective study of nature" but rather a "metaphysical stance" determined by nineteenth-century European ruling-class politics. Facing threats from below, rulers championed the values of "order, harmony and continuity," which then found their way into the dominant models of science, including Darwinian biology. Noting that Karl Marx was a great admirer of Darwin, Gould and Eldredge quoted Marx on how Darwin had translated the social structure of English society into the natural world. Gould and Eldredge were not questioning the fundamentals of Darwinism. But they were suggesting that Darwin was a creature of his "cultural context."[31]

From this vantage point, Gould and Eldredge observed that in other cultural contexts, such as the "socialist nations," a Marxist punctuational theory of change was the norm. That theory encompassed dialectical materialist laws drawn originally, in idealist form, from Hegel. One of these—the law of the transformation of quantity into quality—seemed particularly relevant to PE. To illustrate how this law applied to both society and nature, the authors quoted from the Soviet handbook of Marxism-Leninism: "The transition of a thing, through the accumulation of quantitative modifications, from one qualitative state to a different new state, is a leap in development. . . . We often describe modern Darwinism as a theory of the evolution of the organic world, implying that this evolution covers both qualitative and quantitative changes. . . . The evolutionary development of society is inevitably consumed by leap-like qualitative transformation, by revolutions." Consistent with their philosophical position, explained Gould and Eldredge, Soviet paleontologists had long supported a punctuational perspective. Closing out this section of the article, they offered one last example of the influence of politics on biology that would be quoted often by Gould's opponents: "It may also not be irrelevant to our personal preferences that one of us learned his Marxism, literally at his daddy's knee."[32]

Though the authors did not specify which one received an early Marxist education, we know that it was Gould.[33] The "daddy" on whose knee he sat in 1940s Queens was Leonard Gould. The son of eastern European Jewish immigrants, the elder Gould was a self-taught man who worked as a court stenographer and according to various accounts was "a passionate

Marxist" who was "well-versed in Marxist theory."[34] Like Socialist Party evolutionary lecturer Arthur Lewis decades earlier, Leonard Gould combined a commitment to Marxist politics with a fanatical interest in natural history. The latter passion he bequeathed to his son through walks in city parks and visits to the American Museum of Natural History. Just what political inspiration the son derived from the father is unclear, but by young adulthood, Stephen Jay Gould was a left-leaning political activist. As a student in the early 1960s at Antioch College in Yellow Springs, Ohio, Gould joined the Students for a Democratic Society (SDS), the Student Peace Union, and the Congress of Racial Equality. He "threw himself" into the activities of these groups, including a series of sit-ins to racially integrate the local barbershop.[35]

As a young faculty member at Harvard in the late 1960s and early 1970s, Gould continued to join public protests, this time against the Vietnam War and police brutality.[36] He also became associated with a loosely organized group of left-wing, antiwar scientists called Scientists and Engineers for Social and Political Action, better known by the name SFP, after their semi-regular magazine, *Science for the People*. Early issues took up a wide range of subjects, from nerve gas to the Black Panther Party to "Birth Control in Amerika" to "Helping Science Education in Cuba and Vietnam."[37]

Henry Morris may not have known these details of Gould's political biography, but he knew that before promoting PE, Gould had emerged as a major voice for pro-evolutionism through his column in *Natural History* magazine. Morris also knew that in the mid-1970s, Gould became a major public critic of the new field of sociobiology and of its best-known proponent (and Harvard colleague of Gould's) Edward O. Wilson.[38] In *Sociobiology: The New Synthesis* (1975), Wilson had summarized a body of field and theoretical work done originally on social insects such as ants and honeybees. Using the concept of kin selection, sociobiologists offered an evolutionary explanation for "altruistic" behavior that otherwise seemed to contradict the standard view of Darwinism as a bloody struggle for survival between individuals in a population.[39]

Wilson's efforts seemed reasonable enough, but for Gould and his likeminded colleagues in SFP's Sociobiology Study Group (SSG)—which included Harvard geneticist Richard Lewontin (1929–)—this new field had a dark side. Wilson implied that a number of human behaviors, including rigid sex roles, xenophobia, and genocide, might be best understood as the product of our genes. Gould and his SFP colleagues were alarmed. Wilson appeared to give credence to the idea that genes determined "the conduct of termite colonies" as much as they did "the social behavior of man." If racism, sexism, and war were genetic and not products of cultural evolution, then struggle against them was useless. Although Wilson espoused liberal politics, his sociobiology, in the eyes of Gould and the SSG, became "A Tool for Social Oppression."[40]

Gould's colleague Richard Levins (1930–2016) was thinking along the same lines. A Harvard ecologist who began his career under Wilson's wing but soon joined Gould and Lewontin in their critique of sociobiology, Levins took an openly Marxist political stance. Like Gould, Levins learned Marxism at a young age. The "knee" in his case belonged to his grandfather, who read to Levins from a book by none other than "Bad Bishop" Brown, the maverick pro-evolutionist and Communist who came to prominence in the 1920s. Brown's *Science and History for Girls and Boys* (1932) was a favorite for parents and grandparents of "red diaper babies."[41] As Levins recalled, "My grandfather believed that at a minimum every socialist worker should be familiar with cosmology, evolution, and history."[42] Unlike Gould, Levins joined the Communist Party of both the US and Puerto Rico (where he and his wife, activist Rosario Morales, lived for a time) and also spent years living in and promoting the cause of revolutionary Cuba.[43] Levins and Lewontin collaborated closely and shared a Marxist outlook on science and society, reflected in their coauthored volume *The Dialectical Biologist* (1985). In a similar volume published some twenty years later, Levins and Lewontin claimed that their good friend and "comrade" Gould also identified himself as a "Marxist."[44]

With few exceptions, Gould's public association with Marxism faded over time. In the summer of 1977, when Gould used his *Natural History* column to popularize the ideas of PE, he once again enlisted Engels, dialectics, and Soviet scientists, though without the lengthy quotation from the Stalinist manual (and any mention of his "daddy's knee"). In subsequent public discussion, Gould further distanced himself from that manual by describing it as "silly" and "propagandistic." Much later, he also distinguished between Marxism as a set of ideas—his "intellectual ontogeny"—and his "political beliefs," which, he insisted, were "very different from my father's."[45] When *People* magazine ran a two-page spread on Gould in 1986, focusing on his battle with both creationists and a rare stomach cancer, there was mention neither of his politics nor his father's. The article described PE as "heretical" but omitted its Marxist associations. When Gould departed this earth in 2002, the *New York Times* ran a long obituary. It said not a single word about his politics, Marxist or otherwise.[46]

Henry Morris, however, was onto something. Here, finally, was the fleshly embodiment of a claim that Morris had been making for decades—that evolutionism and Marxism were, as Gould himself later put it, "congenial."[47] Sounding a bit like a Marxist himself, Morris, in *Evolution in Turmoil*, offered a materialist explanation for the rise of PE. It might not be "an accident of history," speculated Morris, that PE emerged during the 1970s, in the wake of the 1960s student protests. Its proponents were younger scientists (Gould was thirty-five in 1977) who had been in graduate school during that previous decade. They were less conservative than their scientific elders and less favorable toward "patriotism" and "western-style evolutionary capitalism." To them, Marxism sounded "noble" and "idealistic." Its proponents had acquired a foothold in the universities, despite "what may have been the excesses of McCarthyism." Under the influence of such people, struggles against racism and imperialism, wrote Morris, were "comingled" with "resistance to fighting Vietnamese Communism," resulting in "campus riots" and "street violence." Whether or not they were directly involved in these "riots"—Morris was not clear on this point—scientists like Gould came to see the existing gradualistic model of evolution as supporting the "establishment evils" they were combating. Thus they "devised" a new evolutionary "mechanism" and supported a "different political and social system" (socialism) that would accompany that new scientific explanation.[48]

Morris reminded readers that the evolution-communism connection had a long pedigree. As an atheist, Karl Marx was "a committed evolutionist" even before Darwin's *Origin*. He was so excited by Darwin's materialist scheme of natural history, wrote Morris, that Marx wanted to dedicate *Das Kapital* to him. Morris's basic point was correct, and the story about Marx and Darwin had been repeated by historians and others for years, including by Gould himself in a 1977 *Natural History* article. But by the publication of *Turmoil* in 1982, Morris could have known what historian Margaret Fay had established in 1978: it was not Marx who wrote to Darwin, but rather Marx's "son-in-law" Edward B. Aveling. He hoped to dedicate to Darwin *The Student's Darwin*, an edited collection of the master's writings. Darwin did, in fact, decline, knowing Aveling's association with atheism, materialism, and socialism.[49]

Morris also offered his readers a historical account of Marxist evolutionism in Soviet Russia. He correctly noted that many pro-evolutionists in the Soviet Union gravitated to a Lamarckian belief in the inheritance of acquired characteristics. The German-born Kammerer and the Russians Pavlov and Lysenko were attracted to Lamarckianism because it offered the possibility of rapid evolutionary change. Under Stalin, Morris explained, Lysenkoist evolutionism was even "imposed on Russian scientists as official state dogma." But ultimately failing the test of science, it was "falsified" and abandoned. In its place, Soviet scientists—and other Marxist evolutionists—sought out a theory of evolutionary change that was not slow and gradualistic.[50]

Even though Henry Morris abhorred Marx and Marxism, the ICR leader viewed Stephen Jay Gould and the newer generation of creationists as tactical allies. For a number of years, Morris had been tying the scourge of racism to Darwinism. Now came Gould and others who were willing to hold the evolution "establishment" responsible for its complicity. Not only did Gould criticize sociobiology on this score, but his *Natural History* articles covered figures in the evolutionary pantheon whose ideas were used to promote racial hierarchy. As quoted by Morris in *Turmoil*, Gould wrote of Haeckel's recapitulation theory that it "provided a convenient focus for the pervasive racism of white scientists."[51] As Gould later noted, this description fit a number of eminent researchers whom Clarence Darrow had invited to provide expert testimony at the 1925 Scopes trial, including Henry Fairfield Osborn, then president of the American Museum of Natural History.[52] Of course, Morris was silent on the long line of antievolutionists who were doctrinaire segregationists.

On evolution's "responsibility" for the "racist" Nazis and Fascists, Morris found less common ground with Gould. Morris explained that former left-wing student radicals like Gould had mixed feelings about Hitler. After all, Morris wrote, "Mussolini and Hitler did call their movements 'socialistic,' and the student movements of the 1960s bore many striking resemblances to the Hitler youth of the 1930s and early 1940s."[53] As with his former melding of socialists and Nazis into a "totalitarian collectivism," Morris obscured the opposite political views held by the SDS and the Nazi Party. But his equation of student radicals, Nazis, evolution, and riots played into the long-standing equation of Darwinism with animalistic social disorder.

Having firmly linked Gould's political biography to Darwin and Marx, Morris treated the "revolutionary evolutionism" of PE. He quoted Gould and Eldredge from the 1977 *Paleobiology* article to show that they openly acknowledged their theory's "Marxist pedigree" and that it ran through Gould's own family. Morris also observed that Gould had "waffled" about whether or not he was a Marxist. Citing British Marxist and Darwin scholar Robert M. Young, who described Gould as a non-Marxist radical, Morris suggested that it might be advantageous for "doctrinaire" Marxists that Gould was not officially identified as one of their tribe, despite "whatever his actual beliefs might be." Gould might be a non-Marxist Darwinist Trojan horse concealing the true Marxist Darwinists inside. Morris's formulation gestured in the direction of an imaginary Marxist-Darwinist conspiracy. But the suggestion gained credibility from Gould's lack of clarity about what being a "Marxist" meant to him. In any event, Morris concluded, PE had captured the hearts and minds of younger evolutionists who were fed up with the "capitalistic establishment." These scientific rebels were "seeking social justice and full egalitarianism not by slow evolutionary change but by rapid and even violent change if need be."[54] Campus riots, indeed.

For all of the attention that Henry Morris lavished on Gould's Marxism in 1982, anticommunism was starting to lose its political punch. As Tim LaHaye predicted the following year, citing an obscure attorney as the source of his information, "The primary battle for the eighties and nineties will not be between Communism and anti-Communism or socialism and antisocialism, but between secular humanism and Christianity." LaHaye played a major role in fulfilling this prophecy about secular humanism.[55] He drew inspiration from the combined work of two figures in the world of Christian conservatism—Francis A. Schaeffer IV (1912–1984), the quirky, world-famous evangelist, and John W. Whitehead (1946–), that obscure attorney and follower of R. J. Rushdoony who did the legal and historical brainwork that made "secular humanism" a viable new political target. That pair produced a vision of conservative Christian activism that spoke a new language but also represented a fundamental continuity with the Christian creationist anticommunism of previous decades.[56]

Hailing from Germantown, Pennsylvania, Francis Schaeffer began his pastoral career as a diehard fundamentalist and anticommunist. Studying with famed Presbyterian fundamentalists J. Gresham Machen and Cornelius Van Til and then briefly allying with ultra-separatist Carl McIntire, Schaeffer and his wife Edith traveled to Europe in 1955 to scout out possibilities for extending the reach of separatist fundamentalism. Soon souring on McIntire's approach, the Schaeffers opened up a Christian community called L'Abri Fellowship in their Swiss home. Situated on the way to a popular ski resort, L'Abri ("the shelter") by the mid-1960s had become famous as a pilgrimage site for evangelical travelers, curious hippies, celebrities, US soldiers stationed in Germany, and lost souls. Schaeffer was a big part of the attraction. Dressed in lederhosen and sporting long hair and a beard, Schaeffer could hold forth on a bewildering variety of topics, from ancient history to the latest trend in rock music. Unlike the old-style fundamentalists like John Rice and Carl McIntire, Schaeffer understood that he needed to meet young people where they were. He convinced many visitors that a Christian worldview was relevant and intellectually defensible.[57]

With his newfound fame, Schaeffer hit the American lecture circuit in the 1960s and published books that cemented his reputation as an original and thought-provoking Christian apologist. While Schaeffer never accepted the full "biblical law" program of Christian Reconstructionism, he did take from R. J. Rushdoony's writings the concept of America as a Christian nation. By the late 1970s, Schaeffer's activist son Francis Schaeffer V (better known as "Franky") persuaded his father to produce a series of films and companion books to spread his worldview to an even larger audience. Those included the panoramic, ten-part *How Should We Then Live? The Rise and Decline of Western Thought and Culture* (1976). In one episode, the elder Schaeffer, perched atop a pile of crushed cars in a junkyard, spoke to viewers about the devastating consequences for Western culture of evolutionary materialism. Evolution meant that all matter, including human beings, had arisen as a result of "chance." We were thus relegated to the status of "machines" that could be easily discarded.[58]

Whereas Francis Schaeffer had become an evangelical celebrity, John Whitehead was an unknown liberal, pot-smoking civil rights lawyer fresh out of law school in Fayetteville, Arkansas, when he found Jesus in 1974. Moving to California to join popular dispensationalist Hal Lindsay's church, Whitehead met R. J. Rushdoony and quickly came under his influence. By the mid-1960s, Rushdoony had begun providing expert testimony in court cases where Christian parents were prosecuted for failing to send their children to public or accredited private schools. In these early years of Christian homeschooling, Rushdoony was pivotal in formulating the necessary legal arguments, and in this effort Whitehead proved a valuable ally. Making use of Rushdoony's vast private library and his Christian publishing connections, Whitehead brought out *The Separation Illusion: A Lawyer Examines the First Amendment* (1977), his first work that aimed to break down the First Amendment barriers to creating a godly America.[59] By that time, the US Supreme Court had issued a clear ruling, in *Epperson v. Arkansas* (1968), against the introduction of openly religious creationist teachings in the science curriculum. This was the ruling that Clarence Darrow had tried and failed to obtain more than forty years earlier as attorney for John Scopes.

Rather than trying to find a way to slip religion into a nonreligious curriculum, Whitehead now argued something completely different. In a 1978 law review article, he contended that the prevailing "secular" science curriculum was a form of "religion" to which courts were granting unconstitutional favor.[60] Whitehead hung his argument on a little-noticed footnote in the 1961 *Torcaso v. Watkins* US Supreme Court decision that referred to secular humanism as a religion.[61] The idea of an explicitly "secular" belief system as religious reinforced claims that R. J. Rushdoony had long been making—no thought was possible without God.[62] From the *Humanist Manifesto* (1933) and the *Humanist Manifesto II* (1973), Whitehead gleaned a definition of the secular humanist faith. Secular humanism, according to Whitehead, "is a religion whose doctrine worships Man as the source of all knowledge and truth." Its most important "tenet" was the "absolutism of evolution."[63]

Whitehead then proceeded to underline the dire practical effects of secular humanism, echoing decades of Red Dynamite political logic. Using an aggressive metaphor, Whitehead noted that evolutionary thinking had "penetrated" all arenas of life. By replacing absolute morality with relative morality, evolutionism had "altered the course of history." The result was "totalitarian regimes" such as Hitler and Mussolini; the "class struggles and atheistic posture" of communism; Freud's "sex-drive philosophy"; as well as racism and "unethical" capitalism. In Whitehead's own field of law, he lamented, evolution had created "sociological jurisprudence" in which the rules changed according to "experience." "The implications of this philosophy are frightening," wrote Whitehead, "when executed by someone with despotic power, such as a Hitler, a Stalin, or a Mussolini." Or as expressed by Russian historian and dissident Vadim Borisov, in yet another violent analogy quoted by Whitehead, Darwin, Marx, Nietzsche, and Freud provided the "humanist" theoretical basis for twentieth-century "totalitarianism, trampling the human personality and all its rights, rhinocerouslike [*sic*], under foot."[64]

By following the footnote in *Torcaso* to its logical legal conclusions, Whitehead had performed a real service for opponents of secular humanism. In 1980, this rising legal star on the Christian Right received an unexpected call from Franky Schaeffer. Schaeffer senior and Whitehead were soon working closely together on a book that would make the sage of L'Abri even more famous and sought-after by conservative Christians: *A Christian Manifesto* (1981). Just as Whitehead had sought to reorient his readers to the dangers secular humanism posed to their constitutional freedoms, Schaeffer also aimed to "connect the dots" in a new way. The problem, as he told readers, was that Christians had been viewing things in "bits and pieces" rather than in "totals." They had failed to see that the moral decline of society was the product of not just a list of specific policies, but an entire "world view."[65] Like Whitehead, Schaeffer made evolution the central component of secular humanism, but repeatedly referred to it, in his idiosyncratic way, as the "material-energy, chance concept of reality." Consonant with his charge that human beings were treated as "machines," Schaeffer used the language of physics, rather than biology. Secular humanists, that is, viewed human beings "as a complex arrangement of molecules." But they also viewed "Man" as "an intrinsically competitive animal."[66]

As much as Schaeffer helped to shift the discourse of Christian conservatism away from the fanatical anticommunism of his former colleague Carl McIntire, the traces of that tradition were evident. In *How Should We Then Live?* Schaeffer spent several pages portraying the repression and relative humanist morality of Marxist communism as a "gigantic contrast" to the absolute God-based morality and freedom produced by the Protestant Reformation.[67] This theme reappeared in *A Christian Manifesto*. At the center of the page following the preface were three lines of text, one stacked on top of the other: "The Communist Manifesto 1848," then, "Humanist Manifesto I 1933," and finally, "Humanist Manifesto II 1973."[68] Humanism was a genealogical descendant of communism. Even Schaeffer's first invocation of those who promoted the "material-energy chance" worldview qualified it by adding, "whether they are Marxist or non-Marxist."[69] The specter of anticommunism thoroughly haunted the new edifice of anti–secular humanism.

If Whitehead and Schaeffer succeeded in laying down the legal and intellectual framework for confronting the forces of secular humanism, the task of popularizing that framework fell to Tim LaHaye. By the late 1970s, LaHaye had gained a wide audience in the Golden State for his distinctive mix of conservative Christian counseling, anticommunist activism, jeremiads on the decline of American morality, and Bible prophecy. In 1979 he took a step more directly into politics by helping to form Californians for Biblical Morality (CBM). LaHaye's group channeled a conservative backlash to the California gay rights movement. In the previous year, gay rights activists and their allies had defeated Proposition 6, which would have made it legal to deny employment to teachers based on their sexual orientation. In response, CBM members rallied and signed petitions in support of politicians who opposed gay rights, supported capital punishment, and called for the return of prayer in the public schools. They sought, as LaHaye described it, to "halt the juggernaut of amoral humanism in our state." They were joined by a number of allies from outside California, including a charismatic preacher from Lynchburg, Virginia, by the name of Jerry Falwell. CBM provided a model for the formation of Moral Majority Inc. later that year.[70]

The centerpieces of LaHaye's effort to publicize the dangers of secular humanism were *The Battle for the Mind* (1980) and *The Battle for the Public Schools* (1983). In both, LaHaye tipped his hat to Whitehead. The 1980 volume came out a year before Schaeffer's *A Christian Manifesto*, but LaHaye had been reading Schaeffer, and he dedicated *Battle for the Mind* to that "renowned philosopher-prophet of the twentieth century."[71]

LaHaye's dystopian vision suggests that the campaign against secular humanism did not break from the anticommunist tradition but adapted it to new circumstances.[72]

As LaHaye told the story, a secular humanist conspiracy was afoot in the land. Christians faced "an invisible enemy." Its name was "humanism," and its "target" was "your mind." The conspirators had "quietly woven" alarming changes into the social fabric. The public schools had fostered "hostility" to morality, religion, creationism, and "moral" sex education. As America marched toward "Sodom and Gomorrah," LaHaye offered both a "shocking, detailed exposé of this regression and a "practical handbook" for fighting back.[73] The final section of the book told readers how they could wage that fight. It included a questionnaire for political candidates to determine their positions on "morals"; a "Key 16" prayer list of the federal, state, and local officials a Christian voter had a hand in choosing and might pray for; a pitch for Moral Majority Inc.; and a list of practical political tips.[74]

Where Francis Schaeffer had made ample use of film to spread his message to the Christian masses, LaHaye used schematic graphics that conveyed the contrasting tenets of the humanist and Christian worldviews. Employing Schaeffer's metaphor of a Christian "base," LaHaye placed the personified rival worldviews (both apparently male) on a three-level platform, which sat atop a row of books. The Christian figure, wearing a large letter *S* for "servant," was perched upon the foundation stones of God, Creation, and Morality. Foundational reading material was labeled "Law," "History," and a variety of books from the Old and New Testaments. A globe, adorned with a cross and labeled "Compassionate World View," hovered above. The Humanist, whose upraised arms rendered his bodily form into the shape of the letter *H*, stood upon Amorality, Evolution, and Atheism (tenets one through three). His globe read "Socialist One World View" (tenet four). His reading list featured early Greek materialists, Enlightenment thinkers (including Illuminati founder Adam Weishaupt), skeptics, and modern philosophers. He wore what looked like a lowercase *s* imprisoned inside a larger *A*, which presumably stood for "autonomous man" (tenet five).[75]

Just as Francis Schaeffer called Christians' attention to "totals" and sought to awaken them from their slumber, LaHaye warned that humanists had cleverly "duped" many Americans into innocently accepting the

tenets of what he termed "the world's greatest evil." By drawing heavily from the writings of socialist and humanist Corliss Lamont as well
as Lamont's old friend Julian Huxley, LaHaye aimed to open the eyes
of duped Christianity. Humanists were atheists; believed in evolution;
viewed morality as subject to change ("amorality"); thought that human
beings were essentially good and capable of solving their own problems
without God's help ("autonomous man"); and were inclined to favor
world government and to disfavor "Americanism, capitalism, and free
enterprise." LaHaye not only cited humanists' leadership of UNESCO
and UNICEF but pointed to Lamont's favorable comments on the Marxist philosophy of dialectical materialism. Evolution provided the foundation for all these tenets, and LaHaye quoted Whitehead and John Conlan
on evolution "shifting the base" from absolute morality to arbitrary
absolutes.[76]

All five tenets of humanism were meant to shock LaHaye's readers into
action, but perhaps none more than evolution-inspired sexual "amorality." "If you believe that man is an animal," he wrote, "you will naturally expect him to live like one." This was an old weapon in the arsenal
of antievolutionism, but LaHaye tied it to new manifestations of sexual
social change: easy divorce, abortion "on-demand," sex education "forc-
ibly taught" to schoolchildren, coed college dormitories, homosexuality
as "an optional life-style," easy access to pornography, and, to boot, the
availability of marijuana and "hard drugs." No matter whether one called
it "permissiveness" or "free love," LaHaye wrote, these were just code
words for "adultery, fornication, perversion, abomination, and just plain
sin."[77] Raising the stakes even higher in his subsequent volume, *The Battle for the Public Schools: Humanism's Threat to Our Children* (1983),
LaHaye focused in on the effects of this moral sea change on the nation's
youth. The threat posed by humanist sex education encompassed five
chapters, starting with "How to Make Sexual Animals Out of a Generation of Children."[78]

In case his picture of "amorality" in *Battle for the Mind* was not sufficiently alarming, LaHaye wrapped up his survey of humanist evil with
a nightmare scenario reminiscent of the Bolshevik Bureau of Free Love.
Humanists were plotting to merge America with Communist countries
and "third-world dictatorships" into a "one-world socialist state." An
elite humanist "ruling class," supported by a military, would rule over

the "masses." Children would be "wards of the state." Men and women would do "the same work." And given humanist "amorality," the masses, both young and old, would apparently be encouraged to engage in promiscuous sex and prostitution, "practice" homosexuality, obtain abortions at will, read pornography, and use drugs. All of this, the humanists hoped, would come to pass by the year 2000.[79]

LaHaye's updated conspiratorial viewpoint had roots stretching back to William Bell Riley of the *Protocols of the Elders of Zion*. This continuity is evident in *The Battle for the Public Schools* where LaHaye confronted detractors who found his claims "bizarre."[80] Most educators, LaHaye acknowledged, "scorn the conspiracy theory." But "many people," he writes, believe that the conspiracy is real and involves the Illuminati, the Rockefeller-funded Council on Foreign Relations, and Trilateral Commission. Riley had argued that even if the *Protocols* were a forgery, they still accurately foretold world events. LaHaye insisted that he "did not know" whether or not the humanist conspiracy theory was valid. But it might explain why, for example, the evolutionist and humanist John Dewey was a "committed world socialist" and why he spent three years in the young Bolshevik republic. Whether he then worked for the "socialists or Marxists" in America "would be difficult to prove," LaHaye admitted. But then again, he asked, would the outcomes—progressive education and the erosion of moral absolutes—have been any different?[81] Whether or not there was an actual communist/humanist conspiracy going back centuries to the Illuminati, assuming that it existed made good practical sense for conservative Christians.

Battle for the Mind was a Christian best seller. By July 1981, Revell had printed 375,000 copies. The term "secular humanism" emerged from obscurity to become common currency in conservative Christian circles. As one study noted, "by the end of 1980, nearly all had adopted it as their enemy." A letter to the *Charlotte Observer* that year reflected the rich mix of ingredients that LaHaye had identified: "abortion, pornography, evolution, sex and values education, socialism, communism, and bureaucratic government are all part of secular humanism."[82] The success of *Battle* testified to LaHaye's ability to make Schaeffer's more historical and theoretical discussion relevant to the masses. But LaHaye also had good timing. As indicated by blurbs on the book's back cover—written by Jerry Falwell (1933–2007) and D. James Kennedy (1930–2007)—LaHaye wrote not as a lone individual but as part of an organized and growing movement that aimed to put a "moral majority" back in charge of America.

As the founder and central spokesman for the Moral Majority, Rev. Jerry Falwell Sr. has gone down in history as the man who made the New Christian Right a potent political force in late twentieth-century America. Less appreciated is the Red Dynamite political genealogy that enabled Falwell to draw inspiration from both J. Frank Norris and John R. Rice. Growing up in Lynchburg, Virginia, Falwell accepted Jesus into his life at the age of nineteen at the Park Avenue Baptist Church in 1952. Park Baptist was one of a small but growing number of churches that chose to affiliate with the Bible Baptist Fellowship (BBF), led by G. Beauchamp Vick, the former right-hand man of J. Frank Norris.[83] Attending Lynchburg's Brookville High School, Falwell had dreams of attending Notre Dame or Virginia Tech to study mechanical engineering—he was both a star athlete and class valedictorian. But with his conversion at Park Avenue, he decided to devote his life to God. When Falwell asked his pastors where he should continue his schooling, they directed him, consistent with his church's independent fundamentalist affiliation, to study with Vick at Baptist Bible College in Springfield, Missouri. If Falwell had gone to nearby Virginia Tech instead, he would have met professor of civil engineering and devout Baptist Henry Morris. But they would soon cross paths. After Morris started up his own non-SBC-affiliated Baptist church in Blacksburg in 1962 he invited Falwell to guest preach there.[84]

By the early 1960s, Falwell was pastoring the growing BBF-affiliated Thomas Road Baptist Church back in Lynchburg, which he had launched in 1956. He combined an orthodox fundamentalist Baptist message with the new technologies of radio and television and a vigorous door-knocking campaign. Falwell rapidly attracted new congregants. His televised *Old Time Gospel Hour* show was consciously modeled on Baptist fundamentalist preacher Robert C. Fuller's *Old Fashioned Revival Hour* radio show, which had mesmerized Falwell as a youngster.[85] Falwell preached a typical Southern Baptist fundamentalist message that revolved around the inerrancy of scripture, the sinfulness of worldly pleasures, the dangers of communism, and the virtues of racial segregation.[86] But he possessed an atypical zeal for building what became one of the first "mega-churches" in the nation. In 1963, the Thomas Road Baptist Church spawned a summer camp for kids (which Henry Morris's children attended), and shortly thereafter a prison program, a Christian school, and Lynchburg Bible College, soon to be Liberty Baptist College and then Liberty University. By the early seventies, Falwell's weekly congregation numbered nearly seven thousand. He reached millions more through the airwaves. And he was now Dr. Jerry Falwell, thanks to an honorary doctoral degree from Lee Roberson's fundamentalist Tennessee Temple University in Chattanooga.[87]

Given Falwell's independent Baptist fundamentalism, his zeal for moral, upright living, and his Norrisite-breakaway heritage, it was almost foreordained that his fate would mix with that of John R. Rice. When Falwell began preaching in Lynchburg, he took out a subscription to *Sword of the Lord* and met Rice soon thereafter. Over the course of the decades, Rice gave Falwell prominent attention in his newspaper. By the late 1970s, Rice viewed the Lynchburg preacher as his protégé and potential successor. In 1971, Rice added Falwell to the *Sword* "Cooperating Board" and invited him to preach at *Sword* conferences. In 1975, Rice paid tribute to Falwell, "our beloved friend," whom "God is wonderfully blessing." Rice ran ads in *Sword* for Falwell's pastoral conferences and for Liberty College.[88] In 1976, Rice even commissioned *The Grim Reaper*, a "Christploitation" film by Nashville-based filmmaker Ron Ormond, in which Falwell portrayed a preacher hammering away at the point that Satan, his demons, and hell are all real.[89] Most importantly, Rice published Falwell's words. Between 1974 and 1979, Falwell's sermons, always accompanied by his photo, appeared on the front page of *Sword of the Lord* twelve times.[90]

Falwell's primary theme was winning souls for Jesus. But he consistently wove into his sermons a concern about the moral "permissiveness" of America, the ongoing dangers of communism, and the obligations of preachers to meet all these issues head-on. As he related in a 1979 sermon reprinted in *Sword*, when a talk show host questioned whether it was appropriate for a preacher to address political issues, Falwell responded by saying that "homosexuality, abortion, pornography are not political issues, they are moral issues that have become political."[91] Falwell rarely addressed evolution explicitly in this period. But in a 1978 sermon on abortion, Falwell did highlight the "evolutionary explanation" as the culprit since it taught Americans that we were descended from animals. "Americans are aborting birth," he charged, "as if the child is an unnecessary animal."[92]

Despite his relative silence on evolution in the 1970s, Falwell had the opportunity to be educated on the subject by his regular reading of *Sword*. Rice reran older pieces that sought to show that evolution was "only" a theory and therefore amounted to a pseudoscientific "hoax." He printed articles showing that evolutionary claims contradicted the Bible, including his own "God's Perfect Creation in Six Literal Days." Most relevant for Falwell's future trajectory, Rice featured commentary that placed evolution in the lens of moral decline as yielding evil "fruits." These included a front-page reprint of Rice's 1954 sermon "Dangerous Triplets" (renamed "Our Triple Enemies"). More accessible for Falwell's generation was a new sermon by Rice on the moral decline of American public education. The schools, Rice revealed, were run by "infidels" who believed in "godless evolution," that man is only an "animal," and that therefore there are "no absolute rights and wrongs." Instead, students get "free sex without marriage," disrespect for the American flag and American heroes, and "filthy, dirty" books like *Catcher in the Rye*. Evolution was the basis of the "humanistic philosophy" of John Dewey, which had brought about this educational nightmare. It was the basis, too, "of Russian communism."[93]

Regardless of what Falwell absorbed from Rice on the subject of evolution, we know that Rice meant a great deal to him. Speaking at the *Sword* editor's funeral in late 1980, Falwell said that Rice was "God's man for the hour. I looked on him as the guardian of fundamentalist truth for this generation." Later that day, Falwell confided a more personal appreciation to Rice's grandson Andrew Himes. "I must tell you that John R. Rice was a father to me," said Falwell. He was, Falwell told Himes, "my mentor, my teacher, my friend, and my prayer partner."[94] As Falwell assumed the mantle of "God's man" for a new age, he owed much to the example and teachings of John R. Rice.

What Falwell did not say is just how indebted he was to Rice for a tangible asset that helped to launch Moral Majority Inc. As Falwell's Liberty Baptist College associate Elmer Towns recalled, the two of them visited Rice in Murfreesboro, Tennessee, in the summer of 1979 as talks about forming the Moral Majority were proceeding. Falwell understood that even if he and Rice shared a similar political and theological outlook, the old man had no interest in forming an explicitly political organization.

With the explanation that they sought to double the enrollment at Liberty Baptist, Falwell asked to borrow something that could reach untold numbers of just the right people: the *Sword of the Lord* mailing list. Rice readily agreed. They left town with the names of two hundred thousand Baptist ministers. As Elmer Towns recalled, "that was magic."[95] Just as William Bell Riley passed the torch to Billy Graham in Minneapolis in 1947, so did John R. Rice become a living link, in more ways than he knew, between the Old and New Christian Right.

And yet, like Henry Morris, who minimized the importance of George McCready Price to his own thinking, Falwell was ambivalent about Rice's legacy. In a 1982 issue of the Liberty Baptist College–sponsored *Fundamentalist Journal*, editor Falwell ran a tribute to John R. Rice. It accurately described Rice as an enemy of "atheism, evolution, modernism, and worldliness." On the other hand, it omitted Rice's anticommunism. And it said nothing about Rice's uncompromising segregationism and opposition to interracial marriage, which he carried to his grave. Already by the early 1970s, Rice was isolated on these issues even among fundamentalists.[96] In response to his continuing opposition to interracial marriage and defense of segregation, Rice received a string of letters from critics, both Black and white, who canceled their *Sword* subscriptions and scored Rice's racism. In 1970, when Rice defended the refusal of Bob Jones University to admit Black students, a self-described "black fundamentalist" wrote Rice from Chicago to say that "Your position is not moral, biblical, Christian or even truly American." Rice replied by calling the man "wholly racist."[97] Falwell had been as pro-segregationist as any Southern Baptist, but by the 1980s he had decisively changed his tune.[98] As Elmer Towns commented, "John R. Rice was to the right of Jerry."[99] The evolution of Falwell's views probably helps explain a remarkable fact. In Falwell's best-selling autobiography of more than four hundred pages published in 1987, the man who was like a "father" to Falwell is nowhere to be found.[100]

As John R. Rice might have expected, the Moral Majority Inc., Falwell's Liberty Baptist College, and Falwell's own televised sermons played important roles in promoting creationist politics in the 1980s. Falwell's various roles often overlapped. A case in point was a "paid ad" from Falwell's *Old Time Gospel Hour* (OTGH) that appeared in the April 20, 1981, issue of the monthly *Moral Majority Report*. In response to the ACLU-backed court challenge to the "balanced-treatment" law in

Arkansas (which would be struck down in federal court the following year in *McLean v. Arkansas*), Falwell asked readers to "Cast Your Vote" on the "Creation or Evolution" "ballot." He could be fairly sure that his readers—many of whom were longtime subscribers to *Sword of the Lord*—would vote "NO" to the question, "Do you agree that public school teachers should be permitted to teach our children *as fact* that they descended from APES?" In return for sending in their ballots, readers were offered free copies of *The Remarkable Birth of Planet Earth* (1972). Moral Majority members might have been forgiven for assuming that Falwell was the book's author, but the tiny, blurred type obscured the name of Henry Morris.[101]

One of the first productions of the ICR's Creation-Life Publishers, *Remarkable Birth* was designed to reach beyond the world of "scientific creationism." Even the title—alluding to Hal Lindsay's apocalyptic best-seller *The Late, Great Planet Earth* (1970)—bespoke those marketing ambitions.[102] Coming in at only ninety-five pages with a three-color cover, this pocket-size paperback was aimed, Morris wrote in his preface, at "busy, but interested readers." Like *Twilight, Troubled Waters,* and *Turmoil,* the book recast arguments from *The Genesis Flood.* But in this volume that Falwell offered to his readers, Morris got right to the political point. In the opening sentence of the preface's second paragraph, he explained why readers should keep on reading: "evolution . . . is largely responsible for our present-day social, political, and moral problems."[103] Morris quoted Jesus on evil fruit and then got more specific about the results of the "root" of evolution. The list included all the usual suspects: "atheism, communism, nazism, behaviorism, racism, economic imperialism, militarism, libertinism, [and] anarchism." If readers of the *Moral Majority Report* had any lingering doubts about whether evolution had to be opposed, Falwell hoped to scotch them with the words of the man whom the back cover of the book identified as "one of America's greatest authorities on scientific creationism."[104]

Which is not to say that Falwell was trying to hide his collaboration with the ICR. The creationist biology department at Liberty Baptist College was headed by Dr. Lane Lester (1938–). With a PhD in genetics from Purdue, Lester had until recently served the ICR as research associate in biology, taught at Christian Heritage College, and was a regular speaker at ICR summer institutes. Lester penned articles for the *Moral Majority*

Report that urged teachers to order ICR materials for their classrooms, insisting that they were "completely scientific with no reference to the Bible." ICR speakers ran workshops in the West, explained Lester, while Liberty Baptist College would provide that service for the eastern half of the country.[105] In 1978, the ICR's Duane Gish, who had recruited Lester to creationism, and Morris himself had spoken at Falwell's church in Lynchburg.[106]

The political and moral concerns that underlay Liberty Baptist College's biology curriculum surfaced in the mandatory "History of Life" course taught by assistant professor of natural science James L. Hall.[107] While the bulk of the course material concerned problems with the "evidences" of evolution, the class began with a meaty section on the "History of Human Thought." Based implicitly on Francis Schaeffer's framework from *How Should We Then Live?* Hall's history traced the development of a non-God-based "worldview" starting with the Renaissance and ending with modern-day "Atheistic evolutionary Humanism." Once again reflecting Richard Weaver's "ideas have consequences," the course outline indicated that "Inner Thought world determines outward action." In the modern period, the "flood tide of humanism" had produced Nazism, the "hippie world of the 1960's," the widespread use of hallucinogenic drugs and rock music, and the horrors of the French and Russian Revolutions. The former was reduced in meaning to "Napoleon Authoritarian rule." The latter was summed up in three words: "Lenin—Bloodbath—Communism."[108] That summary neatly encapsulated the decades of Christian fundamentalist and antievolutionary teaching that boiled the Bolsheviks down to murderous animals.

In his own sermons, on the weekly televised *Old Time Gospel Hour* and elsewhere, Jerry Falwell stuck to themes that he had raised in the initial issue of the *Moral Majority Capitol Report* sent to a choice list of fundamentalist preachers, thanks to the goodwill of John R. Rice. There, in August 1979, he had lamented that "the schools are steeped in humanistic philosophy, guided by atheistic and vulgar textbooks, rotten with drugs, sexual permissiveness and lack of discipline." Evolution was never his main theme. But its undertones reverberated widely.[109] On abortion, he charged on the *OTGH*, young people had become "amoral." They want to "live like animals" and have sex with "no consequences." In contrast, Bible-believing Christians "reject any form of evolutionary teaching."[110] In his sermon on communism, Falwell explained that it was essentially "godless atheism," which "represents and approaches humanity as nothing more than animal creatures." The "real controversy" was between "Christ and Satan," which logically linked the Great Deceiver with communism and evolution both.[111] Preaching on "sexual promiscuity"—which for Falwell included adultery, premarital sex, and homosexuality—he contended that the story of God's creation of Adam was essential for maintaining a true Christian and therefore moral perspective. Life didn't "just happen," Falwell said. "We did not evolve from some lower form of life." Believing in God as creator and rejecting evolution would strengthen Christians' ability to live a sexually clean life. As he put it, "You don't need to live like an animal." Speaking to the Christian Life Commission's "Strengthening Families Seminar" in 1982, Falwell cited Francis Schaeffer on the dangers of secular humanism and explained that "I personally believe that evolution is the cardinal doctrine of secular humanism." Thus, whenever Falwell spoke of the evils of humanism—which he often did—he was implicitly taking aim at evolution.[112]

Like many fundamentalists before him, Falwell possessed a deep conviction that the Christian masses of America had been lulled into complacency and needed to be shocked into recognizing the twin dangers of evolutionary humanism and communism. He expressed this in a variety of ways, including the notion that they needed to get a case of "spiritual heartburn" to awaken them. In frustration, and using another well-worn metaphor, Falwell wrote, "I don't know what it takes to *dynamite* Christians out of their spiritual sleep."[113] Preaching was Falwell's strength, but he also got into the book-writing business. Beyond the 1.5 million who watched the *OTGH* in the early 1980s—who made it the most popular syndicated Christian program in the US—a potentially bigger audience was awakened in August 1980, just months before the nation elected Ronald Reagan the nation's fortieth president, when they read Falwell's *Listen, America!*[114]

Having made headlines with the launching of the Moral Majority the previous year, Falwell wrote *Listen* to bid for public respectability as a "serious" author. Published by Doubleday and then brought out as a Bantam paperback to the tune of 150,000 copies in 1981, the book sported a fifty-title bibliography. As a reporter for the *New York Times* discovered while scouting out the political scene in Oklahoma City shortly before the 1980 elections, Falwell's book was prominently featured in local bookstores. While Oklahomans had gone Republican in the 1976 election, Ford's margin over Carter had been slim, and Falwell hoped to help Republicans' chances. When he spoke in Tulsa at a missions conference in the summer of 1980, he boosted the campaign of State Senator Don Nickles, who would ride in on Reagan's coattails that fall and serve as US senator from Oklahoma until 2005.[115]

Packing his 234-page book with historical specifics, Falwell also hedged his bets, providing a two-page prologue that began right inside the front cover and ended inside the back one. To draw in readers, Falwell retold the story of "Jack and the Beanstalk" and likened America to the sleeping giant, who had been lulled to sleep by indulging in the "good life." A veritable army of Jacks, consisting of "the abortionists, the homosexuals, the pornographers, secular humanists, and Marxists," were making off with the giant's "goods." But all was not lost. Signs that the giant was awakening included the virtual defeat of the ERA, the passage of the Hyde Amendment, and support for balanced-treatment bills in state legislatures. A slightly longer three-page précis of the book—under the heading "Seven Principles That Made America Great"—was confusingly sandwiched in between the last chapter and the end of the prologue. It also highlighted evolution, lamenting under the principle of "God-centered education" that creationism was no longer taught.

For those who persevered and tackled the meat of the book, there was a fair chance they would encounter the issue of evolution and its immoral communist connections. In "Understanding Our Times," which spelled out the wide extent of humanist-inspired sinfulness, Falwell pointed to humanism and "naturalism," the latter term referring to the view of "man as a kind of biological machine." From this standpoint, Falwell explained, "sexual immorality is just another bodily function, as is eating and drinking."[116] Either because of sloppy editing or a desire to reinforce the point, Falwell repeated this paragraph almost word for word in a later chapter. It was followed by a section profiling "sexual anarchy" on the nation's campuses, language that Falwell also used in the pages of the *Fundamentalist Journal*. That phrase nicely captured the political and sexual implications of evolution. Finally, on the first page of the chapter on "The Threat of Communism," Falwell informed readers, accurately, that the Marxist conception of morality denied absolutes and depended, quoting Lenin, on the needs of the class struggle. Moving from social to biological evolution, Falwell indicated (quoting an unnamed source) that another "scientific law of communism" was that "man is simply matter in motion." In his own words, Falwell explained that Marx taught the idea that "man was an evolutionary animal and as such had no eternal life."[117] Falwell was no Fred Schwarz, but he grasped the philosophical essentials that linked Darwin and Marx.

The same can be said of another man who learned from Fred Schwarz and became president of the United States: Ronald Reagan. Even before his election in November 1980 with the help of Falwell's Moral Majority, Reagan had given signs that he would be friendly to creationist politics. To be sure, he had some fences to mend with Jerry Falwell. In 1978, the former California governor had opposed Proposition 6, which banned gays and lesbians from working in the state's public schools. He was divorced and not much of a churchgoer. But Reagan did count himself a born-again Christian. And he was an inveterate anticommunist, having been schooled by his activism in the Screen Actors Guild, his employment as a company spokesman with General Electric, and his exposure to Fred Schwarz, the Harding College program, and similar ventures.[118] In August 1980, Reagan spoke at the Religious Roundtable National Affairs Briefing in Dallas to some fifteen thousand evangelicals. He lamented the moral decline of America. The schools, he told the crowd, "have tried to educate without ethics." The result, Reagan said, was a rise in "crime rates, drug abuse, child abuse, and human suffering." When a reporter asked Reagan about evolution, the candidate responded, "I have a great many questions about it. It is a theory, a scientific theory only. . . . I think recent discoveries down through the years pointed [to] great flaws in it." "Creationist theory," Reagan added, ought to be taught alongside evolution in the public schools.[119] Reagan's comments alarmed pro-evolutionists and even appeared to be a "gaffe" that might sink his election chances. But the ICR happily headlined the candidate's comments in *Acts & Facts*, bolding the quotations from Reagan about evolution for emphasis and fixing on his comments about "ethics."[120]

Once Reagan took office, he rarely said a word about biological evolution. But Reagan's anticommunism and social antievolutionism combined in an intriguing way in one of his most controversial addresses: the "evil empire" speech. Remembered mainly for its militant characterization of the Soviet Union, Reagan's March 8, 1983, speech to the National Association of Evangelicals began, after the obligatory folksy humor, with an attack on young women's reproductive rights. In response to growing access to birth control for teenagers at federally funded clinics, the Reagan administration had proposed a new rule requiring parental notification.[121] For Reagan, like John R. Rice in the 1940s, the growing use of birth control was a sign of spreading sexual sin. The president noted with scorn that the adjective "sexually active" had replaced the more presumably accurate "promiscuous." Reagan then proceeded to decry abortion as an attack on the sacredness of human life. Only after outlining these signs of America's moral decline did he arrive at his "final point," an indictment of the morality of Soviet leaders.[122]

To build his case that the Soviet Union was "the focus of evil in the modern world," Reagan pointed to the Marxist conception of morality, as he had done in his very first presidential press conference in early 1981. At that time, in response to a question about Soviet foreign policy aims and the possibility for détente, Reagan had replied that Soviet leaders had been perfectly open about their aims for world revolution. Their moral code was based on "whatever will further their cause." As opposed to Americans who followed a higher morality, Reagan explained, the Soviets "reserve unto themselves the right to commit any crime, to lie, to cheat," as long as it serves their ultimate goal. Now, two years later, he took the same tack, this time citing a 1920 Lenin speech. As Reagan paraphrased the Bolshevik leader, the Soviets "repudiate all morality that proceeds from supernatural ideas—that's their name for religion—or ideas that are outside class conceptions. Morality is entirely subordinate to the interests of class war. And everything is moral that is necessary for the annihilation of the old, exploiting social order and for uniting the proletariat."[123]

Delivered after a decade of détente and "peaceful coexistence," Reagan's speech caused a storm of protest.[124] But his summary of Lenin's argument was correct. The president implicitly targeted the Marxist materialist and evolutionary conception of morals. Moral standards change, both over time and depending on which class interest they serve. In all of this, Reagan demonstrated himself to be a worthy student of Fred Schwarz. He even quoted, without mentioning names, Pat Boone's comment at Schwarz's Los Angeles rally about letting his little girls die in a nuclear holocaust rather than live under communism.[125] Reagan had arrived at the philosophical heart of the matter.

While Baptists of various kinds—Bible Baptist Fellowship, Southern Baptist Convention, and independent—dominated the Moral Majority executive board that helped land Ronald Reagan in the White House, the one non-Baptist, D. James Kennedy (1930–2007), was no bit player.[126] Even more than Jerry Falwell, Kennedy established a Red Dynamite political legacy that still lives today. Born in Augusta, Georgia, and raised in Chicago and then Tampa, Florida, young Jim Kennedy was a talented clarinet player who won a music scholarship to the University of Tampa. But he soon dropped out to work as an Arthur Murray dance instructor and engage, by his own account, in all manner of sin. It was a seemingly chance encounter with radio preaching, emanating from his alarm clock, that led Kennedy to God. In 1956, he entered Columbia Theological Seminary in suburban Atlanta, affiliated with the conservative Southern Presbyterian Church in the United States (PCUS), formerly the Presbyterian Church in the Confederate States of America.[127] Columbia had been the site of fierce battles over evolution in the late nineteenth century. Their primary casualty was Professor James Woodrow, a theistic evolutionist fired for his heretical views in 1888.[128] Columbia had also been home to Professor James Henry Thornwell (1812–1862), a famed Christian apologist for slavery and an influential exponent of the thesis that the Civil War pitted a devout Christian South against a heretical, atheistic North.[129] Kennedy would absorb elements of both these intellectual traditions as he went forward into the world as an ordained Presbyterian minister in 1959.

Taking the helm at tiny Coral Ridge Presbyterian in Fort Lauderdale, Florida, the following year, Kennedy built it into a mega-church over the next two decades, attaining a membership of five thousand by 1976. In 1978, Kennedy took his church out of the PCUS and into the newly formed and even more conservative Presbyterian Church in America (PCA). Like Jerry Falwell, Kennedy reached a growing body of congregants by radio and television and added a school and seminary to his church. Unlike the Lynchburg preacher, Kennedy earned his doctorate—from New York University in religious education. He was an effective teacher, best known for his methodical, door-to-door patient explanations, which he trade-marked in a popular handbook, *Evangelism Explosion*.[130] He also relied on the force of his personality. Once speaking at a creationist conference with Kennedy, the ICR's John Morris heard a comment about Kennedy that rang true: "That's the only person I've ever seen who can scream in body language."[131] Starting in 1978, millions of viewers watched Kennedy weekly on the *Coral Ridge Hour*, as the pastor expressed that body language with a rich, baritone voice, boyish good looks, his signature "royal blue gown," and an authoritative professorial air.[132]

Kennedy conveyed the same basic message about moral decline that fellow Moral Majority leaders LaHaye and Falwell were spreading, but with a Reformed "Reconstructionist" flavor. In *Why I Believe* (1980), Kennedy armed his followers against doubts that Satan planted in the minds of those who had failed to put in the "intellectual effort" to ground their Christian faith securely in knowledge. On the topic of evolution—on which Kennedy repeatedly preached over the decade—such knowledge included the laws of probability, which meant that a human cell could not have evolved "by chance." There was also the compelling political and historical knowledge that evolution, as Dan Gilbert had explained, was the root of every "ism." Nazism drew upon the "evolutionary platitudes" of Nietzsche. In a unique twist on the Marx-Darwin legend, Kennedy told readers not that Marx had offered to dedicate *Das Kapital* to Darwin, but that it was "well known" that Marx had asked Darwin to write the book's introduction. However badly Kennedy had mangled his facts, his concluding thought would have resonated with many readers: the same people pushing the "Communist conspiracy" were also pushing "an evolutionary, imperialistic, naturalistic view of life," which, in a poetic phrase, would "crowd the Creator right out of the cosmos."[133] Summing up the godly alternative to that worldview, Kennedy provided a clue to his emerging Reconstructionist theology. The system of God-given ethics, wrote Kennedy, using the distinctive language of R. J. Rushdoony, was "theonomous," or based on God law.[134]

Kennedy's next book was more forthright on this point, starting with its title: *Reconstruction: Biblical Guidelines for a Nation in Peril* (1982).[135] Focusing his fire on the "morally bankrupt leadership" of the country, Kennedy lamented the "evils" that this had imposed on the American people: liberal theology, relativistic ethics, socialist economics, statism, corruption, and cultural immorality. The task before us, Kennedy believed, was to *"reconstruct America with the Biblical guidelines which God has given us."*[136] Starting with the subject of ethics, which he considered the "seed-plot" of his plan, Kennedy surveyed the variety of non-godly ethical systems. Since the goal, in Jeremy Bentham's famous formulation, was the "greatest good for the greatest number," anything done to perpetuate the human "species" could be justified, even at the expense of those with "bad genes." The Nazis took this logic to its natural conclusion. Then the Communists applied it, killing tens of millions in the name of a promised earthly "paradise."[137]

Kennedy's Reconstructionist-tinged discussion of education also led him to the evolutionary-communist nexus. He asked, What is the purpose of education? Was it to produce good citizens? To enable self-realization? To graduate young people who were "well adjusted" to their environment? None of these was adequate. Citing Calvin, Kennedy explained that the purpose of education was, instead, to "know God and glorify Him as God." Citing R. J. Rushdoony, and taking special aim at the "adjustment" theory, Kennedy pointed out the false "naturalistic assumption" (or presupposition) underlying secular education. What if the environment to which students were adjusting was immoral, such as in Communist Russia where the naturalistic atheistic rulers viewed "men as little more than complicated mice" who could be killed "by the tens of millions" without consequences? During his own college education, Kennedy recalled that a professor had told him that "matter in motion" was all that existed. If so, then the products of American secular education would be well adjusted to the "natural" environment but "maladjusted" to God.[138]

Even Kennedy's pro-capitalist, antisocialist discussion of economics was based on a rejection of evolution. Referring to the biblical basis in Genesis for Rushdoony's "Dominionism," Kennedy reminded readers that as human beings created in the image of God, they were God's "vice-regents" who had "dominion" over the earth. As "steward" over the things we possess, thanks to God's creation, we are obligated to use that wealth in a manner consistent with biblical teachings, which favor "capitalism" (evident by virtue of the Eighth Commandment—"Thou shalt not steal"). Kennedy contrasted capitalism with socialism, which he defined as "government ownership." Despite claims that Nazi Germany was opposed to socialistic Russia, Kennedy wrote, they were just "two forms of the socialistic concept." His exposition of the biblical basis of free enterprise drew on familiar arguments developed by fellow conservative Presbyterian Carl McIntire, Baptist John R. Rice, and others during the early Cold War years.[139]

What distinguished Kennedy's Christian pro-capitalism from that of John R. Rice was his Reformed slant. In developing the pro–free enterprise argument, Kennedy leaned heavily on John Calvin. The great Geneva theologian was the "prime mover" of capitalism, who freed money from the "bondage" imposed on it by the medieval Catholic Church. The source of Kennedy's information on Calvin was mentioned only once in the text, but it was likely significant. This source was historian C. Gregg Singer (1910–1999), a PhD from the University of Pennsylvania and an elder in the Southern Presbyterian Church (PCUS). Unlike his fellow aspiring historian and contemporary R. J. Rushdoony, Singer had a successful academic career, teaching history in a string of Christian colleges and publishing books that reflected his conservative Reformed perspective. Most influential was *A Theological Interpretation of American History* (1964; rev. 1981 and 1994), which surveyed the impact of theology on the main currents of American history from the Puritans to the Moral Majority. By the mid-1980s, Singer's book was required reading at Liberty University.[140] Baptist and Reformed theological strains intertwined on the Liberty campus to convey the twin dangers of evolution and communism.

By the late 1980s, thanks to the combined efforts of Kennedy, Falwell, LaHaye, Rushdoony, Reagan, and others, the significance of the evolution question had become clearer for millions of evangelical Christians. The cross-fertilization between the work of scientific creationists and fundamentalist activists bore plentiful fruit. Henry Morris had now spent four decades writing about the evils of evolution. In 1989, Baker House published Morris's crowning achievement, *The Long War against God*. The book contained more than three hundred footnoted pages. But it was accessibly written. And it conveniently summed up in one volume the lessons that Morris wanted the nation's Christians to learn before it was too late.

Even as the ICR was gaining momentum in the broader stream of conservative Christianity, the challenges of making evolution relevant emerged in the book's introduction penned by Rev. David Jeremiah. Succeeding Tim LaHaye at Scott Memorial Baptist in 1981, Jeremiah served as president of Christian Heritage College and preached on the radio every week through his Turning Point Ministries. When he first met Henry Morris, Jeremiah was not sure Morris was on the right track in placing evolution at the center of the ICR's ministry. "I wondered if perhaps his perspective had been clouded by a narrow focus of study over the years," Jeremiah wrote. One can imagine Jeremiah making this way through a technical article in *Acts & Facts* and scratching his head in puzzlement. But having learned the full scope of evolution's consequences from Morris, Jeremiah had seen the light. "I am now convinced," he wrote, "that all significant problems of society are the children of an ignorant or indifferent attitude toward creationism." Even though the ordinary person "neither knows nor cares much about the error of evolution," Jeremiah acknowledged, it affects his life profoundly. The evil fruits of evolution included "pornography, adultery, divorce, homosexuality, premarital sex, [and] the destruction of the nuclear family." Jeremiah also embraced Morris's conclusion, reflected in his book's title, that the true origin of evolutionary thinking went back to ancient times when Satan first uttered his "big lie about the universe."[141]

In *Long War*, after documenting the undeniable impact of evolution on a wide spectrum of academic theory, Morris proceeded to deal with the social and political practice that had resulted, what he called "Political Evolutionism—Right and Left." Morris's chapter title promised political evenhandedness, and his approach did yield some valuable insights. In tracing conservative social Darwinism, of the Carnegie and Rockefeller type, for instance, Morris noted that this tradition persisted in the world of latter-twentieth-century conservatives who did not identify strongly enough, in his opinion, with evangelical Christianity. He correctly observed that Robert Welch of the John Birch Society, a nominal Christian, was a "strong evolutionist." He bemoaned the fact that conservative evangelical hopes in Ronald Reagan's administration had been edged out by "economic measures designed to restore a greater degree of Darwinist *laissez-faire* capitalism."[142] These observations supported Morris's contention that his opposition to evolution was not political, since both left- and right-wingers were, in practice, evolutionists.

The same seeming impartiality appears in Morris's substantial sections on Hitler and communism. Citing and quoting from scholarship that traced the impact of the ideas of Spencer, Darwin, Nietzsche, and Haeckel on Hitler, Morris made the case that the *Führer* was the "ultimate fruit of the evolutionary tree." Even the title of *Mein Kampf*, "my struggle," displayed Hitler's evolutionary bent, his debt to the "Darwin-Spencer-Haeckel emphasis" on the "struggle for existence." To back his claims about left-wing evolutionism, Morris quoted amply from scholarship showing the support Darwin received from Marx and Engels. He pointed to the Marxist aspects of Gould and Eldredge's work on punctuated equilibrium and their opposition to Wilson's sociobiology. He trotted out Marxist historian Robert Young on the inevitably political nature of science. He even quoted from a 1980 analysis of creationism by Socialist Workers Party (SWP) leader Cliff Connor published in the *Militant* newspaper that accurately reported, "Defending Darwin is nothing new for socialists." Like their political predecessors in the era of the Scopes trial, the SWP continued to argue that the "cultural" issues of social and biological evolution were critical for the working-class fight for power.[143]

Despite his seemingly undifferentiated hostility to evolutionists of the Right and Left, a closer examination brings Morris's own politics more sharply into focus. In an attempt to distance his own worldview from Hitler's, Morris refers to "certain attempts to depict Hitler as a right-winger, or even as a Christian." Morris's response: "Even though he was an anti-Communist (except when it suited his devious thinking to unite with Russia in the pact that precipitated World War II!), Hitler was certainly not a Christian, in any sense whatever."[144] Yet Morris offers no additional information to his readers about Hitler's anticommunism—an essential ingredient in Nazi ideology. In effect, Morris admits that Hitler was "one of us" in his militant anti-communism but fails to explore the damaging implications. It was precisely that common ground that had led William Bell Riley—who had chosen Morris as his heir apparent at Northwestern—to praise Hitler and led others to sympathize more quietly with him. Though Riley was long gone, Morris's ties to Northwestern lived on, as the ICR regularly held summer institutes on that St. Paul, Minnesota, campus.[145]

Morris's political perspective comes out more clearly in his coverage of left-wing evolutionism. Modern evolutionists may try to dismiss early cases of right-wing social Darwinism as "irrelevant today," notes Morris. But they cannot deny that socialism and communism represent a "current problem in every sense of the word!" Governments promoting some form of Marxist ideology "reign over most of the world's nations"; Marxist theory is taught in colleges as "the wave of the future"; even liberals in the Democratic Party promote "Marxist policies." If one compares the human toll of communism and Nazism, the latter pales before the former. The number of people slaughtered in the "class struggle" exceed "by a factor of ten or more" those who died from Hitler's "genocidal aggressions." Hitler was bad, but the communists were at least ten times worse.

Not that Morris's comparative judgment on Marxist and Nazi evolutionism was based on a mathematical calculation. The heart of his argument could be found, instead, in a chapter called the "The Dark Nursery of Darwinism." Here Morris's readers learned once again about Darwin's predecessors and their participation in a shadowy worldwide political conspiracy linked to the French Revolution. The plotters included geologist Charles Lyell, Erasmus Darwin, who anticipated his grandson's evolutionary ideas, and Jean-Baptiste Lamarck. Morris reviewed the attraction of Lamarckism for Soviet science, restated once again the demonstrably false fact that Marx wanted to dedicate *Capital* to Darwin, and reminded readers about Stephen Jay Gould's Marxist ties. The main point was that Marx and Lamarck were somehow linked in a worldwide secret plot.[146]

To the charges of mass murder and secret plotting, Morris now added the whopper that gave his book its distinctive punch: the key player behind the revival of evolutionary ideas during the nineteenth century was Satan. He worked directly through two key figures, who represented both biological and social evolution: Alfred Russel Wallace and Karl Marx. As Morris tells his readers, Wallace was a religious skeptic who "was long enamored of socialism, Marxism, and even anarchism." He also developed a fascination with spiritualism based on a pantheistic notion of disembodied spirits in nature. But Wallace is best known as the codiscoverer of the idea of natural selection, which came to him while enduring a "rather severe attack of intermittent fever" in the Molucca islands of Malaysia. By Wallace's account, the idea of incorporating the population theories of Thomas Malthus—a key piece of the natural selection concept—"suddenly flashed upon me." For Wallace, this inspiration must have come from the natural spirit world. But the only spirits Morris recognized were the demons mentioned in the Bible. From this perspective, Morris concludes, "it is not naïve fundamentalism but essential realism to recognize that Satan . . . would somehow be very directly involved in this watershed development of 1858–59."[147]

Whereas Wallace appears in Morris's account as an unwitting tool of Satan, Karl Marx was an active accomplice. Here Morris relied on a book by Richard Wurmbrand, the former imprisoned Romanian pastor and anticommunist confederate of Billy James Hargis. In *Marx and Satan* (1986), Wurmbrand probed the writings of the young Karl Marx to argue that he not only became an atheist after his years in a Lutheran high school but "became a Satan worshipper who regularly participated in occult practices and habits." Evidence from Marx's early work included the drama *Oulanem* (in Wurmbrand's estimation, a satanically inspired anagram of Emmanuel, a biblical name of Jesus) and the poem "The Player," both written during Marx's high school years. Morris quotes a stanza from the latter, including the lines, "See this sword? / The prince of darkness / Sold it to me." According to Morris, Wurmbrand explains that these lines reference "rites of initiation in the Satanist cult, in which an enchanted sword ensuring success in life is sold to the initiate for the price of a blood covenant with Satan for his soul in death."[148] Marx thus appeared worse than Hitler not only in the number of his minions' victims, but because Marx, apparently unlike Hitler, was a practicing satanist.

Following the satanic trail backward through history, Morris arrived in ancient times, where he moved through ancient evolutionary-inspired pantheism to the early Greek materialists and finally to Babel. Morris had trod this territory since his collegiate poem on King Nimrod in the 1940s. But *Long War* was his definitive treatment. He summed up the results of his historical investigation for readers: "This means, finally, that the very first evolutionist was not Charles Darwin or Lucretius or Thales or Nimrod, but Satan himself!" Morris seemed less than 100 percent certain on his sub-points. On the possible connection between Marx's satanism and the millions murdered by Communists in the twentieth century, Morris ventured that "one cannot help sensing some kind of occult cause-and-effect relation." But the ICR's publicity for *The Long War against God* was unequivocal. The cover story on the book in *Acts & Facts* noted that while a satanic origin for evolutionism might be "ridiculed" by some, Henry Morris had demonstrated that it was "the only viable explanation."[149] Creation science had proved Satan's role in evolution once and for all.

The ICR was flying high by the 1990s. One reason was the addition of a new Australian-born ICR staffer with a gift for gab. Born in 1951 into a fundamentalist family in Queensland, Australia, Ken Ham spent five years as a high school science teacher in the late 1970s. Under the impact of Henry Morris's writings, he became a full-time creationist activist. After spending a number of years in Australian creationist groups, specializing in providing educational materials to Queensland public schools—where teaching creationism along with evolution was legal until 1987—Ham came to the US in that year to offer his services to the ICR. An article in *Acts & Facts* introduced Ham to the ICR faithful as "a skilled communicator, with a keen sense of humor and an entertaining accent."[150]

Over the next seven years, Ham carved out a prominent place in the ICR landscape. He wrote a regular "Back to Genesis" column, adorned with cartoons, and taught "Back to Genesis" seminars to large crowds around the country. Ham's populist style contrasted sharply with Morris's more academic approach, but he was building on the older man's fruitistic foundation. In his first column for *Acts & Facts*, Ham described evolution as Darwin's "repopularization" of "an ancient pagan belief" that today gave people license to practice "humanistic morality" and act as if there were no "absolutes." Evolutionary thinking explained the homosexual "lifestyle," abortion, and "easy divorce." A cartoon that accompanied a Ham column—which he used often in the coming decades—depicted a medieval battle between knights guarding rival castles, "Evolution" and "Christianity." The latter fired canons at balloons flying from the former, which included Divorce, Homosexuality, Pornography, Racism, and Abortion. In a sign of things to come, Ham's "fruits" pointedly did not include communism.[151]

The ICR reached an expanding popular audience in the 1980s and '90s through its Museum of Creation and Earth History. Founded in 1976 and originally housed on the Christian Heritage College campus in El Cajon, the museum attracted five thousand visitors by 1980. Its exhibits included the Origin of Mankind, the Origin of Birds, the Case for Creation, the Origin of Horses, and Creation: Life before Birth. In 1981, the ICR added a new display, Dinosaurs and Man in History, based on the soon-to-be discredited "Paluxy Man Tracks" near Glen Rose, Texas, which supposedly provided evidence that humans and dinosaurs lived at the same time. In 1986, the ICR opened new headquarters, including a new museum building, in nearby Santee, California. Perhaps thanks to the arrival of Ken Ham the next year, the ICR began to market a range of dinosaur-themed products designed for children. Visitors to the museum gift store could buy T-shirts, mugs, posters, and a *Dinosaur ABC's Activity Book*.[152] Speaking at the new building's dedication ceremony, an aging John Whitcomb Jr., coauthor with Morris of *The Genesis Flood*, said that the completion of the new building could only have happened thanks to the power of an "infinite God," who could counter the influence of Satan and his "vast armies of evil spirits."[153]

The final incarnation of the museum, in which rival creation and evolution trees displayed the good and evil fruits of Darwinism, took shape in 1992. The ICR moved the museum up a floor, where it occupied some four thousand square feet. Henry and John Morris wrote the bulk of the text for a new set of exhibits, relying heavily on *The Long War against God*. Visitors learned about the origins of evolutionary thinking in the ancient world, from Babel through the Greek materialists through "evolutionary" non-Christian world religions. To underline why Christians needed to understand the evolution issue, Morris devoted a panel to explaining why it was a "dangerous error" to ignore the question of origins. "The tree of evolution bears only corrupt fruits," the panel read. "Creationism bears good fruits." By 1997, five years after the renovation, some one hundred thousand people, from nearly every US state and at least twenty other countries, had visited the new museum. To a degree that Henry Morris and John C. Whitcomb Jr. could never have imagined, young-earth creationism had arrived.[154]

It was only the beginning. In 1994, Ken Ham struck off on his own, and in 2007 his organization, Answers in Genesis, opened a multimillion-dollar creation museum in Petersburg, Kentucky, that put the Santee museum to shame. The fruits of evolution in the new facility's exhibits did not explicitly include Karl Marx or communism. In the new political context of the early twenty-first century, anticommunism did not carry the same weight as it did even in the early nineties, when the Soviet Union was crumbling. That did not mean, however, that Red Dynamite had forever faded. A whole cast of characters, both old and new, kept the tradition alive into the Trump era.

8

THE NIGHTCRAWLER, THE WEDGE, AND THE BLOODIEST RELIGION

On New Year's Day 2011, the online conservative news outlet WorldNet-Daily (WND) blared this headline: "From Darwin to Marx to Kinsey to Obama." The author was Dave Welch (1961–), head of the Houston Area Pastors Council and a former national field director for Pat Robertson's Christian Coalition. Welch was reacting to the news that President Obama was "evolving" toward approval of gay marriage. According to Welch, Obama and Vice President Biden had a secret "agenda," which was to carry out a "dialectic Marxist strategy" to "overturn traditional morals." There was a "direct linkage" between Marx and Darwin—"Marx and Engels based their communistic philosophy squarely on the foundation of evolutionism." Sex researcher Alfred Kinsey was a devoted Darwinist. As the result of this godless conspiracy, the economy was becoming social-ist, and sexual morality and marriage were headed in the direction that "Darwin, Kinsey and their ilk intended." Invoking the Frankish Catholic hero of the battle of Tours, Welch bemoaned the failure of American con-servatives to serve as "spiritual, moral, and intellectual Charles Martels against the hordes of 'Moors' sweeping through our land." (Welch also referred to the Marxist movement as a "jihad.") Welch then harked back to the biblical prophet Nehemiah, who acknowledged that the Jewish people had abandoned the ways of the true God and that He had justly punished them for their "evil deeds." It was time, Welch wrote, for God's pastors and people to "restore His righteousness to our land."[1]

Neither historians nor pro-evolutionist academics in any field were likely to take Welch's plot involving Marx, Darwin, Kinsey, and Obama seriously. The anonymous pro-evolutionist blogger who goes by the name of "the Sensuous Curmudgeon"—and is widely respected by academic pro-evolutionists—described Welch's article as a "towering monument of foolishness."[2] He called WND (which received the Curmudgeon's "Buffoon Award" in 2008) "an absolutely execrable, moronic and incurably crazed publication."[3] Founded in 1997 by journalist Joseph Farah, a "birther" activist who helped publicize the charge that Obama was not a US-born citizen, WND is far enough to the right that first-time liberal visitors to its website have trouble believing it is not a hoax.[4]

But as "crazed" as Welch's ideas might seem to some, they surely struck a chord for others. Conservative Christian leaders had been teaching about the tie between Marxist philosophy, Darwinian evolution, and sexual immorality for at least a century. To be sure, the context was changing rapidly. The Soviet Union was long gone. A decade of the "war on terror" in Iraq, Afghanistan, and elsewhere had made "jihad" a household word. Welch contended with America's first African American president. Gay rights were ascendant. The Internet vastly multiplied the potential sources of political and moral authority. Due in part to the grassroots conservative activism pioneered by Welch's former employer Pat Robertson, the Republican Party had become "God's Own Party." Conservative evangelicals had a seat at the political table like never before. At the same time, opposition to evolution was increasingly likely to skirt the realm of "creation science" and instead take the form of a seemingly nonreligious "intelligent design" (ID). For all of the ways in which the world had evolved, however, the end of the Cold War did not quash the persistent appeal of Red Dynamite political rhetoric in the new millennium. That rhetoric came in varied forms—the writings of Summit Ministries' David Noebel; the pro-capitalist politics of the Discovery Institute, the new ID think tank; and the continuing creationist campaigns of the ICR and of the new kid on the block, Answers in Genesis.

Given the pivotal role of the Christian Coalition in transforming the Republican Party into a viable vehicle for Christian conservatism, we begin with a man who was no stranger to the allied evils of evolution and communism: Marion Gordon "Pat" Robertson (1930–). The son of southern segregationist US Senator Willis Robertson (R–VA), Pat grew up in Lexington, Virginia, and attended the McCallie prep school in Chattanooga, close on the heels of *Genesis Flood* coauthor John C. Whitcomb Jr. Robertson graduated magna cum laude from his hometown college of Washington and Lee and then received his JD from Yale Law School. Ordained a Baptist minister after attending the Biblical Seminary in New York, he began to imbibe a strong dose of the Christian charismatic tradition, with its strong emphasis on faith healing and speaking in tongues. Robertson founded the Christian Broadcasting Network in 1960. By the late 1970s he had become a major Christian media mogul and TV personality, seen by millions weekly on *The 700 Club*.[5]

Until the mid-1970s, Robertson hewed to his father's conservative Democratic sympathies and had high hopes for the newly elected Southern Baptist president, Jimmy Carter. But Carter's liberal politics pushed Robertson to the right. Robertson spoke out against abortion on *The 700 Club*. His newsletter, *Pat Robertson's Perspective*, aired his support for Anita Bryant's campaign against gay rights. In 1980, Robertson campaigned for Ronald Reagan and organized a giant "Washington for Jesus" rally in the nation's capital. After Reagan's election, Robertson lobbied hard for a proposed constitutional amendment to allow prayer in public school. Though the amendment failed to secure passage in 1984, Robertson gained increasing attention from Christian conservatives. A televised speech Robertson gave in 1986 effectively launched his 1988 presidential campaign and, through it, the Christian Coalition. Speaking at Constitution Hall in Philadelphia, on the 199th anniversary of the document's adoption by the delegates to the Constitutional Convention, Robertson made a promise: if three million registered voters would sign petitions in the coming year pledging their support to Robertson, he would run for president.[6]

Robertson hailed the wisdom of the nation's founders who created a country "under God." He bemoaned the state of America, which in the previous quarter-century had strayed from "our historic Judeo-Christian faith." Public schools had replaced moral absolutes with "values clarification" and "situation ethics." They had replaced the "Holy Bible" with the familiar pantheon of communist and evolutionary evil: Charles Darwin, Karl Marx, Sigmund Freud, and John Dewey. Young people were learning that "if it feels good, do it." For conservative Christians paying attention to the warnings of Francis Schaeffer, Tim LaHaye, and the like, the "whirlwind" of immoral consequences was also familiar: one million teenage pregnancies and four hundred thousand abortions each year; a massive number of sexual assaults; and an epidemic of sexually transmitted diseases, including AIDS.

To save the children, Robertson called for tougher measures on drugs and alcohol in the schools; removal of control of education from the "leftist" teachers' unions; the return of God to the nation's classrooms; and the replacement of "progressive education" with a focus on the "facts" of history, geography, and science.[7] Robertson got his three million signatures. But his campaign quickly spiraled downward. The marginality of Robertson's Christian charismatic leanings and a series of sex scandals involving former supporters and associates of Robertson doomed his campaign. Still, the essential message—that the Christian "worldview" was under attack and needed to be aggressively defended—had serious staying power for conservative Christians in post–Cold War America.

Dave Welch, the Houston pastor who linked Darwin, Marx, Kinsey, and Obama, found Robertson's message compelling. Welch's story helps us appreciate the worldview of a twenty-first-century conservative Christian activist. Born in Seattle, Welch grew up in coastal Hoquiam, Washington, where his mother took care of the large family, and his father, Herb Welch, a fourth-generation logger, worked as an independent small "gyppo" operator. That label originated as a derisive term for strikebreakers by organizers for the radical Industrial Workers of the World (IWW), who built a logging local of their union in Hoquiam in the early twentieth century.[8] Herb Welch's politics were far to the right of the IWW. He was active in the local chapter of the John Birch Society, and Dave Welch recalls reading Birch material as a teenager. It included *None Dare Call It*

Conspiracy (1972) by Gary Allen, a conservative journalist, Birch Society member, and former writer for George Wallace's 1968 presidential campaign.[9] In a slim paperback complete with explanatory diagrams, Allen argued that communism was just a front for diabolical conspiracy by a group of "power-mad billionaires" called the "Insiders" whose history reached back to Adam Weishaupt and the Illuminati. One contingent of them were Jewish bankers, though Allen ambiguously cautioned against imputing Insider status to all Jews. The victims of the conspiracy were the middle class—including small businessmen like Herb Welch—who were squeezed from below by "schoolboy Lenins and teenage Trotskys" and oppressed from above by the Rothschilds, the Rockefellers, and the Council on Foreign Relations.[10]

Dave Welch entered electoral politics at age nineteen when he voted for Ronald Reagan in 1980. Over the next fourteen years, Welch became a small businessman like his father, taking over his father-in-law's service station and auto repair business. In 1984, he attended his first Republican gathering and was elected as a county delegate to the state convention. The issue of abortion—"the sanctity of life"—drew him into Republican politics. In 1988, he joined the Pat Robertson campaign. As Welch recalls, Robertson "linked the godlessness, the atheism of communism very much to the policies and the tyranny of communism that go hand in hand." Robertson shared Gary Allen's conspiratorial views. Railing against George H. W. Bush's "globalist" politics in the aftermath of the Gulf War, Robertson explained history as a plot masterminded by the Illuminati, Jewish bankers, and other shadowy figures in *The New World Order* (1991).[11] Dave Welch's 2011 piece on Darwin, Marx, Kinsey, and Obama had a similar conspiratorial tone. His allegation that Marxists aimed to overturn morals, "create chaos and then serve as savior," suggests, with echoes of the *Protocols*, that they were secretly plotting behind the scenes.

The Christian Coalition emerged out of the promise and failure of the Robertson campaign. Welch served as state director and then national field director starting in 1997. He worked closely with history PhD Ralph Reed, whom Welch regarded as "a masterful strategist." Welch tried to follow Reed's methodical model when he formed his own organization in Houston in 2003, the US Pastor Council (USPC). Welch concentrated on local politics and building up strength from the grass roots. As he described his approach to local pastors, "'We're going to teach you how to organize a precinct,' but ultimately, 'We will give you the information historically, biblically, politically as to how to be well prepared to do that.'" The information packet Welch put together included step-by-step instructions, an organizational chart, and a summary of the AMERICA plan. "A" stood for "Articulate a Biblical position on important issues of the day": abortion, gay marriage, sexual immorality, creation/evolution, and "Socialism/Marxism vs. Constitutional Republic." To these Welch added "Race Relations and Equal Justice," in an attempt to broaden the base of the US Pastor Council's operation in Houston.[12]

While Welch had no scientific training, his political education had taught him that evolution was *"the* question" underlying the war of the worldviews. Pointing to the writings of Francis Schaeffer, David Noebel, Tim LaHaye, and D. James Kennedy as resources for his recruits, Welch took up the challenge of explaining to his local preacher allies why the seemingly academic issue of evolution was so important.[13] "If we assume that we're an animal," he explains, "then the truth is that we don't have a real case to stand on for developing any kind of framework of laws" based on true, Christian morality. This is where "we connect the dots," said Welch: the evolutionary assumption "can produce things like communism which are based on the state being the supreme authority, not God."[14] Any pastor who embraced any form of evolution "is no more a Christian than the chimpanzees from which he or she claims to have evolved."[15] In his own distinctive way, Welch carried forth the ideas linking Darwin and Marx that George McCready Price outlined a century earlier.

When it came to the topic of evolution and communism in the early twenty-first century, there was no greater dot connector than the old anticommunist warrior David Noebel. By the 1980s, Noebel's Anti-Communist Youth University had reinvented itself as Summit Ministries. Summit received a boost in 1987 when a young man named Ryan Dobson attended its summer camp. His father was James Dobson, the doctor who headed up the prominent conservative Christian group Focus on the Family, headquartered a few miles from Summit, in neighboring Colorado Springs. When Dobson promoted the value of his son's experience at Summit on his *Focus on the Family* radio show, the Summit staff received nearly fourteen thousand new applications. The "explosion" ignited by

Dobson's promotion resulted in a major expansion of the Summit property in Manitou Springs. It brought welcome publicity for what had been begun by "Doc" Noebel as a modest undertaking more than two decades earlier.[16]

Summit's new success also motivated Noebel to consolidate his nearly thirty years of teaching material into a beefy 891-page textbook, *Understanding the Times: The Religious Worldviews of Our Day and the Search for Truth* (1991). The book eventually sold more than half a million copies. It carried back-cover endorsements from James Dobson, D. James Kennedy, and Tim LaHaye. The original edition compared three worldviews: Christian, secular humanist, and Marxist/Leninist. Noebel added "cosmic humanist" (New Age–ism) for the 1995 edition and then Islam and postmodernism starting in 2006. In all versions, readers learned how competing worldviews taught about theology, philosophy, ethics, biology, psychology, sociology, law, politics, economics, and history. For its breadth and depth of coverage it was an impressive piece of work. Noebel's confidence in his own Christian worldview was admirable. In the first edition, he encouraged students to prepare for absorbing the material in *Understanding the Times* (UTT) by reading key texts from all three viewpoints: the *Communist Manifesto*, the *Christian Manifesto*, and the *Humanist Manifestos* I and II. Noebel's confidence was also reflected in the fact that he consciously adopted an evenhanded tone, refusing to portray Marxists or humanists as "either stupid or insane" despite what he claimed, with some accuracy, was "their tendencies to describe Christians in such unflattering terms."[17]

In his opening chapter, "The Battle for Hearts and Minds," Noebel made it clear why it was crucial for America's young Christians to grasp the essentials of competing worldviews. Quoting from *Children at Risk*, by James Dobson and Gary Bauer, Noebel affirmed that Americans were in the midst of fighting "a great Civil War of Values," what Patrick Buchanan would label a "cultural war" the following year at the Republican National Convention. The widespread ignorance among young Christians of worldview fundamentals meant, in Noebel's evocative phrase, that they "stood intellectually naked before left-wing professors." The Cold War might have ended, Noebel acknowledged, but as Fred Schwarz had taught, the campuses continued to be the "nurseries of communism." For all the similarities between secular humanism and Marxism, the latter, in

Noebel's estimation, was more consistent and stood on a stronger philosophical foundation. As Noebel had been explaining to legions of young Christians in Manitou Springs, if they followed the example of the sons of Issachar from 1 Chronicles 12 and understood the world around them, they would "know what Israel ought to do." The book was both an intellectual exercise and a call to action.[18]

If some youngsters at a Summit summer session might be forgiven for missing the connections between evolutionism and Marxism, no one who persevered through *UTT* could fail to understand what George McCready Price had meant in 1925 when he called Darwinism "Red Dynamite." Even one of the handful of political cartoons that adorned the text of the original edition emphasized the point: public school students enjoying a field trip to the "Land of Ideas" are free to climb the trees labeled "Karl Marx," "Charles Darwin," "John Dewey," and "Voltaire" but are prevented from visiting "Religion," which is "Off Limits." In every one of the ten chapters on the Marxist worldview, evolutionary theory made a prominent appearance. In the opening summary of his treatment of Marxist philosophy, Noebel concluded that "for better or worse, the Marxist's philosophy of dialectical materialism is built primarily on the 'science' of Darwinian evolution." In outlining Marxist ethics, Noebel devoted an entire section to "The Evolution of Morality." His discussion of Marxist psychology leaned heavily on the ideas of Ivan Pavlov, "an avowed evolutionist." On Marxist sociology, Noebel quoted Lenin on society as "a living organism" in a section titled "Society as an Evolving Entity." The "Marxist evolutionary perspective" provided a framework for analyzing history and law. Most extensively, a twenty-six-page chapter on Marxist-Leninist biology wove multiple threads between Marx, Engels, Lenin, and evolutionary science. Six of those pages were devoted to punctuated equilibrium and Stephen Jay Gould, with an obligatory comment about his "daddy's knee."[19]

No one knows how many teenagers powered through all forty-two chapters of the abridged edition of *UTT*. But Noebel and Summit did make a considerable effort to reach young people. The 1995 abridged edition omitted endnotes and featured large, easy-to-read pull quotes, photos, and concise timelines of the lives of important figures. The 2006 edition was still in black and white but included many more text boxes with quotes and definitions of key terms, a more attractive layout, and a

Figure 13. "Land of Ideas" cartoon, 1984. Appearing in anticommunist David Noebel's pioneering worldview textbook *Understanding the Times* (1991), the cartoon helped readers understand the twin threats of Marxism and Darwinism. Originally published in the *Colorado Springs Sun*, December 3, 1984. Cartoon by Chuck Asay. Courtesy of Special Collections, Pikes Peak Library District, Colorado Springs, CO, MSS 0448.

regular feature called "The Pop Culture Connection." In the chapter on Marxist-Leninist biology, Noebel supplemented his discussion of punctuated equilibrium by means of a scene from *X-Men 2*. Whereas evolution usually moves slowly, explains character Dr. Jean Grey, sometimes it "leaps forward," producing a range of "mutant" characters. Young viewers of the film might remember characters like Nightcrawler. He has two-toed feet, three-fingered hands, yellow eyes, and a prehensile tail—not exactly what Gould and Eldredge had in mind, but surely fodder for a lively conversation.[20]

In many cases, that conversation would have taken place outside of the nation's public schools. The publication of *UTT* coincided with a meteoric rise in both private Christian schools and conservative Christian homeschooling. The Christian school movement started by Tim LaHaye and others in the 1960s and '70s had gone mainstream. The copublisher of the 1995 edition of *UTT* was the Colorado Springs–based Association of Christian Schools International (ASCI), the largest national accrediting organization in the field. Between 1983 and 2005, the number of ASCI schools jumped from 1,900 to 3,957, with a consequent rise in student enrollment from 270,000 to 746,681.[21]

Christian homeschooling provided a ready market for Noebel's book. Based on the pioneering ideas of R. J. Rushdoony and the legal activism of John Whitehead and Michael Farris, homeschooling became legal in all fifty states by the early 1990s.[22] Summit produced a *UTT* homeschooling package, which included the 2006 revised second edition, teacher and student manuals, and eight DVDs. In her review of the new package, popular conservative homeschooling book reviewer Cathy Duffy opined that she was "so impressed with the course that I think it or a similar course should be offered to (or maybe even required of) all Christian teens and parents."[23]

Noebel's efforts to reach the Christian masses drew strength from a continuing collaboration with Henry Morris and the ICR. During the 1990s, Manitou Springs was a regular stop for the ICR's traveling lecturers in the Summer Institute on Scientific Creationism.[24] Like Noebel, the ICR continued to emphasize the political and moral "fruit" of evolutionary thinking, including its communist component. Most prominently, this took the form of the third volume of the *Modern Creation Trilogy* (1996), coauthored by the father-and-son team of Henry and John Morris. Titled "Society and Creation," it recast the material from *Long War against God* in a shorter, updated form. After tracing evolution to Babel and Satan, the Morrises offered by-now familiar material in two key chapters on "The Corrupt Fruits of Evolutionism" and "Evolution and Its Deadly Social Philosophies." Marx's supposed Satanism was now relegated to a footnote, but the twelve-page section on Marx and evolution unfortunately repeated the thoroughly discredited claim that Marx wanted to dedicate *Das Kapital* to Darwin.[25]

As a front-page article in *Acts & Facts* noted, the appearance of the *Trilogy* as a boxed set was "an unusual publishing event."[26] There is no evidence to suggest that it sold well. But as before, the ICR was not content only to write about evolution's fruits. In the summer of 1995, John Morris visited fundamentalist Baptist Pensacola Christian College (PCC) to give a series of *Trilogy*-themed talks. Founded by Bob Jones University graduates Arlin and Beka Horton in 1974, PCC was committed to a young-earth creationist perspective. Its publishing arm A Beka Books was a leading supplier of conservative Christian curriculum materials. John Morris's third lecture, given to some three hundred graduate students at PCC, concerned "how evolution has brought a host of evils upon society."[27] *Trilogy* may have been a marginal contributor to the ICR's revenue stream. But Morris's visit to PCC reminds us that the ICR was part of a network of conservative activists who made a significant impact on the education that millions of young Christians would receive.

As the ICR continued to spread its fruitistic worldview, creationism was evolving. Under the leadership of the Discovery Institute, founded in Seattle in 1990, the new buzzword for those opposing evolutionary science was no longer the Morris-spawned "scientific creationism," but rather "intelligent design." ID seemed worlds away from the young-earth perspective. Its promoters denied that it relied on biblical authority for its conclusions or harbored any political aims. Both claims were deceptive. In its own distinctive way, the Discovery Institute and its campaign for ID are a twenty-first century continuation of the Red Dynamite tradition.

In the hands of biochemist Michael Behe, mathematician William Dembski, philosopher of science Stephen C. Meyer, and law professor Philip Johnson, ID offered a scientific-sounding challenge to Darwinian evolution. The poster child for ID was the bacterial flagellum—a whiplike structure whose functioning depended on a set of intricately interdependent parts—which, Behe argued, could not have evolved gradually, as Darwin claimed. The flagellum, in Behe's terms, was "irreducibly complex" and so must have an intelligent designer. ID's approach relied on the argument from design that William Paley had proposed almost two hundred years earlier. But ID proponents' credentials and careful excision of explicitly religious references gave the impression of a bold new direction.[28]

The nerve-center for ID was a far cry from Henry Morris's ICR. Formed initially as a branch office of the Indianapolis-based Hudson Institute, a conservative think tank formed in Croton-on-Hudson, New York, by RAND Corporation employees, the Discovery Institute was the brainchild of liberal Republicans Bruce Chapman and George Gilder. Their coauthored book, *The Party That Lost Its Head* (1966), bemoaned the ultraconservatism of Barry Goldwater and his supporters in the John

Birch Society. Later the two moved to the right, Chapman as a Reagan appointee and Gilder as one of the intellectual inspirations for Reagan's supply-side economics policies.[29] In its early years, the Discovery Institute took on a range of projects, including one on transportation networks in the Pacific Northwest funded heavily by the Bill and Melinda Gates Foundation. Not until 1993, when Chapman read a pro-ID piece by Cambridge PhD Stephen Meyer, did it wade into the waters of the evolutionary controversy.

Meyer's article came to the defense of young-earth creationist Dean Kenyon, a professor at San Francisco State University with a PhD in biophysics. Won over to young-earth creationism in 1970s, Kenyon offered expert creationist testimony in the *McLean v. Arkansas* and *Edwards v. Aguillard* cases. During the latter case, he authored a creationist textbook, *Of Pandas and People*, published two years later in 1989. Kenyon routinely included creationist material in his San Francisco State biology class, and in 1992 was ordered to cease and desist by his department chair. Kenyon appealed on the grounds of academic freedom and won the right to keep teaching. Meyer branded the whole affair a "Scopes Trial for the '90s."[30]

Championing the ID cause, Meyer was at pains to emphasize that intelligent design was not a form of creationism. As he told neoconservative talk show host Ben Wattenberg on the PBS show *Think Tank* in 2006, intelligent design involved an "inference from biological data, not a deduction from religious authority."[31] Despite Meyer's protestations, the 2005 *Kitzmiller* case in Dover, Pennsylvania, had already blown a large hole in the claim that Bible-based creationism was not implicated in ID. The Dover school board had required biology teachers to introduce intelligent design through Kenyon's *Of Pandas and People*. Doing research for the plaintiffs in the original proofs of *Pandas*, pro-evolutionist philosopher Barbara Forrest discovered an intriguing pattern. After the *Edwards* ruling made teaching creationism unconstitutional, Kenyon's team had searched and replaced every instance of "creationists" with the seemingly nonreligious phrase "design proponents." The smoking gun that tipped her off was the one instance where the transformation misfired, rendering "creationists" as "cdesignproponentists." This "missing link" was one of many facts introduced at trial that led Judge John Jones III, a Republican appointed by President George W. Bush, to comment in his ruling that the school board had acted with "breathtaking inanity."[32]

Even before *Kitzmiller*, it was clear the Discovery Institute had not only religious inspiration but also a political vision straight out of the ICR creation museum. True, chief ID ideologist Philip Johnson distinguished his conception of "creationism" from young-earth "creation-science." Johnson thought the earth might be ancient, and he even held the door open to a God-directed evolution. But he joined Henry Morris in pinning a range of moral evils on the dominant philosophy of "evolutionary naturalism and materialism." These included easy divorce, single parenting, the "sexual revolution," radical feminism, and gay liberation.[33] While Johnson consciously steered away from explicit theological arguments, he did employ a tree metaphor that meshed with the "fruits" of Matthew 7:15 and was remarkably similar to the Santee museum exhibit. Johnson used the allied metaphor of the "wedge" to describe the way in which ID could "split" the apparently "solid log" of "scientific materialism."[34]

Johnson's political logic also led to the demonization of the unholy materialist trio identified by his conservative predecessors: Darwin, Marx, and Freud. In his first book, *Darwin on Trial* (1991), Johnson gave ample space to Karl Popper's critique of Marx and Freud; they were unscientific since their systems of thought were not falsifiable. Johnson suggested that Darwin should be added to the mix as yet another proponent of "pseudo-science."[35] In *Defeating Darwinism* (1997), Johnson applauded the fall of Marx and Freud and predicted that "Darwin is next on the block." Unlike Henry Morris, Johnson did not take the trouble to show how intimately Marx and Darwin were connected. But he did celebrate the fall of the Soviet Union as an example of how "a cultural tower built on a materialist foundation can look extremely powerful one day and yet collapse in ruins the next."[36]

To accelerate the collapse of Darwinism, Johnson and others organized a conference in 1995 called "The Death of Materialism and the Renewal of Culture." There they refined their "wedge" strategy to split the Darwinist materialist tree. As Bruce Chapman explained—and as Dan Gilbert would have appreciated—the conference prepared the group to take on "the ideology of materialism and the host of social 'isms' that attend it." In 1996, Johnson and others established the Center for the Renewal of Science and Culture (CRSC), under the auspices of the Discovery Institute, to carry out this grand goal. One of the primary funders of the CRSC, to the tune of $1.5 million, was multimillionaire Howard Ahmanson Jr., to whom Philip Johnson dedicated *Defeating Darwinism* (1997). Heir to the Home Savings of America fortune, Ahmanson was a Christian Reconstructionist who until 1995 had sat on the board of R.J. Rushdoony's Chalcedon Foundation and provided key funding to conservative Christian attorney John Whitehead. As Ahmanson explained his philanthropic goal, "My purpose is total integration of Biblical law into our lives."[37] Aware of Ahmanson's Reconstructionist leanings, Johnson was careful to disclaim any interest in a Christian "political party."[38] These qualms, however, did not prevent Johnson from speaking in 1999 at fellow conservative Presbyterian and Reconstructionist D. James Kennedy's Reclaiming America conference.[39]

The "spiritual and intellectual movement" preferred by Johnson was inevitably political. This fact emerged clearly from an initially secret CRSC five-year public relations plan leaked online in 1999—the "Wedge Document."[40] It began by affirming that the major elements of "western Civilization"—"representative democracy, human rights, free enterprise, and progress in the arts and sciences"—were based on the idea that human beings are created in the image of God. It then claimed that this idea had been undermined for more than a century by the chief advocates of the "materialist conception of reality," Darwin, Marx, and Freud. They viewed people not as "moral and spiritual beings" but rather "as animals or machines who inhabited a universe ruled by purely impersonal forces and whose behavior and very thoughts were dictated by the unbending forces of biology, chemistry, and environment." This worldview created "cultural consequences" that were "devastating" in a number of arenas: the teaching of social sciences, which held the environment responsible for individual behavior, thus removing any sense of objective moral standards; modern approaches to criminal justice, product liability, and welfare, all of which promoted a "victim" mentality; and "utopian" and "coercive" political projects pushed by "materialist reformers" who "promised to create heaven on earth."[41]

To fell the tree of materialism with the wedge of intelligent design, the document's authors proposed three phases of activity: research/writing, publicity, and "Cultural Confrontation and Renewal." Commenting on their document after it became public, Discovery Institute leaders denied any political aim. The institute was not seeking to establish a "theocracy" but rather to influence "science and culture" with "ideas." And yet, the

chief "governing goal" listed in the document is "to defeat scientific materialism *and its destructive moral, cultural and political legacies.*" Since
those "legacies" included not just ideas but social and political practices—
such as welfare or product liability law—"renewal" necessarily meant
changing power relations between groups of people. It was not just that
the Discovery Institute would go on to lobby for changes in state laws to
allow for the teaching of intelligent design. It was taking part in the "culture war" over who ruled America.[42]

The document's emphasis on "free enterprise" as a product of belief
in God, paired with its anti-Marxism, was an important—and often
unrecognized—component of the Discovery Institute's political agenda.
In 2009, Jay Richards, a Discovery Institute senior fellow and early leader
of the CRSC, published *Money, Greed, and God: Why Capitalism Is the
Solution and Not the Problem.* As he noted in his introduction, the United
States may have won the Cold War, but "Just turn on the TV and you'll
see capitalism blamed for almost every social problem."[43] Given a decade
of corporate scandals and then the Great Recession of 2008, this was no
surprise. In the face of growing anticapitalist sentiment, Richards offered
an updated Carl McIntire-esque paean to free enterprise—complete with
biblical quotations—for the twenty-first century. Defending intelligent design and critiquing evolutionary science, as it turns out, also meant defending capitalism.

If intelligent design creationism seemed well adapted to the political and
legal environment of the new millennium, it hardly spelled the end of ICR-
style "scientific creationism." It, too, evolved with the times and starting in 1994 took the form of a new organization spawned from the very
loins of the ICR. Within the next two decades, Ken Ham's Answers in
Genesis (AiG) left its creationist parent in the dust. Having started his
American creationist career in the 1980s leading the ICR's Back to Genesis seminars, Ham sharpened his populist focus, established a highly successful website (over forty million page views by 2014), produced a shiny
new magazine, *Creation* (later renamed *Answers*), set up shop in northern Kentucky, and in 2007 welcomed the first visitors to AiG's $27 million
Creation Museum. By 2019, more than four million people had walked
through the museum's doors.[44] In the summer of 2016, AiG opened the
first installment of a planned Ark Encounter theme park, within an hour's

drive of the museum. Costing over $100 million and featuring a 510-foot ark, the park exhibit was expected to draw two million visitors in its first year of operation alone.[45] AiG's essential young-earth creationist message was the same one that Henry Morris and John Whitcomb Jr. had purveyed in *The Genesis Flood* more than thirty years earlier: the earth was about ten thousand years old. Evolutionary ideas had devastating consequences. And one of these satanic consequences was communism, though its role was considerably reduced in Ken Ham's reformulation.

An early indication that Ham would be applying Morris's "fruit test" arrived in the wake of the tragic events at Columbine High School in Littleton, Colorado. On April 20, 1999, students Dylan Klebold and Eric Harris—the latter wearing under his trench coat a T-shirt bearing the words "NATURAL SELECTION"—gunned down twelve students and one teacher. Leading up to the shootings, Harris had ranted online about the need to cull the unfit from the human population.[46] In a statement on the AiG website nine days later, Ken Ham suggested that evolutionary education was the "missing link" that could explain the murderous Colorado attacks. Since schools had been teaching that students are "just evolved animals, and that there is no absolute authority," such attacks would continue until the nation "allows God to be the absolute authority."[47] Ham was probably nudged by a broadcast by conservative radio commentator Paul Harvey, who read onto the air a letter to the editor from Addison L. Dawson, a right-wing resident of San Angelo, Texas. Dawson believed that evolution helped to cause the shootings. "Children are taught," Dawson wrote in a style reminiscent of fellow Texan J. Frank Norris, "that they are nothing but glorified apes who have evolutionized out of some primordial soup of mud." In June of that year, US Representative Tom DeLay (R–TX), the House majority whip, read Dawson's letter into the *Congressional Record* and immortalized the Texas man's words for a national audience.[48]

AiG's fruitistic perspective was also in evidence in Creation Museum exhibits. While nearly all of the text in the museum concerned either direct biblical apologetics or attacks on the inadequacies of evolutionary science, AiG found other ways to direct visitors' attention to the social and political impact of evolutionary thinking. As millions of people walked through "Graffiti Alley," they saw a hodgepodge of magazine and newspaper clippings pasted to the faux-brick walls. School shootings, gay marriage, assisted suicide, cloning, removal of the Ten Commandments from the public square—these all testified to a nation in moral decline. They resulted from the shift from a God-centered absolute standard of morality and truth to a relativistic one where, as a nearby sign declared, "Today Man Decides Truth." To hammer home the relativism, the word "truth" is crossed out with red spray paint and replaced with an artistically graffitied "Whatever." The same political logic was at work in the "Culture in Crisis" room, where visitors could peep, in voyeuristic fashion, into the windows of suburban American homes. In a video playing on a flat screen, wayward youth are seen and heard exhibiting the presumed evil effects of the godless, Darwinian worldview: violent video games, pornography, drugs, teen pregnancy, abortion, parental neglect, and alcoholism.[49] These exhibits may have made up a small portion of the Creation Museum, but they stood at the center of its mission. Ultimately, AiG aimed to mobilize visitors for right-wing political action.[50]

If AiG underplayed its worldview politics at the Creation Museum, this was not the case in other venues, where Ken Ham and his compatriots were clear about the worldly stakes in the debate over "origins." At the Answers in Genesis Mega-Conference in July 2013, a key focus was the recent pair of US Supreme Court decisions (*United States v. Windsor* and *Hollingsworth v. Perry*) upholding the right of gay marriage. In his talk, Tony Perkins of the Family Research Council urged his audience not to be discouraged, because "that's what our adversary the Devil wants."[51] Even when the explicit conference focus was evolutionary science, politics always lurked nearby. One afternoon, the roughly one thousand attendees were treated to the world premiere of *Evolution v. God* (2013). The film uses man-on-the-street interviews on a secular college campus to "expose" the fact that none of the science students or professors interviewed, despite their belief in evolution, could say that they actually had "seen" evidence of "macro-evolution" taking place, from, say, dogs to whales.[52] The real meaning of the film, however, emerged in the introduction by filmmaker Ray Comfort. "If atheistic evolution is true," Comfort told the crowd, "and we are past primates with no moral accountability—then fornication is nothing but our species following our instinct to procreate. If Darwinian evolution is true, then adultery, pornography, homosexuality, lust, lying, and blasphemy are culturally acceptable. If there is no God, then anything goes." The film was yet another device to emotionally prepare the AiG faithful—in part through the bonding effect of group laughter—to take on these serious "culture war" issues in the political realm.[53]

The same fruitistic dynamic was on display less than a year later, this time before an online audience of millions who watched Ken Ham debate Bill Nye "the Science Guy."[54] In his opening thirty-minute presentation, Ham began by appearing to talk about science, but by the end it was clear that he was really talking about politics. Ham focused on the long-standing distinction the ICR and AiG made between "observational" and "historical" science. As he explained, the former treats events taking place in the present so that we presumably can reach solid, reliable conclusions. The latter, however, as it concerns events taking place without human witnesses, cannot lead to objective conclusions.[55] This then deprives evolutionary science of any privileged status. It is, in effect, another religion, a point resting on the ideological work that John Whitehead, Francis Schaeffer, and Tim LaHaye had done decades before.

Having freed himself from the need to address any details of evolutionary science, Ham now got to the heart of the matter with a slide that illustrated the foundation stones of both evolution and creationism. On the left side was a stack of evolution-inspired bricks labeled "abortion," "euthanasia, "marriage ???," "moral relativism," and "man's ideas/naturalism." On the right side, appropriately, was the creationist stack: "life begins at fertilization," "sanctity of life," "biblical marriage," "moral absolutes," and "God's word." "See," Ham commented, "the battle is really about authority. It's more than just science or evolution or creation; it's about who's the authority in this world. Man or God? If you start with naturalism, then what about morals? Who decides right and wrong? . . . Abortion—get rid of spare cats, get rid of spare kids, we're all animals."[56] For Ham's audience at the Creation Museum and for his supporters worldwide, this was the political red meat of the debate.

For his part, Bill Nye talked knowledgeably about the Grand Canyon; snow-ice cores in Antarctica; tree rings; *Tiktaalik*, the amazing half-fish and half-amphibian; and cosmic background radiation. But Nye failed to recognize that he was firing blanks in a political battle. His condescending terms of address—"my Kentucky friends"—did not help. Nye concluded with a vain attempt to connect the battle over evolution to a political issue he thought might appeal to his audience: economic nationalism. If we devalue science education, Nye warned, America would fail to compete with its rivals in the world. "We need to innovate to keep the United States where it is in the world," he said.[57]

For all of Ken Ham's relentless focus on the politics of the culture war, he rarely ever brought communism into the picture. In the early twenty-first century, Ham understood that anticommunism did not carry the same punch. A two-part political cartoon published by AiG in the wake of the Columbine shootings serves as an index of this shift. The first panel, "How to Build a Bomb in the United States Public School System," features a scowling, alienated teenager whose head is a ticking, black dynamite bomb. The accompanying text offers three steps to produce the "bomb" in question: teach evolutionary ideas, take away the Bible, and then "stand back and wait!" To the Sixth Commandment injunction, "Thou Shalt Not Kill," the bomb-headed figure responds, "Why not?!!" In the second panel, a happy, clean-cut Bible-carrying teenager responds to the same question, "I understand!" He has benefited from learning that "people are *not* evolved animals." Given the long history of associating bomb throwers with anarchists and communists, it is striking that in 1999, the bomb no longer carried explicit political connotations. The link between George McCready Price's "red" and "dynamite" seems to have been severed.

At AiG's Creation Museum—in contrast with the ICR museum in Santee—one searches in vain for any textual reference to Karl Marx (the satanist or not), communism, or socialism. The only explicit political reference even close to this vein—and then a visual and auditory one—comes in the Cave of Sorrows. Visitors pass by a projected still image of a Nazi parade with jarringly loud audio of Hitler speaking.[58] And yet, the reference to Hitler—given the creationist movement's constitutional inability to distinguish Nazism from communism—turns out to be a clue that AiG had not dropped anticommunism from its creationist arsenal.

Decades after the end of the Cold War, Ken Ham and colleagues continued to find ways to take potshots at the communists. Take, for instance, AiG's response to the 2007 killings at Jokela High School in Finland.[59] Shooter Pekka-Eric Auvinen was a deeply alienated admirer of Finnish ecofascist Pentti Linkola, who advocates radical human depopulation to save the earth. Like Columbine killer Eric Harris, Auvinen viewed himself as a heroic "natural selector" personally eliminating the "unfit" from the human population.[60] Calling Auvinen a "self-proclaimed Social Darwinist," AiG writer Bodie Hodge repeated the arguments Ken Ham made about Columbine but with a twist. In his evolutionary morality, Hodge argued, Auvinen resembled others who had borne violent "fruits of evolution." Darwin's arguments had influenced not only mass murderer Adolf Hitler, Hodge wrote, but also "Karl Marx, Pol Pot, Leon Trotsky, and Joseph Stalin." Can we afford more "fruit," Hodge asked, like that produced by the "Marxist teachings of Hitler and Stalin"?[61]

Not only was Hitler outnumbered by violent Communist criminals, but he vanished into the Communist woodwork. The vehement opponent of Bolshevism, in Bodie Hodge's hands, miraculously becomes a Bolshevik. In a sense, it was the same logic, driven by political necessity, that led Gerald Winrod to conclude that Franklin Roosevelt was Jewish. But it also drew on a long conservative tradition that equated communism and fascism. Readers of Hodge's article who wanted to know more could click on the hyperlinked names of Marx and associates and read another AiG piece, "Darwin's Impact—the Bloodstained Legacy of Evolution." It damned the communist Darwinians in more detail, though it did distinguish between Marxist class struggle and Nazi "racial" struggle.[62]

In 2009, Hodge followed up with "The Results of Evolution: Could It Be the Bloodiest Religion Ever?" The centerpiece was a table listing the number of deaths caused by the by-now-familiar list of "leaders with evolutionary worldviews": Hitler, Trotsky and Lenin, Stalin, Mao, and Pol Pot. A loose notion of causation led Hodge to attribute all deaths from the following conflicts to evolution: World War I and World War II, including the Holocaust; the Russian Revolution and Civil War; the entire Chinese Revolution; and the Cambodian revolution. (A later edition of this article added the Korean and Vietnam Wars.) To this Hodge added estimates of total abortions—also evolutionary casualties—from a select group of countries over decades—twenty-six million for the US alone from 1928 to 2007. In the twentieth century, the "consequence" of the "idea" of evolution—as lived out by communists, Nazis, and abortionists—was a pile of 778,000,000 corpses.[63]

Weird as Bodie Hodge's statistical "research" might seem to historians of the twentieth century, the table of casualties was consistent with a long-standing line of argument that held up communism as an illustration of evolution's amorality and "culture of death." In the 1920s, William Jennings Bryan pinned the origins of World War I on the evolutionary proclivities of the German General Staff. In the 1950s, John R. Rice pilloried communists for their willingness to cheat and murder. Now came Hodge and AiG to reinforce this idea with a scientific-sounding set of "data." The fact that his first four endnotes cited nothing more than Wikipedia did not matter to the AiG faithful. Hodge had confirmed what they already knew. Anticommunism was a minor weapon in AiG's arsenal. But it was a live one. Hodge's revised article was included in AiG's *A Pocket Guide to Atheism* (2014, with an introduction by Ken Ham), still offered for sale at the Creation Museum bookstore today.[64]

For a scholarly version of Hodge's anticommunist arguments, AiG collaborated with Jerry Bergman, a Jehovah's Witness turned atheist turned young-earth creationist scholar. The possessor of nine degrees in fields ranging from biology to public health to social psychology, Bergman is a prolific writer, best known as the champion of those who, he claims, have been fired from their jobs or otherwise victimized for challenging the evolution "dogma."[65] Bergman counts himself among these modern-day reverse John Scopeses (Bergman was denied tenure at Bowling Green State University). He has compared the supposed persecution of creationists in the US to the early stages of Nazi persecution of German Jews.[66]

In "The Darwinian Foundation of Communism," originally published in AiG's *Creation* magazine in 2001, Bergman argued that the "communist holocaust" of the twentieth century was due in large part to Darwin's influence over Marx, Engels, Lenin, Stalin, and Mao. Bergman was more restrained with his statistics—he estimated that the victims numbered one hundred million, barely one-eighth of Hodge's figure. His writing was also more grounded in research. He accurately traced the impact of Darwin on the early Russian Marxists. He appropriately quoted from Richard Hofstadter on the Darwinian presence in the output of the socialist Kerr publishing house in Chicago. At the same time, Bergman provided a strangely distorted version of Marxist revolutionary theory, in which "the strong overthrow the weak." Fond of loose Hitler analogies, Bergman did little to distinguish Nazis from communists. In one block quote, meant to illustrate the impact of Darwin on Marx, most of the text addressed Hitler's politics.[67]

Bergman has also given scholarly cover to the claim, made most loudly by the ICR in its Santee museum, that Darwin is responsible for what Bergman calls "ruthless laissez-faire capitalism." As in the museum, the poster boys are Carnegie and Rockefeller, the classic "social Darwinists" whose choice quotations presumably "prove" Darwin's evil impact on society.[68] In one incarnation of this argument in his book *The Darwin Effect* (2014), Bergman describes poor working conditions, low pay, and profiteering, citing as one authority Howard Zinn's *A People's History of the United States*.[69] To the uninitiated, Bergman seems to be taking the side of workers and raising the flag of revolution. But as creationists have done since George McCready Price, Bergman instead provides a defense of capitalism using anticapitalist-sounding language. The key is that Bergman never specifies what un-ruthless capitalism looks like. His work suggests, without ever explicitly saying so, that there exists a kinder, gentler, capitalism, compatible with the biblical creationist ideas. Bergman's argument complements the openly pro-capitalist work of Jay Richards over at the Discovery Institute.

Where Bergman leaves the magical realm of ideal creationist capitalism to the imagination, another AiG ally, pastor Chad Hovind, has spelled it all out. The nephew of young-earth creationist Kent Hovind, the former owner of Dinosaur Adventure Land who is now serving time in federal prison for tax evasion, Chad Hovind pastors a church in nearby Cincinnati.[70] His cousin Eric (Kent's son), a graduate of the Pensacola Christian schools of A Beka Books fame, was one of the featured speakers at AiG's 2013 Mega-Conference. Right outside the main hall, Chad Hovind staffed a booth promoting his new book, *Godonomics: How to Save Our Country—and Protect Your Wallet—through Biblical Principles of Finance* (2013). With blurbs from former Arkansas governor and presidential hopeful Mike Huckabee, TV and radio personality Glenn Beck, and Christian conservative historian-celebrity David Barton, the book was selling like hotcakes. Featuring chapters like "What Would God Say to FDR about Liberty?" and "What Would God Say to Karl Marx about America?" Hovind's book makes a blindingly positive case for the free enterprise system.

Like Jay Richards, Hovind wrote his book because he perceived that an increasing number of young Christians were rejecting capitalism. In the name of "social justice," with references to Jesus's Sermon on the Mount, young people mistakenly "consider socialism to be the economic system that God endorses." Bemoaning excessive government taxation and debt, Hovind cites neoclassical economist Milton Friedman and preaches the values of personal responsibility, generosity, hard work, incentive, and reward. Instead of buying his son a new Xbox, Hovind insisted that the boy earn the money to buy it. For a model employer, Hovind ironically chose the prototype of the ICR's anti–social Darwinist position: Andrew Carnegie. Asked by a reporter how he fostered the personal growth of his employees, Carnegie "said people are developed the same way gold is mined. Tons of dirt need to be moved to find an ounce of gold." But if you focus on the positive potential, your employees will thank you and reach their goals.[71]

When Hovind directly confronted Karl Marx, he not only drew on arguments that Carl McIntire, Raymond Edman, and John R. Rice had made decades before, but he explicitly brought evolution into the picture. In a table titled "*Communist Manifesto* vs. the Bible," Hovind highlighted the differences between the way Marx and God would answer a series of fundamental questions, such as "Who are we?" Marx: "Humans are a highly evolved animal and are inherently good; in need of a good environment to stimulate good behavior." God: "Humans are spiritual and physical beings made in the good image of God, but possessing a sinful heart that makes us individually responsible for our evil behavior." For readers wanting more, Hovind referred them to *Understanding the Times* by David Noebel, who is indeed the leading worldview warrior on this issue.[72]

At the 2013 AiG Mega-Conference booth, Hovind not only hawked his book and six-DVD *Godonomics* course, but handed out, for free, a clever Marx-themed faux credit card to reinforce his message. Adorned with a portrait of Marx, the "Government Master Card" (subtitled the "Spend-Like-There's-No-Tomorrow Card") is issued to "I. M. Enslaved." By refusing God's warnings and spending more than they had, cardholders would find themselves in a state of "serfdom," echoing economist F. A. Hayek's classic anticommunist tract.[73] While the card made no reference to evolution, Hovind's presence at the conference and the clear references in his book to Marx's evolutionary thinking help to demonstrate how AiG was not merely poking holes in a scientific theory or upholding a particular interpretation of the Bible, but engaging in a long tradition of Red Dynamite politics.

As much as David Noebel, Jerry Bergman, and Chad Hovind might seem to be lone voices in the twenty-first-century wilderness, the persistence of creationist anticommunism decades after the end of the Cold War is significant. For a century, Red Dynamite warriors, from George McCready Price to Aimee Semple McPherson to John Rice to Dave Welch, were concerned above all about changing social, political, and moral standards that relied on human rather than divine dictates. The fall of Stalinist regimes in the former Soviet Union and Eastern Europe in the early 1990s put an end to one model of social change but did little to resolve fundamental issues of inequality, exploitation, and injustice worldwide. By 2016, conflicts over these issues shook up the American political system and gave Red Dynamite politics a new lease on life into the Trump era.

EPILOGUE

The Baby Christian and the Dark Place

It is July 7, 2016, opening day at Ark Encounter. Some four thousand of us have descended on tiny Williamstown, Kentucky, where Amish construction workers hired by Answers in Genesis have built a 510-foot-long, multilevel wooden replica of Noah's Ark. Up on level three we learn about the limits of natural selection, rival interpretations of the fossil record, and how many ice ages there were (one, in contradiction to modern science). But the main point is on level one, where visitors are reminded of why Noah's family built the ark. God created a "Perfect World"; humanity became "extremely wicked"; and in his "Divine Judgement," God slaughtered nineteen billion people with a worldwide flood. Quotes from scripture underline the point that we are once again living in the "days of Noah." Our current sins signal the coming end times, when Jesus will return and redeem the Christian faithful. AiG's population figures may contravene the conclusions of demographers, but they make a fundamentally political point, not a scientific one. If you dare to rebel against God's word, the cost will be unthinkably high.

Compared to the Creation Museum, where the Graffiti Alley and Culture in Crisis exhibits were specific about "extremely wicked" behavior, Ark Encounter is more circumspect, with only indirect references to gay marriage and abortion.[1] If the treatment of general culture-war politics was muted at Ark Encounter, the specific connection that George McCready Price had originally drawn nearly a century earlier between evolutionary thought and communism was absent. In the Ark Encounter bookstore, you could find AiG's *Pocket Guide* to Noah's Ark, but not Hodge's volume on atheism that mathematically tallies up communism's victims. It might seem as if the Red Dynamite tradition were dead and buried forever. But such a pronouncement is premature in light of the broader political currents in which Answers in Genesis swims. Ark Encounter opened its door less than two weeks before the Republican Party convention in Cleveland nominated Donald Trump for president. The 2016 presidential race and its aftermath revealed that socialism, the immorality of rebellion, and evolutionary ideas can still be linked in ways that resonate with American evangelicals.

The persistence of Red Dynamite politics emerged from a strategic alliance between conservative Christians and Donald Trump's presidential campaign. Months before Ark Encounter opened its doors, AiG's Ken Ham penned a positive appraisal of the "Donald Trump Phenomenon." While he did not formally endorse the real estate mogul, Ham was "intrigued" by the mass support for Trump and felt impelled to offer his thoughts on how the candidate's success offered positive lessons for Answers in Genesis. Acknowledging that Trump "does not promote the Christian worldview," Ham still admired the way that Trump "comes across as genuine" and speaks "with conviction and authority." Many Christians like Trump, Ham believed, because they "are sick of the political correctness in this nation, as well as the liberal, humanistic agenda of much of the secular media!"[2]

Other prominent evangelicals jumped on the Trump bandwagon early. Like Ken Ham, they recognized that millions of rank-and-file evangelicals were drawn to Trump's plainspoken calls for barring immigration from Mexico and the Middle East, his denunciation of trade deals, his "outsider" status, his nostalgia for a mythical American past, and his willingness to tell the truth about the miserable economic conditions facing working people. The first to do so was Jerry Falwell Jr., then president of Liberty University and son of its esteemed founder. Trump addressed conservative Christian fears in his January 18, 2016, invited speech at Liberty, telling students they were "under siege" from secular political correctness.[3]

With Falwell's blessing, the candidate gained entrée into the larger evangelical world. At a June 2016 summit meeting with several hundred nationally known evangelical leaders in New York City, failed Republican presidential hopeful Governor Mike Huckabee of Arkansas and Focus on the Family founder James Dobson played prominent roles. Huckabee set the tone by making it clear the group would not challenge Trump's Christian credentials. "I don't think anybody here expects you to be theological today," he said. What they did expect were answers to some key questions: if elected, would Trump appoint "pro-life" US Supreme Court justices? Would Trump protect the Second Amendment? Would Trump revoke the Johnson Amendment, prohibiting tax-exempt nonprofits, including churches, from endorsing candidates for public office?[4] In an interview after the meeting, James Dobson put his theological seal of approval on the presidential hopeful. Trump was a "baby Christian" on the road to salvation.[5]

It may seem inexplicable that antievolutionist Ken Ham and his conservative Christian counterparts positively entertained the prospect of a Trump presidency. Trump is "social Darwinism" incarnate. He built a career on ruthless competition. He boasted that he would recruit "Wall Street killers" to conduct trade negotiations. He was catapulted to reality TV fame by *The Apprentice*, in which job applicants engage in a battle for survival of the fittest (to work for Donald Trump).[6] Trump grew up attending Norman Vincent Peale's Presbyterian church.[7] But from Trump's reference to the Eucharist wafer as the "little cracker," to his rendering of 2 Corinthians as "two Corinthians," to his admission that he never asks forgiveness of God for his sins, he has difficulty convincing anyone that he is part of any Christian faith community.[8]

Yet it would be a mistake to imagine that conservative evangelicals' embrace of Trump is a radical departure from the norm. Politics—in the broad sense of who has power over whom in society—has always been intertwined with religious faith commitments.[9] This fact explains how a freely cursing, nominally Christian Barry Goldwater, and Ronald Reagan, the only divorced president, could become the darlings of the Christian Right. It also explains how antievolutionist and anticommunist J. Frank Norris could ever have become a hero to rank-and-file Baptists from Fort Worth to Detroit. Unlike Trump, Norris spoke the language of Christian faith fluently. But he also shot and killed a man in his office under questionable circumstances. He stood trial for murder, was suspected of arson, and was so unscrupulous and dictatorial in his dealings with his church associates that longtime Norris loyalist G. Beauchamp Vick left First Baptist and started his own non-Norris denomination. Norris reveled in his bad-boy persona and mixed it with a good dose of populism. Typical of his era's Southern Baptists, Norris was a staunch segregationist and friend of the Klan. In his own way, he was as Trumpian as they come.

In the wake of the June 2016 New York meeting, despite Trump's lowly "baby Christian" status, he created a twenty-six-member Evangelical Executive Advisory Board whose members included those with a live connection to Red Dynamite politics. Take former US Representative Michele Bachmann (R–MN). A graduate of the Oral Roberts University Law School, Bachmann learned about the law from John Eidsmoe, whose book *God and Caesar* continued the hallowed tradition of fabricated Bolshevik quotes to prove the immorality of communism.[10] Oral Roberts employed John Whitehead, the Rushdoonian collaborator of Francis Schaeffer, whose *How Should We Then Live?* made a major impact on the young Bachmann. After she and her husband went through the experience of founding a charter school that taught creationism (but then lost a legal battle on this point), Bachmann became acquainted with the writings of David Noebel and soon joined the Summit board of directors. She spoke at Summit summer workshops, and as a Minnesota state senator listed on her website Noebel's *Understanding the Times* as one of her nine favorite books. When D. James Kennedy's Coral Ridge Ministries issued a new antisocialist (and anti-Obama) "documentary" film, *Socialism: A Clear and Present Danger* (2010), Bachmann appeared as one of the talking heads, along with David Noebel and Jay Richards, the pro-capitalist Discovery Institute staffer.[11] Bachmann's presence on Trump's advisory board linked him to a deep legacy of conservative Christian activism.

That legacy runs even deeper in evangelical Trump adviser Rev. Robert Jeffress Jr. (1955–). Jeffress grew up attending First Baptist Church in

Dallas, Texas, where pastor W. A. Criswell (1909–2002) held forth every Sunday. A contemporary and admirer of J. Frank Norris, Criswell was a longtime fundamentalist Baptist leader (and fierce segregationist) who inherited Norris's role after his death in 1952.[12] Jeffress, now at the helm at First Baptist in Dallas, with a congregation of some 110,000, channels Norris by associating evolution with moral decline. In *Countdown to the Apocalypse* (2015), Jeffress decried the idea that human civilization is progressing—"the whole thesis of evolution." Rather, the country has descended into "moral disorder" and "an orgy of self-gratification."[13] Jeffress has also lent his voice to "scientific" creationism. At the groundbreaking ceremony for the new ICR Discovery Center (a Dallas-based rival to AiG's Creation Museum), Jeffress was a featured speaker.[14]

Not only is he outspoken about politics, appearing regularly on Fox News to denounce gay marriage and Islam, but Jeffress also channels the Shooting Parson's combativeness. Asked why he supported Trump, given the candidate's abrasive tone, Jeffress replied that when a president is negotiating with Iran or at war with ISIS, "I couldn't care less about that leader's temperament or his tone or his vocabulary. Frankly, I want the meanest, toughest son of a gun I can find. And I think that's the feeling of a lot of evangelicals. They don't want a Caspar Milquetoast as the leader of the free world."[15] When it comes to US foreign policy, the ends justify the means. Anticommunists from Dan Gilbert to Ronald Reagan had consistently called attention to what Gilbert termed Marxism's "wolf pack" ethics as a detestable alternative to Christian moralism. Reagan had paraphrased Lenin on this—that whatever is necessary to advance the march to socialism is moral. Today's conservative Christians are no different. They are also determined to achieve their ends—to resist reformist and revolutionary social change—by any means necessary.

As Donald Trump moved toward Republican front-runner status, he began to gather around him a number of conservative Christians with solid antievolution credentials. They included Trump's running mate Mike Pence (1959–), a Christian evangelical who had publicly attacked the validity of evolutionary science in 2002 as a Republican US representative from Indiana.[16] Reacting to a newly published article in *Nature* that suggested that ideas about human evolution would need to be revised, Pence used the occasion to lambaste evolution as a "sincere theory" but not a "fact," and called for schools to teach it as such. He affirmed his own belief in creationism, backed up by the Declaration of Independence and the Bible, both of which spoke of a "creator."[17]

Whereas Mike Pence questioned evolution but stopped short of pinning social and political evils on Darwin, Ben Carson (1955–), the man Trump would pick as housing secretary, pointedly connected socialism and evolution with Satan. Born and raised in Detroit in a Seventh-day Adventist family, Carson became a world-famous Johns Hopkins University pediatric surgeon and entered national politics when he delivered a blistering attack on President Barack Obama at the 2013 National Prayer Breakfast. *America the Beautiful* (2012), coauthored with his wife, Candy Carson, gave a glimpse of the politics Ben Carson brought to the table. The authors paid tribute to the "Tea Party" revolt within the Republican Party, slammed "political correctness," called for a return to traditional Judeo-Christian values, and, like Chad Hovind and Jay Richards, took on the challenge to defend "capitalism" from its critics.[18]

Satan came into the picture in late 2015 when Carson was still considered a viable Republican front-runner. As attempts mounted to discredit Carson in the eyes of potential Republican voters, a video surfaced of a talk he gave to an Adventist audience in 2012. Asked about the status of evolution as a "theory" and not a fact, Carson answered, "I personally believe that this theory that Darwin came up with was something that was encouraged by the adversary, and it has become what is scientifically, politically correct."[19] As that audience knew, the adversary was Satan. A writer for the *New Yorker* called Carson's ideas "wild delusions."[20] But no one should have been surprised. Satan plays a particularly active part in Adventist theology, as illustrated by Ellen G. White's *The Great Controversy*, the book that the original flood geologist George McCready Price sold in the backwoods of Prince Edward Island a century earlier. ICR founder Henry Morris had explained for decades that the origin of evolutionary thought goes back to the Great Deceiver.

At the 2016 Republican National Convention, now firmly in the Trump camp, Carson elaborated further on the satanic connection. As he went off script before millions of television viewers, Carson pointed to the disturbing association between then-presumptive Democratic presidential nominee Hillary Clinton and socialist "community organizer" Saul Alinsky. Carson accurately related that the young Hillary Rodham considered Alinsky one of her "heroes" and "mentors." Then Carson dropped his bomb: the epigraph to Alinsky's book *Rules for Radicals* (1971) included the following line: "Lest we forget at least an over-the-shoulder acknowledgment to the very first radical . . . who rebelled against the establishment and did it so effectively that he at least won his own kingdom—Lucifer." Reeling off examples of America's public expressions of indebtedness to "God" and our "Creator," Carson asked, "So, are we willing to elect someone as president, who has as their role model, somebody who acknowledged Lucifer?" "No!" the crowd thundered back.[21] For the thousands who had passed through the ICR museum in Santee, California, and believed the authoritative-looking panel informing them that the pro-evolutionist Karl Marx was "(according to some) a Satanist in college," Carson's news about Alinsky and Clinton added credence to the charge.

Although she had not spoken publicly of Satan, the Trump cabinet appointee who most reliably could provide a transmission belt for creationist anticommunism was his choice for secretary of education, Betsy DeVos. A native of Holland, Michigan, who grew up in the conservative Christian Reformed Church, DeVos had made headlines campaigning for charter schools in Michigan. Her father-in-law was Richard DeVos Sr., the founder of the Amway empire and funder of many conservative causes, including D. James Kennedy's Coral Ridge Ministries. The foundation he ran with his wife Helen, a onetime member of Kennedy's congregation, gave Coral Ridge more than $15 million from 1998 to 2009.[22] While Betsy DeVos generally kept mum in public on hot-button culture-war issues, she did share with a high-level Christian philanthropic meeting her Reformed desire to "advance God's kingdom" on earth.[23] That goal apparently included support for teaching intelligent design. Through the Dick and Betsy DeVos foundation, she helped fund the Thomas More Law Center just a few years before it represented the Dover, Pennsylvania, school board and its plan to introduce ID in local public schools.[24]

There is no sign that DeVos was publicly boosting the creationist cause or denouncing socialism during the 2016 campaign. But conservative Christians saw their opportunity to lobby her, and they took it. Before DeVos was confirmed, she heard from leaders of the Council for National Policy (CNP). A high-level Christian conservative group founded by Tim

LaHaye in 1981, CNP has strong links to the DeVos family. The CNP Education Committee sent Betsy DeVos a report urging the "restoration" of education based on "Judeo-Christian principles." Among their suggestions for the secretary nominee was that the White House forge relationships with "key pastoral networks"—prominently including Dave Welch's US Pastor Council—to help Christian believers understand the stakes involved. The ultimate goal was to promote a "gradual, voluntary" shift from public "secular" education to private, church-based, and home schools.[25]

A clue to what the content of that nonpublic education might look like came from Dan Smithwick, the chair of the CNP Education Committee. A former AT&T executive who worked for Pat Robertson's campaign (as did Welch), Smithwick comes out of the Reformed tradition and has been strongly influenced by the Reconstructionism of both R. J. Rushdoony and D. James Kennedy.[26] In 1986, he founded the Nehemiah Institute, a conservative Christian worldview training organization that has worked closely with both the ICR and David Noebel's Summit Ministries.[27] Nehemiah is best known for a worldview opinion survey called PEERS (Politics, Economics, Education, Religion, Social Issues) that it has administered to more than 110,000 students since 1987. Based on ratings of a battery of seventy true-false statements—for example, "Human life came into existence less than 10 thousand years ago" (true)—Nehemiah classifies young people as holding worldviews ranging from "Christian Theism" (best) to "Moderate Christian" to "Secular Humanism" to "Socialism" (worst).

The results have been discouraging for Smithwick's group. Based on responses to a different worldview survey, one group of Atlanta students were split into creationist and evolutionist cohorts. After administering the PEERS test to them, Smithwick found that a sizable minority of the creationists were in the secular humanist or socialist categories. As he reported to ICR members in *Acts & Facts*, the scores of the evolutionists were even "lower . . . well into the Socialism worldview category."[28] In 2001, Smithwick warned that if the decline continued, evangelical public school students would be lost to socialism by 2014.[29] This nightmare prediction converged with a similar recognition by Answers in Genesis that they are losing young people on the critical culture war issues of gay marriage, abortion rights, and religious "freedom" laws. Thus Ken Ham's jeremiad,

Already Gone: Why Your Kids Will Quit Church and What You Can Do to Stop It (2009).[30] Donald Trump's planned appointment of Betsy DeVos gave Smithwick and his fellow worldview warriors renewed hope.

In November 2016, conservative Christians celebrated as Donald Trump was elected forty-fifth president of the United States. Exit polls suggested that some 80 percent of white evangelicals voted for Trump.[31] Ken Ham wasted no time using Trumpian themes to boost the creationist cause. On Inauguration Day, in a blog post titled "How to Make America Great Again," Ham put a friendly twist on Trump's campaign slogan. The only way to make America great, Ham explained, was to rely on God's word as the "absolute authority in all areas."[32] Soon after, Ham took up the subject of "fake news," originally a term referring to deliberately fabricated online news stories. Now embracing Donald Trump's expanded definition and deploying it for creationist purposes, Ham wrote, "The two greatest fake news items permeating Western cultures are molecules-to-man evolution and millions of years." Ham identified with Trump as a fellow fake news victim: "The left-wing secular media is doing to President Trump what they've done to us for years—spreading false accusations, lies, and misinformation."[33] In the wake of negative publicity about smaller-than-expected crowds at Ark Encounter, Ham warned Creation Museum visitors that the "media" was not to be trusted. "You can't believe a thing they say," Ham said.[34] For visitors who accepted AiG's claim that evolution was a satanic plot responsible for catastrophic moral decline, it made sense not to trust that plot's promoters.

Four months after Trump's inauguration, Ham sent a letter to AiG supporters spelling out the connection between evolution and moral decline, with a dash of Red Dynamite politics. For more than forty years, Ham wrote, he had been teaching about the link between "atheistic evolution and morality." In a rare acknowledgment of the nonscientific character of ICR founder Henry Morris's work, Ham noted that Morris, too, had connected evolution and morals. Invoking abortion and gay marriage as examples of growing immorality, Ham insisted that the evolution controversy was not just about science, but a clash of worldviews. In his debate with Bill Nye, he had highlighted the moral aspect of the subject and had been ridiculed. Now Bill Nye had proved Ham's point, with his new Netflix show, *Bill Nye Saves the World*. As Ham related, in a recent episode of this educational and scientific show, *Crazy Ex-Girlfriend* star Rachel Bloom performed a "lewd" song called "My Sex Junk" to promote transgender values. To drive home the point, Ham pointed to yet another example of how evolutionary thinking threatened the moral, social, and political order: the novel *Brave New World* by Aldous Huxley, onetime socialist, eugenicist, and grandson of Charles Darwin's "bulldog." Ham quoted Huxley about his youthful rebellion, which aimed at restraints on both "sexual freedom" and the "political and economic system." As Huxley described it, he was engaged in "political and erotic revolt."[35] Jerry Falwell Sr. could not have put it better.

In the summer of 2019, a group of people in Cumberland, Kentucky, engaged in a different kind of revolt and unwittingly illuminated the real stakes in the culture war over evolutionary science. Located about a four-hour drive southeast from the Creation Museum and about the same distance northeast from Dayton, Tennessee, of Scopes trial fame, Cumberland sits in Harlan County, also known as "Bloody Harlan," after the famed coal-mining labor battles going back to the 1930s.[36] Mining jobs are scarce today, and Harlan is Trump country, with nearly 85 percent of voters backing him in 2016. Faithful Christian visitors to tiny Cumberland (population 2,237) have their choice of at least four evangelical churches and can pay a visit to nearby Kingdom Come State Park, which draws hikers from miles around hoping to see a black bear.[37]

What made Cumberland nationally known in July–August 2019 was a coal miners' rebellion. Earlier that summer, the giant Blackjewel mining company had stopped production. Company owners declared bankruptcy, refused to pay miners, blocked access to their retirement funds, and even received permission from a judge to retroactively take back funds already deposited in employees' bank accounts. Starting on July 29, a small group of Blackjewel miners set up a protest camp square in the middle of the railroad tracks, blocking a trainload of coal they had dug worth $1 million from leaving the Cloverlick number 3 mine. Their slogan: "No pay, we stay."[38] A month later, the miners, now joined by hundreds of community members and supporters from near and far, were still there waiting for their money.[39]

Six days after the Cumberland miners began their protest, Ken Ham posted to the AiG blog. He did not mention the miners' action, but his message was all about labor activism and fit squarely in the Red Dynamite tradition of creationist politics. In "Teachers Union Endorses Killing Unborn Children," Ham expressed alarm that delegates at the recent annual meeting of the National Education Association (NEA) in Houston had approved a resolution supporting a woman's "fundamental" right to choose abortion. Ham devoted most of his post to answering the NEA's prochoice arguments with anti-choice talking points. But the key to Ham's underlying worldview appeared in his summary of comments by fellow AiG speaker Bryan Osborne, who attended the Houston conference. As Ham summarized Osborne, "It's a dark place. . . . [Osborne said that] a spirit of rebellion and the idea of 'we won't take this anymore' was everywhere in the expo hall, in the imagery (such as the closed, raised fist) that was on display, and in many of the presentations given. Bryan also shared that *socialist, Marxist philosophy—stemming from a secular, evolutionary worldview*—permeated the convention." The photo following these lines featured a large banner in the foreground that depicts a teacher, an African American woman, speaking into a megaphone, as a large crowd of people in the background hold aloft their closed fists.[40]

As Ken Ham knew, over the previous three years, public school teachers and allied school workers had carried out strikes and protests in primarily "red" states like West Virginia, Arizona, Colorado, Oklahoma, North Carolina, and Kentucky.[41] The "Red for Ed" teachers perfectly embodied the "spirit of rebellion," including the raised fists, animating the 2019 NEA gathering. In the name of adequate funding for public education, the welfare of their students, and dignity on the job, teachers were saying, precisely, "we won't take this anymore." But as John R. Rice had written decades earlier, "the heart of all sin is rebellion against authority."[42] Ham did not need to tell his readers why Osborne had described the teachers' convention as a "dark place." Teachers who went on strike—and presumably coal miners who blocked coal trains—served Satan. Rebellious workers, channeling an evolutionary and communistic worldview, acted against the wishes of God.

More trouble was on the way. In 2020, the ranks of rebellious working people widened under the impact of the COVID-19 pandemic, the deepest economic crisis since the 1930s, and a giant movement against police brutality sparked by the police killing of George Floyd in Minneapolis.

Although Senator Bernie Sanders failed to secure the Democratic presidential nomination and agreed to stump for the Democratic Joseph Biden–Kamala Harris ticket, his 2016 democratic socialist political campaign had irrevocably fractured the party and shifted its political center of gravity to the left. In mass mobilizations during the summer of 2020 in Portland, Chicago, Kenosha, Seattle, and elsewhere, prominent roles were played by activists associated with Antifa, a loose grouping of anarchist-influenced radicals who promoted "direct action" tactics under cover of dark that included setting fires, looting, and smashing windows.[43]

The improbable result thirty years after the end of the Cold War: red-baiting, antisocialist, anti-Marxist politics stood at the center of a Republican presidential campaign. As the 2020 election approached, Fox News warned about "modern-day Bolsheviks" like democratic socialist US Representative Alexandria Ocasio-Cortez (D–NY) and her growing influence in the Democratic Party; President Trump tweeted denunciations of "Radical Left anarchists" and the "Marxist group" Black Lives Matter.[44] Conspiracy theory has also made a significant comeback, with a nod from President Trump.[45] QAnon conspiracy promoters do not invoke Darwin as part of the "deep state" plot they claim is unfolding. But their focus on a secret cabal of satanic pedophiles resonates with past imagined plots involving the devil, evolution, and sexual immorality going back a century.[46] Adding billionaire investor George Soros to the mix lends credence to time-honored claims of Jewish conspiracy as well.[47]

While President Trump did not connect the communist threat with the dangers of Darwinism, D. James Kennedy Ministries was pleased to supply the missing link in *The Coming Communist Wave: What Happens If the Left Captures All Three Branches of Government* (2020). On the first page of this pamphlet, veteran journalist and conservative activist Robert H. Knight warned that socialism (embodied in the Democratic Party) was part of Marx's evolutionary theory. Like Darwin, Marx believed that societies "naturally evolve toward higher stages" without any "divine direction." Knight cited polling data showing that millennials increasingly embraced socialism and rejected capitalism. If the Democrats won the White House in 2020, Knight explained, America would become an evolved, socialist, atheist state dictating "sinister" social norms at odds with the Ten Commandments. Widespread acceptance of abortion rights, gay rights, and transgender rights signaled to Knight that a satanic social evolution was already under way. These movements and their socialist promoters sinned by violating "God's instructions."[48]

Nearly 130 years earlier, shortly before the birth of John Scopes, rebel coal miners in the hills of East Tennessee refused the follow the instructions of state political leaders and instead took up arms against the convict lease system. Defenders of the established order called them communists and anarchists. Within a few decades, conservative Christians pinned similar charges on John Scopes and his allies. Whether or not the Tennessee rebels or today's Blackjewel miners or the "Red for Ed" teachers or fighters for gay and transgender rights or the millions of protesters against police brutality in the nation's streets have given any thought to Charles Darwin, they have acted on the principle of social evolution. Their deeds convey the contention that social norms, morals, and institutions can change so that the world might become a better place for human beings to live and flourish. That contention lies at the heart of the political controversy over evolutionary science. It is through such battles over the state of this world that the conversation about both biological and social evolution can advance.

ACKNOWLEDGMENTS

At the end of a process that took longer than I ever imagined it could, I am pleased to tip my hat to those who have helped make this book possible. It is the product of our collective labor.

Over the past decade, I benefited from conversations with a wide range of friends, colleagues, and subject matter experts. They include Garland Allen, Roger Barnes, Tim Bayly, John Bodnar, Paul Boyer, Scott Burgins, Ben Burlingham, L. D. Burnett, Jon Butler, Joe Cain, Steve Clark, Dennis Drake, Ben Eklof, Nahyan Fancy, Tyler Ferguson, Wendy Gamber, David Gellman, Luke Gillespie, Sandy Gliboff, Andrew Hartman, Victoria Hilkevitch, Sylvester Johnson, Chris Judge, Terry Koger, Charles Latshaw, Michael McVicar, Esther Moudy-Gummere, Naomi Oreskes, Dawn Oyedipe, Mark Pittenger, Leo Ribuffo, Corey Robin, Tatiana Saburova, Sandy Shapshay, Rita Stephens, Rev. Bradley Tharpe, Samrat Upadhyay, Richard Valdez, Dror Wahrman, Barbara Whitehead, Molly Worthen, and Nikos Zirogiannis.

A bevy of archivists and librarians helped me gain access to valuable research material at the Asbury Theological Seminary, Cedarville University, the Center for Adventist Research at Andrews University, the General Conference of Seventh-day Adventists, the Indiana State Library, Liberty University, the Minnesota Historical Society, the Presbyterian Church in America Historical Center at Covenant Theological Seminary, the Rhea County (Tennessee) Archives, the Southern Baptist Convention archives, the Southwestern Baptist Theological Seminary, the University of Arkansas, the University of Northwestern–St. Paul, the Wheaton College Billy Graham Center, Wichita State University, and the Wells Library at Indiana University–Bloomington.

I am grateful to John Morris of the Institute for Creation Research and Tom Cantor of the Light and Life Foundation for granting me interviews. David Noebel, Dave Welch, and the late John C. Whitcomb Jr. all provided valuable information through our correspondence.

Research assistants Gloria Lopez, Paula Tarankow, and Anushka Mansukhani provided invaluable assistance.

My work on this book developed in tandem with my teaching. I first had the opportunity to teach about the politics of evolutionary science thanks to my department head at the University of North Georgia, Christopher Jespersen. He gambled that a junior colleague who had never studied the history of religion or science or philosophy in graduate school could teach a course called the History of Evolutionary Science. I did so only with a professional biologist, Terry Schwaner, sitting through every class. Richard Miller played that role at Butler University. Finally on my own at DePauw University and then Indiana University, I nonetheless learned much from biology colleagues Wade Hazel, Ellen Ketterson (as well as Ellen's graduate students Sam Slowinski and Abby Kimmitt), and Mike Wade.

Students in classes at all four of those institutions asked great questions, enabled me to return to the Answers in Genesis Creation Museum numerous times on field trips, helped me to see my subject from a variety of perspectives, and conducted their own research that enriched my knowledge.

Colleagues who gave feedback on the book manuscript at various stages were enormously helpful. They include Steve Andrews, Constance Clark, Rachel Coleman, Jane Dailey, Janine Giordano Drake, Michael Grossberg, Michael Kazin, Ed Linenthal, Kathryn Gin Lum, Michael McGerr, Jeffrey

Moran, Roberta Pergher, Rebecca Polen, Mark Roseman, Lana Spendl, Dan Williams, and the anonymous reviewers at *Church History*.

Feedback I received from presentations on the book research was also crucial. Venues included the Commons Project: Science and Religion Dialogue and the Faculty Research Colloquium, both at DePauw University; the History and Philosophy of Science Colloquium and US History Workshop, both at Indiana University; the Organization of American Historians; the Society for US Intellectual History; and the Ohio Academy of History.

Participation in an Indiana University Faculty Writing Group directed by Alisha Lola Jones and Laura Plummer enabled me to make substantial progress on the manuscript.

Special thanks go to two friends and fellow historians who read the entire manuscript of the book: Bill Trollinger and Ron Numbers. When I started this project, I knew Bill only as the author of a biography of William Bell Riley. I imagined him as a stern, conservative Baptist like Riley, with whom I would have little in common. On both counts, I could not have been more wrong. Based on our common interest in creationism—he is the coauthor with Susan Trollinger of a book on the Creation Museum—Bill enthusiastically agreed to read my book manuscript. His positive appraisal, insightful comments, and friendship helped me to sustain the energy to finish the project.

Ron Numbers is a rare bird. A world-renowned expert on the history of both science and religion—who coincidentally directed Bill's dissertation—he literally wrote *the* book on the history of creationism. I am eternally grateful that Ron took an early interest in my work, has supported the project through the years, and read the whole book through more than once. His generous spirit, great sense of humor, and encyclopedic knowledge of the scholarly literature has enriched my own work in so many ways.

I am grateful to Darryl Hart and Larry Moore, the editors of Cornell's Religion and American Public Life series, for seeing value in my work. Their detailed commentary and suggestions have made this a far better book, as did comments from historian and reviewer Matthew Avery Sutton, whose work I greatly admire. As I finished up the manuscript, it was a real pleasure working with editor Michael McGandy, acquisitions assistant Clare Jones, and the whole Cornell staff.

Since my children Kevin and Anna were in their mid- to late teens when I started work on this book and thus had sufficient control over their lives to avoid forced conversations about creationism and anticommunism, I am grateful for their willingness occasionally to listen, discuss, and to read short portions of text that I thought they might tolerate and even like. They always make me think and remind me of why this all matters.

Saving the best for last, I want to express my deepest gratitude to Beth Gazley, my wife and life partner, to whom this book is dedicated. She has been my biggest champion and toughest critic. She has endured a lifetime of elevator speeches, disturbing anecdotes, and conspiracy theories that might otherwise be the occasion for comic relief but that I am determined instead to take seriously. I could not have done this without her.

NOTES

Introduction

1. This sketch includes authentic narrative details as well as aspects of what historian Edward Larson called "the modern Scopes legend." Edward Larson, *Summer for the Gods: The Scopes Trial and America's Continuing Debate over Science and Religion* (Cambridge, MA: Harvard University Press, 1997), 3–6, 88–91, 191–92, 225–46; "7 Sleepy Small Towns in Tennessee Where Things Never Seem to Change," Only in Your State, February 13, 2018, https://www.onlyinyourstate.com/tennessee/sleepy-tn-towns/.

2. Larson mentions, though he does not explore, the implications of Dayton's emergence as an industrial town and the Socialist politics of Thomas Scopes. Larson, *Summer for the Gods*, 87–88, 91.

3. Karin Shapiro, *A New South Rebellion: The Battle against Convict Labor in the Tennessee Coalfields, 1871–1896* (Chapel Hill: University of North Carolina Press, 1998), 105.

4. "Honored. A Memorial Meeting for the Dead," *Knoxville Journal and Tribune*, August 21, 1892, 1.

5. Bettye J. Broyles, comp., *History of Rhea County, Tennessee* (Dayton, TN: Rhea County Historical and Genealogical Society), 103–20; James B. Jones Jr., "Coal Mining in the Cumberland Plateau, 1880–1930," Appalachian Cultural Workshop Papers, https://www.nps.gov/parkhistory/online_books/sero/appalachian/sec9.htm; Jack Reynolds, *The Great Paternalist: Titus Salt and the Growth of Nineteenth-Century Bradford* (New York: St. Martin's, 1983).

6. "The Dayton Explosion," *Chattanooga Daily Times*, October 17, 1896, 8; "Situation at Dayton," *Chattanooga Daily Times*, July 17, 1897, 1; "Miners Lay Down Picks," *Chattanooga Daily Times*, December 22, 1897, 1; "Miners' Strike in Tennessee," *New York Times*, December 26, 1897, 2; "Mine Drivers Strike," *Chattanooga Daily Times*, July 6, 1898, 3; "Burned: Dayton Coal and Iron Company's Plant Destroyed," *Knoxville Sentinel*, September 19, 1899, 1; "All Union Men Called Out of the Dayton Coal and Iron Company's Mines," *Chattanooga News*, October 22, 1903, 10; "Strike Battle in Tennessee," *New York Times*, March 26, 1904, 1; "Protection for Miners; Dayton Coal and Iron Company Asks Injunction," *Chattanooga Daily Times*, March 30, 1904, 4; *Seventh Annual Report of the Bureau of Labor, Statistics, and Mines of the State of Tennessee* (Nashville, 1898), 175–77.

7. State of Tennessee Mining Department, *Fourteenth Annual Report of the Mining Department* (Nashville, 1904), 130–31.

8. "Appalling Disaster," *Chattanooga Daily Times*, December 21, 1895, 1.

9. "Tennessee Mine Disaster," *New York Times*, May 28, 1901, 3; "Dayton Mine Horror," *Knoxville Sentinel*, April 1, 1902, 1.

10. "Mistrial in Dayton Mine Case," *Knoxville Sentinel*, August 18, 1902, 1.

11. "Neighborhood News, Tennessee," *Chattanooga Daily Times*, June 4, 1886, 7.

12. Shapiro, *New South Rebellion*, 43. A lively scholarly debate has raged over the racial politics of the UMWA. See Herbert Gutman, "The Negro and the United Mine Workers of America: The Career and Letters of Richard L. Davis and Something of Their Meaning, 1800–1900," in *The Negro and the American Labor Movement*, ed. Julius Jacobsen (Garden City, NJ: Anchor Books, 1968), 49–127; Herbert Hill, "Myth-Making as Labor History: Herbert Gutman and the United Mine Workers of America," *Journal of Politics, Culture, and Society* 2 (Winter 1988): 132–99; and Daniel Letwin, *The Challenge of Interracial Unionism: Alabama Coal Miners, 1878–1921* (Chapel Hill: University of North Carolina Press, 1998).

13. John T. Scopes and James Presley, *Center of the Storm: Memoirs of John T. Scopes* (New York: Holt, Rinehart and Winston, 1967), 5; Todd Hatton and Matt Markgraf, "I Am John Scopes," WKMS Radio, Murray State University, Murray, KY, 2015, https://www.wkms.org/post/audio-wkms-documentary-i-am-john-scopes#stream/0.

14. Scopes and Presley, *Center of the Storm*, 5–6.

15. Richard Schneirov, Shelton Stromquist, and Nick Salvatore, eds., *The Pullman Strike and the Crisis of the 1890s: Essays on Labor and Politics* (Urbana: University of Illinois Press, 1999).

16. Scopes and Presley, *Center of the Storm*, 8–10.

17. *Machinists Monthly Journal* 11 (January 1899): 54; "Socialist Nominations," *Paducah (KY) Sun*, May 28, 1901; "Socialist Labor Ticket," *Crittenden (KY) Record-Press*, November 4, 1904, 8; "Eugene Debs Speaks to a Good Crowd," *Paducah News-Democrat*, May 24, 1910, 5.

18. Scopes and Presley, *Center of the Storm*, 22. For more on the controversy over World War I among working people in Illinois see Carl R. Weinberg, *Labor, Loyalty, and Rebellion: Southwestern Illinois Coal Miners and World War I* (Carbondale: Southern Illinois University Press, 2005), 29–62.

19. Scopes and Presley, *Center of the Storm*, 22, 24–5, 27.

20. Broyles, *History of Rhea County, Tennessee*, 109–10.

21. Broyles, 36–37.

22. "Scopes of 'Monkey Trial' Is Dead at 70," *New York Times*, October 23, 1970; Scopes and Presley, *Center of the Storm*, 30.

23. S. George Pemberton and James A. MacEachern, "History of Ichnology—Carroll Lane Fenton and Mildred Adams Fenton: Pioneers of North American Neoichnology," *Ichnos* 3 (1993): 145–53.

24. Carroll Lane Fenton, *Darwin and the Theory of Evolution* (Girard, KS: Haldeman-Julius Co., 1924), 29.

25. E. Haldeman-Julius, *Studies in Rationalism* (Girard, KS: Haldeman-Julius Co., 1925), 69.

26. "Debs Says the Word: To the Wealthy Folk," *Appeal to Reason*, April 29, 1899, 2; Shapiro, *New South Rebellion*, 244–46.

27. T. W. Callaway, "Father of Scopes Renounced Church; Was Staunch Socialist and Follower of Debs," *Chattanooga Daily Times*, July 10, 1925.

28. "Fight on Anti-Darwin Law to Go On; Verdict Expected, Says Scopes," *Daily Worker*, July 22, 1925.

29. Michel Lienesch, "Abandoning Evolution: The Forgotten History of Antievolutionism Activism and the Transformation of American Social Science," *Isis* 103 (December 2012): 687–709

30. Matthew 7:15–17.

31. George H. Nash, *The Conservative Intellectual Movement in America* (1976; Wilmington, DE: Intercollegiate Studies Institute, 1996), 30–42; Frank S. Meyer, "Richard M. Weaver: An Appreciation," *Modern Age* 14 (Summer 1970): 243.

32. Except when I refer to George McCready Price and his coinage of the term "Red Dynamite," I use it otherwise without quotation marks to denote more broadly this claim, expressed by many other conservative Christians, about evolution, communism, and immorality.

33. Henry Morris, "Reflections on a Legacy: Four Decades of Creation Ministry," *Acts and Facts* 39 (2010): 10–13 (based on two articles previously published by Morris in 2001 and 2005), http://www.icr.org/article/5121/.

34. Dan Gilbert, *Evolution: The Root of All Isms* (San Diego, CA: Danielle, 1935), 6.

35. Robert Pennock, *Tower of Babel: The Evidence against the New Creationism* (Cambridge, MA: MIT Press, 1999), 343.

36. For a contemporary popular version of this conception of Satan see Kenneth Copley, *The Great Deceiver: Unmasking the Lies of Satan* (Chicago: Moody, 2001). See also Henry Ansgar Kelly, *Satan: A Biography* (Cambridge: Cambridge University Press, 2006); Neil Forsyth, *The Old Enemy: Satan and the Combat Myth* (Princeton, NJ: Princeton University Press, 1987); and Elaine Pagels, *The Origin of Satan* (New York: Random House, 1995).

37. Paul Boyer, *When Time Shall Be No More: Prophecy Belief in Modern American Culture* (Cambridge, MA: Harvard University Press, 1992), 279–81.

38. Ronald Numbers, *The Creationists: From Scientific Creationism to Intelligent Design*, 2nd ed. (Cambridge, MA: Harvard University Press, 2006); Ronald Numbers, *Darwin Comes to America* (Cambridge, MA: Harvard University Press, 1998); Ronald Numbers and John Stenhouse, eds., *Disseminating Darwinism: The Role of Place, Race, Religion, and Gender* (Cambridge: Cambridge University Press, 1999); Charles Israel, *Before Scopes: Evangelicalism, Education, and Evolution in Tennessee, 1870–1925* (Athens: University of Georgia Press, 2004); Larson, *Summer for the Gods*; Constance Areson Clark, *God—or Gorilla: Images of Evolution in the Jazz Age* (Baltimore: Johns Hopkins University Press, 2010); Jeffrey P. Moran, "Reading Race into the Scopes Trial: African American Elites, Science, and Fundamentalism," *Journal of American History* 90 (December 2003): 891–911; Jeffrey Moran, *American Genesis: The Evolution Controversies from Scopes to Creation Science* (New York: Oxford University Press, 2012); Adam Shapiro, *Trying Biology: The Scopes*

Trial, Textbooks, and the Antievolution Movement in American Schools (Chicago: University of Chicago Press, 2013); Kimberly Hamlin, *From Eve to Evolution: Darwin, Science, and Women's Rights in Gilded Age America* (Chicago: University of Chicago Press, 2014); Susan Trollinger and William Trollinger, *Righting America at the Creation Museum* (Baltimore: Johns Hopkins University Press, 2016). For studies that briefly touch on creationist anticommunism see George Marsden, *Fundamentalism and American Culture: The Shaping of Twentieth-Century Evangelicalism* (New York: Oxford University Press, 1982), 126, 153–56; Paul Boyer, *When Time Shall Be No More: Prophecy Belief in Modern American Culture* (Cambridge, MA: Harvard University Press, 1992), 152–80; Timothy P. Weber, *Living in the Shadow of the Second Coming: American Premillennialism, 1875–1925* (New York: Oxford University Press, 1979), 87; Michael Lienesch, *In the Beginning: Fundamentalism, the Scopes Trial, and the Making of the Antievolution Movement* (Chapel Hill: University of North Carolina Press, 2007), 90–91; and Matthew Avery Sutton, *American Apocalypse: A History of Modern Evangelicalism* (Cambridge, MA: Harvard University Press, 2014), 164–65, 185–90. For a study that gives sustained attention to creationist opposition to materialism see John Bellamy Foster, Brett Clark, and Richard York, *Critique of Intelligent Design: Materialism versus Creationism from Antiquity to the Present* (New York: Monthly Review, 2008). For works that take creationism's moral consequentialism seriously see Christopher P. Toumey, *God's Own Scientists: Creationists in a Secular World* (New Brunswick, NJ: Rutgers University Press, 1994); Pennock, *Tower of Babel*; and Karl W. Giberson and Donald A. Yerxa, *Species of Origins: America's Search for a Creation Story* (Lanham, MD: Rowman & Littlefield, 2002).

39. On the history of this publication see Joe Cain, "Publication History for *Evolution: A Journal of Nature*," *Archives of Natural History* 30, no. 1 (2003): 168–71.

40. Richard Hofstadter, *Social Darwinism in American Thought* (1944; Boston: Beacon, 1992); Robert Bannister, *Social Darwinism: Science and Myth in Anglo-American Thought* (Philadelphia: Temple University Press, 1989), 136. Ronald L. Numbers, "Myth 17: That Social Darwinism Has Had a Profound Influence on Social Thought and Policy, Especially in the United States of America," in *Newton's Apple and Other Myths about Science*, ed. Ronald L. Numbers and Kostas Kampourakis (Cambridge, MA: Harvard University Press, 2015), 139–46; Hamlin, *From Eve to Evolution*.

41. Nikolai Krementsov, "Darwinism, Marxism, and Genetics in the Soviet Union," in *Biology and Ideology: From Descartes to Dawkins*, ed. Denis R. Alexander and Ronald L. Numbers (Chicago: University of Chicago Press, 2010), 215–46.

42. Richard Hofstadter, *The Paranoid Style in American Politics and Other Essays* (New York: Alfred A. Knopf, 1966), 29.

43. Leo Ribuffo, "Policy Series: Donald Trump and the 'Paranoid Style' in American (Intellectual) Politics," issforum.org, June 13, 2017, https://issforum.org/roundtables/policy/1-5an-paranoid; Brian P. Bennett, "Hermetic Histories: Divine Providence and Conspiracy Theory," *Numen* 54 (2007): 174–209. For an insightful recent attempt to make sense of conspiracy theorizing see Thomas Milan Konda, *Conspiracies of Conspiracies: How Delusions Have Overrun America* (Chicago: University of Chicago Press, 2019).

44. Karl Popper, *The Open Society and Its Enemies*, vol. 2 (London: George Routledge & Sons, 1945), 94; Karl Popper, "The Conspiracy Theory of Society," in *Conspiracy Theories: The Philosophical Debate*, ed. David Coady (Aldershot, UK: Ashgate, 2006), 14n1.

45. Richard Dawkins, "Review of *Blueprints: Solving the Mystery of Evolution*," *New York Times*, April 9, 1989.

46. For studies of how the contested politics of gender played out during the Cold War see Elaine Tyler May, *Homeward Bound: American Families in the Cold War Era* (New York: Basic Books, 1988); Mary C. Brennan, *Wives, Mothers, and the Red Menace: Conservative*

Women and the Crusade against Communism (Boulder: University Press of Colorado, 2008); Donald Critchlow, *Phyllis Schlafly and Grassroots Conservatism: A Woman's Crusade* (Princeton, NJ: Princeton University Press, 2007); K.A. Cuordileone, *Manhood and American Political Culture in the Cold War* (New York: Routledge, 2004). On the intersection of anticommunism and resistance to the civil rights movement see Jeffrey Woods, *Black Struggle, Red Scare: Segregation and Anticommunism in the South, 1948–1968* (Baton Rouge: LSU Press, 2004); George Lewis, *The White South and the Red Menace: Segregationist Anticommunism and Massive Resistance, 1945–1965* (Gainesville: University Press of Florida, 2004). For an analysis of how debates over racial equality played into the controversy over evolution in the era of the Scopes trial see Moran, *American Genesis*, 72–90. For the deployment of a creationist framework to defend racial segregation see Horace C. Wilkinson, "Noted Jurist Discusses This Matter of Segregation," February 5, 1948, *Alabama Baptist*, box 35, F1595, Race Problems, 1948, Papers of J. Frank Norris, Southern Baptist Historical Library and Archives, Nashville, TN (hereafter cited as Norris Papers).

47. Daniel Horowitz, *Betty Friedan and the Making of "The Feminist Mystique": The American Left, the Cold War, and Modern Feminism* (Amherst: University of Massachusetts Press, 1998); Hamlin, *From Eve to Evolution*, 149–65.

48. For a valuable study of grassroots conservatism that takes conspiracy ideas seriously see Erin Kempker, *Big Sister: Feminism, Conservatism, and Conspiracy in the Heartland* (Urbana: University of Illinois Press, 2018).

49. See "Birtherism of a Nation," *Atlantic*, May 13, 2020, https://www.theatlantic.com/ideas/archive/2020/05/birtherism-and-trump/610978/, and "The Prophecies of Q," *Atlantic*, June 2020, https://www.theatlantic.com/magazine/archive/2020/06/qanon-nothing-can-stop-what-is-coming/610567/.

50. F.E. Round, R.M. Crawford, and D.G. Mann, *The Diatoms: Biology and Morphology of the Genera* (Cambridge: Cambridge University Press, 1990), 123–24.

51. Kenne Fant, *Alfred Nobel: A Biography*, trans. Marianne Ruuth (1991; New York: Arcade, 1993), 60–63, 94, 207; OED Online, June 2017, Oxford University Press, s.v. "dynamite, n.," http://www.oed.com/view/Entry/58829?rskey=Mn2Dlz&result=1&isAdvanced=false (November 21, 2017). Thanks to Nikos Zirogiannis for sharing his insights on the original Greek. On how the word in Greek has been interpreted in a biblical context see, for instance, http://biblehub.com/greek/1411.htm.

52. On the predominance of sexual morality as a theme in intelligent design creationism see Pennock, *Tower of Babel*, 318–19.

53. Jerry Falwell, *Listen, America!* (New York: Doubleday, 1980), 185–87.

54. "Les Mis Red & Black," https://www.youtube.com/watch?v=me7DMpMaKI0.

55. Mark 12:17.

56. Jerry Falwell, "The Role of the Churches in the Last Quarter of the Twentieth Century," *Sword of the Lord*, August 24, 1979, 14.

57. William E. Arnal and Russell T. McCutcheon, *The Sacred Is the Profane: The Political Nature of "Religion"* (New York: Oxford University Press, 2013).

58. Darren Dochuk, *From Bible Belt to Sun Belt: Plain-Folk Religion, Grassroots Politics, and the Rise of Evangelical Conservatism* (New York: W.W. Norton, 2011); Darren Dochuk, *Anointed with Oil: How Christianity and Crude Made Modern America* (New York: Basic Books, 2019); Matthew Avery Sutton, *Aimee Semple McPherson and the Resurrection of Christian America* (Cambridge, MA: Harvard University Press, 2007); Matthew Avery Sutton, *American Apocalypse: A History of Modern Evangelicalism* (Cambridge, MA: Harvard University Press, 2014); Daniel K. Williams, *God's Own Party: The Making of the Christian Right* (New York: Oxford University Press, 2010); Molly Worthen, *Apostles of Reason: The Crisis of Authority in American Evangelicalism* (New York: Oxford University Press, 2014).

59. Sutton, *American Apocalypse*, xiii.

60. Naomi Oreskes and Erik M. Conway, *Merchants of Doubt: How a Handful of Scientists Obscured the Truth on Issues from Tobacco Smoke to Global Warming* (New York: Bloomsbury, 2010), 248–49, 254, 262.

61. CNN, "Anger Erupts over New Mask Rule While Experts Say They Are Key to Contain the Virus," June 24, 2020, https://www.youtube.com/watch?v=cc4qgvXgLkc.

62. "Teachers in Colorado and Arizona Walk Out over Education Funding," *New York Times*, April 26, 2018, https://www.nytimes.com/2018/04/26/us/teacher-walkout-arizona-colorado.html?searchResultPosition=1.

1. Lighting the Darwin Fuse

1. C. Allyn Russell, "William Bell Riley: Architect of Fundamentalism," *Minnesota History* 43 (Spring 1972): 20.

2. "Evolution or Sovietizing the State through Its Schools," in *Inspiration or Evolution*, 2nd ed. (Cleveland: Union Gospel, 1926), 91, 97–98.

3. Nikolai Krementsov, "Darwinism, Marxism, and Genetics in the Soviet Union," in *Biology and Ideology: From Descartes to Dawkins*, ed. Denis R. Alexander and Ronald L. Numbers (Chicago: University of Chicago Press, 2010), 215–46.

4. Engels to Marx, December 11 or 12, 1859, in Karl Marx and Frederick Engels, *Marx and Engels: 1860–1864*, vol. 40, *Karl Marx, Frederick Engels: Collected Works*, trans. Peter Ross and Betty Ross (New York: International, 1985), 551.

5. Karl Marx and Frederick Engels, *Correspondence, 1846–1895*, trans. Dona Torr (New York: International, 1936), 125–26.

6. John Bellamy Foster, *Marx's Ecology: Materialism and Nature* (New York: Monthly Review, 2000), 105–10.

7. Marx to Engels, June 18, 1862, in Karl Marx and Frederick Engels, *Marx and Engels: 1860–1864*, vol. 41, *Karl Marx, Frederick Engels: Collected Works*, trans. Ross and Ross, 381.

8. For a study that emphasizes Marx's "ambivalence" about Darwin see Richard Weikart, *Socialist Darwinism: Evolution in German Socialist Thought from Marx to Bernstein* (San Francisco: International Scholars, 1999).

9. Margaret A. Fay, "Did Marx Offer to Dedicate *Capital* to Darwin? A Reassessment of the Evidence," *Journal of the History of Ideas* 39 (January–March 1978): 134.

10. Friedrich Engels, *Herr Eugen Dühring's Revolution in Science* (New York: International, 1939, 78–88.

11. Scholarly interpretations of Engels's linking of Marx and Darwin have varied widely. For a sympathetic account that stresses the "complementary" nature of the two figures see Paul Heyer, *Nature, Human Nature, and Society: Marx, Darwin, Biology, and the Human Sciences* (Westport, CT: Greenwood, 1982), 11. For more skeptical views see David Stack, *The First Darwinian Left: Socialism and Darwinism, 1859–1914* (Cheltenham, UK: New Clarion, 2003), 4, which argues that Engels was actuated by "tactical" and "polemical" motives rather than a sincere belief that Marx and Darwin were complementary figures; and Paul Thomas, *Marxism and Scientific Socialism: From Engels to Althusser* (New York: Routledge, 2008), 65, which argues that Engels misrepresented Darwin, with "destructive" consequences.

12. Frederick Engels, *The Origin of the Family, Private Property, and the State* (1884; New York: Pathfinder, 1972), 48.

13. Mark Pittenger, *American Socialists and Evolutionary Thought, 1870–1920* (Madison: University of Wisconsin Press, 1993), 8.

14. Herbert Spencer, *First Principles*, 2nd ed. (London: Williams and Norgate, 1867), 517.

15. Enrico Ferri, *Socialism and Modern Science (Darwin-Spencer-Marx)*, trans. Robert Rives La Monte (New York: International, 1900), 88.

16. Ernest Untermann, *Marxian Economics: A Popular Introduction to the Three Volumes of Marx's "Capital"* (Chicago: Charles H. Kerr, 1913), 36, 40. On this point see Pittenger, *American Socialists*, 138–39.

17. George Cotkin, "'They All Talk like Goddam Bourgeois': Scientism and the Socialist Discourse of Arthur M. Lewis," *ETC: A Review of General Semantics* 38 (1981): 274; "Arthur M. Lewis Dies at Midlothian Club," *Chicago Daily Tribune*, August 23, 1922, 19. See also Pittenger, *American Socialists*, 140–45.

18. George Cotkin, "The Socialist Popularization of Science in America, 1901 to the First World War," *History of Higher Education Quarterly* 24 (Summer 1984): 209–10.

19. Allen Ruff, *"We Called Each Other Comrade": Charles H. Kerr & Company, Radical Publishers* (Urbana: University of Illinois Press, 1997), 167, 263.

20. Arthur M. Lewis, *Evolution: Social and Organic* (Chicago: Charles H. Kerr, 1908), 6.

21. Lewis, 7, 21.

22. On challenges to Darwin see Peter J. Bowler, *The Eclipse of Darwinism: Anti-Darwinian Evolution Theories in the Decades around 1900* (Baltimore: Johns Hopkins University Press, 1983).

23. Peter J. Bowler, *Evolution: The History of an Idea*, 3rd ed. (Berkeley: University of California Press, 2003), 266–73.

24. Bowler, 253–56.

25. Bowler, 95.

26. Bowler, 79.

27. Mari Jo Buhle, *Women and American Socialism, 1870–1920* (Urbana: University of Illinois Press, 1981).

28. August Bebel, *Woman and Socialism*, trans. Meta Stern (New York: Socialist Literature, 1910), 238.

29. Bebel, 242.

30. Bebel, 254. In addition to lauding women's capacities, Bebel subscribed to a melding of biological and social evolution, common in socialist and non-socialist evolutionary thinkers. See Pittenger, *American Socialists*, 186–97.

31. Buhle, *Women and American Socialism*, 246–87.

32. Lena Morrow, "The Sex and Woman Questions," *Masses*, December 1911.

33. Morrow, 162–65. On Barnes see Ira Kipnis, *The American Socialist Movement, 1897–1912* (New York: Columbia University Press, 1952), 379–80.

34. "Rockefeller's Foes Invade His Church," *New York Times*, May 11, 1914, 1; "Bouck White's Band Guilty; Locked Up," *New York Times*, May 13, 1914; "Bouck White Gets 30 Days," *New York Times*, March 16, 1917; "Social Rebels Burn All Flags," *New York Times*, June 2, 1916; David Shannon, *The Socialist Party of America: A History* (New York: Macmillan, 1955), 60–61.

35. White's first hit was *The Book of Daniel Drew: A Glimpse of the Fisk-Gould-Tweed Regime from the Inside* (New York: Doubleday, Page, 1910), a biting portrait of the Wall Street financier and railroad magnate. It became the basis, along with Matthew Josephson's *The Robber Barons* (1934), of a Hollywood movie, *The Toast of New York* (1937), starring Cary Grant, Edward Arnold, and Frances Farmer.

36. Bouck White, *The Call of the Carpenter* (Garden City, NY: Doubleday, Page, 1911); David Burns, *The Life and Death of the Radical Historical Jesus* (New York: Oxford University Press, 2013), 83–125.

37. "Divorce Key at the Altar," *New York Times*, November 23, 1914. On the broader context see Rebecca Davis, "'Not Marriage at All, but Simple Harlotry': The Companionate Marriage Controversy," *Journal of American History* 94 (March 2008): 1146–48.

38. "Bouck White Tarred and Feathered on Wife's Complaint," *New York Times*, May 30, 1921, 1.

39. White, *Call of the Carpenter*, 297. Bouck White did not cite Andrew Dickson White, but he did make use of three quotations (of Carlyle, Ségur, and Rev. James Tait) that appear within pages of each other in White's work. See Andrew Dickson White, *A History of the Warfare of Science with Theology in Christendom*, vol. 1 (New York: D. Appleton, 1896), 73–86. For a critical appraisal of Andrew Dickson White's work see Ronald Numbers, "Science and Religion," *Osiris*, 2nd series, vol. 1, *Historical Writing on American Science* (1985): 59–80.

40. Krementsov, "Darwinism, Marxism, and Genetics," 217–18; Alexander Vucinich, *Darwin in Russian Thought* (Berkeley: University of California Press, 1989), 308–10; George Windholz, "Pavlov's Religious Orientation," *Journal for the Scientific Study of Religion* 25 (September 1986): 322.

41. Vucinich, *Darwin in Russian Thought*, 356.

42. Vucinich, 368.

43. Vucinich, 363.

44. V.I. Lenin, "What the 'Friends of the People' Are and How They Fight the Social-Democrats (a Reply to Articles in *Russkoye Bogatstvo* Opposing the Marxists)," in *Collected Works of V.I. Lenin*, vol. 1, *1893–94* (Moscow: Progress, 1960), 142.

45. As Paul Pojman notes, "the word 'Mach' has bizarrely entered into popular culture as an icon for razors, sound systems, fighter pilots, and high speed fuels." Paul Pojman, "Ernst Mach," in *The Stanford Encyclopedia of Philosophy (Winter 2011 Edition)*, ed. Edward N. Zalta, http://plato.stanford.edu/archives/win2011/entries/ernst-mach/. Indeed, despite my sympathies for Lenin, I have been shaving, unreflectively, for a good number of years with Gillette's Mach 3 Turbo razor.

46. P. M. Harman, *Energy, Force, and Matter: The Conceptual Development of Nineteenth-Century Physics* (Cambridge: Cambridge University Press, 1982), 146–47; C. Hakfoort, "Science Deified: Wilhelm Ostwald's Energeticist World-View and the History of Scientism," *Annals of Science* 49 (November 1992): 525–44.

47. Marina F. Bykova, "Lenin and the Crisis of Russian Marxism," *Studies in East European Thought* 70 (November 2018): 235–47.

48. V.I. Lenin, *Materialism and Empirio-Criticism* (1909; Beijing: Foreign Languages Publishing House, 1972), 396.

49. V.I. Lenin, "The Attitude of the Workers' Party to Religion," originally published in *Proletary* (May 13, 1909), in Lenin, *Collected Works*, vol. 15, 402–13.

50. K. Marx and F. Engels, *On Religion* (Moscow: Foreign Languages Publishing House, 1957), 42.

51. Isaac Deutscher, *The Prophet Armed: Trotsky, 1879–1921* (Oxford: Oxford University Press, 1954), 7–36.

52. Max Eastman, *Leon Trotsky: The Portrait of a Youth* (New York: Greenberg, 1925), 129.

53. Krementsov, "Darwinism, Marxism, and Genetics," 221–23.

54. V.I. Lenin, "On the Significance of Militant Materialism," *LCW* 33 (Moscow: Progress, 1972), 227–36.

55. Leon Trotsky, "The Tasks of Communist Education," *Communist Review* 4 (December 1922), Marxists Internet Archive, http://www.marxists.org/history/international/comintern/sections/britain/periodicals/communist_review/1923/7/com_ed.htm.

56. Philip Pomper, ed., *Trotsky's Notebooks, 1933–1935: Writings on Lenin, Dialectics, and Evolutionism* (New York: Columbia University Press, 1986), 48.

57. Dmitri V. Pospielovsky, *A History of Marxist-Leninist Atheism and Soviet Antireligious Policies* (New York: St. Martin's, 1987), 132–34.

58. William B. Husband, *"Godless Communists": Atheism and Society in Soviet Russia, 1917–1932* (De Kalb: Northern Illinois University Press, 2000), 65.

59. Kirill Rossiianov, "Beyond Species: Il'ya Ivanov and His Experiments on Cross-Breeding Humans with Anthropoid Apes," *Science in Context* 15 (2002): 277–316; Alexander Etkind, "Beyond Eugenics: The Forgotten Scandal of Hybridizing Humans and Apes," *Studies in History and Philosophy of Biological and Biomedical Sciences* 39 (2008): 205–10.

60. Monte Reel, *Between Man and Beast: An Unlikely Explorer, the Evolution Debates, and the African Adventure That Took the Victorian World by Storm* (New York: Doubleday, 2013).

61. Laura Davidow Hirshbein, "The Glandular Solution: Sex, Masculinity, and Aging in the 1920s," *Journal of the History of Sexuality* 9 (July 2000): 277–304.

62. Rossiianov, "Beyond Species," 286–89.

63. Rossiianov, 293–308; Etkind, "Beyond Eugenics," 206–7.

64. "Men and Apes," *Time*, June 28, 1926, 24. On Smith's activism see R. Laurence Moore and Isaac Kramnick, *Godless Citizens in a Godly Republic* (W.W. Norton, 2018), 38–40.

65. "Russian Admits Ape Experiments," *New York Times*, June 19, 1926, 17; "Soviet Backs Plan to Test Evolution," *New York Times*, June 17, 1926, 2.

66. Elizabeth A. Wood, *The Baba and the Comrade: Gender and Politics in Revolutionary Russia* (Bloomington: Indiana University Press, 1997); Richard Stites, *The Women's Liberation Movement in Russia: Feminism, Nihilism, and Bolshevism, 1860–1930* (Princeton, NJ: Princeton University Press, 1978); Wendy Z. Goldman, *Women, the State, and Revolution: Soviet Family Policy and Social Life, 1917–1936* (New York: Cambridge University Press, 1995).

67. Beatrice Farnsworth, *Alexandra Kollantai: Socialism, Feminism, and the Bolshevik Revolution* (Stanford, CA: Stanford University Press, 1980).

68. Julia L. Mickenberg, "Suffragettes and Soviets: American Feminists and the Specter of Revolutionary Russia," *Journal of American History* 100 (March 2014): 1043. On the way that antiradicalism and antifeminism were intertwined see Kim Nielsen, *Un-American Womanhood: Antiradicalism, Antifeminism, and the First Red Scare* (Columbus: Ohio State University Press, 2001), and Kirsten Marie Delegard, *Battling Miss Bolsheviki: The Origins of Female Conservatism in the United States* (Philadelphia: University of Pennsylvania Press, 2012).

69. Mickenberg, "Suffragettes and Soviets," 1045.

70. Nielsen, *Un-American Womanhood*, 30–31; Delegard, *Battling Miss Bolsheviki*, 28–30.

71. "Decree Provides Maidens Become Property of State," *Huntington (IN) Press*, October 26, 1918, 1; "Soviets Make Girls Property of State," *New York Times*, October 26, 1918, 5. For representative coverage see, e.g., "Maidens Become Property of State 'Bureau of Free Love,' a Bolsheviki Plan," *San Bernardino County (CA) Sun*, October 26, 1918, 1; "All Russian Girls Shall Become State Property, Slav Soviets Decree," *Topeka Daily Capital*, October 26, 1918; "Girls Become State Chattels; Bolshevik Order Requires Them to Register at Eighteen with 'Free Love Bureau,'" *Cincinnati Enquirer*, October 26, 1918, 7.

72. Alex Goodall, *Loyalty and Liberty: American Countersubversion from World War I to the McCarthy Era* (Urbana: University of Illinois Press, 2013), 45–54.

73. "Bolshevism Bared by R. E. Simmons," *New York Times*, February 18, 1919, 4.

74. Stephen Eric Bronner, *A Rumor about the Jews: Reflections on Antisemitism and the "Protocols of the Learned Elders of Zion"* (New York: St. Martin's, 2000); Lucien Wolf, *The Myth of the Jewish Menace in World Affairs or, The Truth about the Forged Protocols of the Elders of Zion* (New York: Macmillan, 1921).

75. Barbara Evans Clements, "The Effects of the Civil War on Women and Family Relations," in *Party, State, and Society in the Russian Civil War: Explorations in Social History* (Bloomington: Indiana University Press, 1989), 105–6. Clements accepts as authentic the evidence that young "enthusiasts" for a Marxist vision of sexual liberation did create a "Bureau of Fee Love" in the city of Vladimir. Her source is A. G. Kharchev, *Brak i sem'ia v SSSR*, 2nd ed. (Moscow, 1979), 133.

76. Quoted in Nick Paton Walsh, "'Nationalisation' of Wives Made Lenin See Red," *Guardian*, April 19, 2006; Patrick Wright, *Iron Curtain: From Stage to Cold War* (Oxford: Oxford University Press, 2007), 439.

77. Krementsov, "Darwinism, Marxism, and Genetics," 238–39.

78. Theodore Draper, *The Roots of American Communism* (1957; New Brunswick, NJ: Transaction, 2003); Irving Howe, *The American Communist Party: A Critical History* (New York: Praeger, 1962); Bryan Palmer, *James P. Cannon and the Origins of the American Revolutionary Left, 1890–1928* (Urbana: University of Illinois Press, 2007); James P. Cannon, *The First Ten Years of American Communism: Report of a Participant* (New York: Pathfinder, 1973).

79. Draper, *Roots of American Communism*, 360–61. According to Draper, the proponents of undergroundism were dubbed the "Goose Caucus" by Jay Lovestone, because of how much panicked noise they made about saving the party, an apparent reference to a legend about the sacred geese of Juno having alerted Roman consul Marcus Manlius Capitolinus to an imminent Gallic invasion of Rome in 387 BCE.

80. Palmer, *James P. Cannon*, 241–45.

81. "Anti-Evolution Law Branded Unconstitutional in Fight for Freedom of Education," *Daily Worker*, July 14, 1925, 1. For other *Daily Worker* trial coverage see, for example, "Bryan Seeks to Bar Sciences from Schools," July 9, 1925; "Scopes Trial Underway with New Indictment," July 11, 1925; "Judge in Scopes Trial Asked to Rule on Constitutionality of Tenn. Anti-evolution Law," July 15, 1925; "Bryan Seeks to Bar Scientists in Scopes Case," July 13, 1925; "Dayton Judge Rules Ape Law Constitutional," July 15, 1925; "State Moves to Bar Science, Scopes' Defense Afraid to Expose Bosses' Bible Dope," July 17, 1925; "Closing Arguments in Scopes Trial Begin Today"; and "Darrow and Bryan Clash over Evolution," July 20, 1925.

82. "Darrow Cited for Contempt by Tenn. Judge," *Daily Worker*, July 14, 1925, 1, 3.

83. "Fight on Anti-Darwin Law to Go On; Verdict Expected, Says Scopes," *Daily Worker*, July 22, 1925. For an example of the story appearing as an INS exclusive see "Scopes Says Science Will Gain Victory," *Logansport (IN) Pharos-Tribune*, July 22, 1925, 1.

84. "Darrow Sidesteps in Dayton," *Daily Worker*, July 17, 1925; "Evolution, Capitalism and the Workers," *Daily Worker*, July 9, 1925.

85. Alan Wald, *Exiles from a Future Time: The Forging of the Mid-Twentieth Century Literary Left* (Chapel Hill: University of North Carolina Press, 2002), 39–70.

86. "'Monkey or Man' Mike's Play at Defense Picnic," *Daily Worker*, August 7, 1925; "'Monkey or Man?' Screaming Farce by Michael Gold," advertisement, *Daily Worker*, August 20, 1925.

87. "The K.K.K. in Washington," *Daily Worker*, August [7 or 8], 1925; "Klan Declares War on Science and Foreigners," *Daily Worker*, August 22, 1925.

88. Michael Kazin, *A Godly Hero: The Life of William Jennings Bryan* (New York: Alfred A. Knopf, 2006), 278–84.

89. William Schneiderman, "What about Evolution?," *Daily Worker*, August 15, 1925. On *Komsomolskaya Pravda* see *Closer to the Masses: Stalinist Culture, Social Revolution, and Soviet Newspapers* (Cambridge, MA: Harvard University Press, 2004), 106–8.

90. H. L. Mencken, *Religious Orgy in Tennessee: A Reporter's Account of the Scopes Monkey Trial* (New York: Melville House 2006), 15; Marion Elizabeth Rodgers, *Mencken: The American Iconoclast* (New York: Oxford University Press, 2005), 287–90.

91. "Discuss Evolution at Minneapolis Open Air Meeting Sunday," *Daily Worker*, July 16, 1925; "Scopes Trial, Religion versus Evolution, Free Masonry and Communist Membership Discussed," *Daily Worker*, July 20, 1925.

92. "Release Cicero Communists in Speech Fight," *Daily Worker*, July 20, 1925.

93. Palmer, *James P. Cannon*, 269.

94. Ralph Lord Roy, *Communism and the Churches* (New York: Harcourt Brace, 1960), 21; Ron Carden, "The Bolshevik Bishop: William Montgomery Brown's Path to Heresy, 1906–1920," *Anglican and Episcopal History* 72 (June 2003): 197–228; Ron Carden, *William Montgomery Brown (1855–1937): The Southern Episcopal Bishop Who Became a Communist* (Lewiston, NY: Edwin Mellen, 2007). Since Brown was the only bishop ever to be deposed in the US for heresy, the church's action was front-page news around the nation. See, for instance, "Church Ousts Bishop Brown as a Heretic," *Brooklyn (NY) Daily Eagle*, October 12, 1925, 1.

95. William Montgomery Brown, *Communism and Christianism: Analyzed and Contrasted from the Marxian and Darwinian Point of View*, 4th ed. (Galion, OH: Bradford-Brown Educational, 1921), 20, 180; illustrations on 83–84, 155–56.

96. On the history of this publication see Joe Cain, "Publication History for *Evolution: A Journal of Nature*," *Archives of Natural History* 30 (2003) (1): 168–71. For downloadable PDFs of the magazine as well as an index by author and subject see https://profjoecain.net/category/projects/evolution-journal-nature-ejn/.

97. Branko Lazitch and Milorad M. Drachkovich, *Biographical Dictionary of the Comintern: New, Revised, and Expanded Edition* (Stanford, CA: Hoover Institution, 1986), 212; "Crusader," *Time*, June 28, 1937, 44; "Katterfeld Indicted; Former Topeka Man Accused of Conspiracy against U.S.," *Topeka Daily Capital*, January 22, 1920; "Kansan a Communist," *Topeka Daily Capital*, September 6, 1919; "Millionaire 'Red' Must Go to Jail; Illinois Supreme Court Rejects Appeal of William Bross Lloyd and Four Associates," *New York Times*, November 17, 1922.

98. Whittaker Chambers, *Witness* (1952; Washington, DC: Gateway Editions, 2002), 210.

99. "Interview with Ludwig E. Katterfeld by Theodore Draper [conducted September 8, 1956]," edited by Tim Davenport, document in the Hoover Institution Archives, Theodore Draper Papers, box 30, http://www.marxists.org/history/usa/parties/cpusa/1956/09/0908-draper-katterfeldint.pdf, 5.

100. "Crusader," *Time*, June 28, 1937, 44.

101. Constance Areson Clark, *God—or Gorilla: Images of Evolution in the Jazz Age* (Baltimore: Johns Hopkins University Press, 2008), 3.

102. Quoted in Constance Areson Clark, "'You Are Here': Missing Links, Chains of Being, and the Language of Cartoons," *Isis* 100 (September 2009): 578–81.

103. For a discussion of those implications in the late nineteenth and early twentieth centuries see Moore and Kramnick, *Godless Citizens*, 26–40.

104. Masthead, *Evolution: A Journal of Nature*, April 1928: 8; "Our Policy," *Evolution: A Journal of Nature*, December 1927: 8; for the inclusion of Brown see, for instance, "Some Good Books," *Evolution: A Journal of Nature*, March 1928: 13, where his work appears at the top of the list.

105. Letter from John Dewey, *Evolution: A Journal of Nature*, April 1928: 8.

106. "Funnymentals," *Evolution: A Journal of Nature*, February 1928: 11.

107. For articles he wrote for Katterfeld see Henshaw Ward, "What Is Evolution?," *Evolution: A Journal of Nature*, December 1927: 5; "The Nebraska Tooth," *Evolution: A Journal of Nature*, July 1928: 3; Clark, *God—or Gorilla*, 101.

108. "Funnymentals," *Evolution: A Journal of Nature*, October 1928: 12.

109. Elof Axel Carlson, *Genes, Radiation, and Society: The Life and Work of H.J. Muller* (Ithaca, NY: Cornell University Press, 1981), 91–108.

110. H.J. Muller, "Observations of Biological Science in Russia," *Scientific Monthly* 16 (May 1923): 549.

111. Carlson, *Genes, Radiation, and Society*, 204–34.

2. The Lamb-Dragon and the Devil's Poison

1. Clarence Darrow, *Attorney for the Damned: Clarence Darrow in the Courtroom* (1957; Chicago: University of Chicago Press, 2012), 213–14.

2. "Letter to the editor of Science from the principal scientific authority of the fundamentalists," *Science* 63 (March 5, 1926): 259.

3. See, for example, Albert Einstein to George McCready Price, February 6, 1954, box 2, folder 2, Papers of George McCready Price, Adventist Heritage Center, Andrews University, Berrien Springs, MI (hereafter cited as Price Papers).

4. Ronald L. Numbers, *The Creationists: From Scientific Creationism to Intelligent Design* (Cambridge, MA: Harvard University Press, 2006), 103.

5. Ronald L. Numbers and Jonathan M. Butler, eds., *The Disappointed: Millerism and Millenarianism in the Nineteenth Century* (Bloomington: Indiana University Press, 1987).

6. On Ellen White see Ronald Numbers, *Prophetess of Health: A Study of Ellen G. White* (New York: Harper & Row, 1976); and Terrie Dopp Aamodt, Gary Land, and Ronald L. Numbers, eds., *Ellen Harmon White: American Prophet* (New York: Oxford University Press, 2014). For an insightful analysis of White's emergence as a prophet in Portland see Jonathan M. Butler, "Prophecy, Gender, and Culture: Ellen Gould Harmon [White] and the Roots of Seventh-day Adventism," *Religion and American Culture: A Journal of Interpretation* 1 (Winter 1991): 3–29.

7. Paul Boyer, *When Time Shall Be No More: Prophecy Belief in Modern American Culture* (Cambridge, MA: Harvard University Press, 1992).

8. Douglas Morgan, *Adventism and the American Republic: The Public Involvement of a Major Apocalyptic Movement* (Knoxville: University of Tennessee Press, 2001), 15–17.

9. Numbers, *Creationists*, 7.

10. Ellen G. White, *Spiritual Gifts: Important Acts of Faith, in Connection with the History of Holy Men of Old*, vol. 3 (Battle Creek, MI: Seventh-day Adventist Publishing Assn., 1864), 90–92; Ronald Numbers, "'Sciences of Satanic Origin': Adventist Attitudes toward Evolutionary Biology," in *Darwinism Comes to America* (Cambridge, MA: Harvard University Press, 1998), 92–110.

11. White, *Spiritual Gifts*, 75.

12. Ronald Osborn, "True Blood: Race, Science, and Early Adventist Amalgamation Theory Revisited," *Spectrum*, Fall 2010, 16–29; Gordon Shigley, "Amalgamation of Man and Beast: What Did Ellen White Mean?," *Spectrum*, June 1982, 10–19.

13. For a leading twentieth-century Adventist apologist's defense of the "of man and *of* beast" interpretation, which also sees a role for Satan in amalgamation, see Francis D. Nichol, "Amalgamation: Ellen G. White Statements regarding Conditions at the Time of the Flood," 1951, http://www.whiteestate.org/issues/amalg.html.

14. Clifton L. Taylor, "Pioneer Days," *Eastern Canadian Messenger*, April 16, 1918.

15. George McCready Price, "If I Were Twenty Again," *These Times*, September 1, 1960, 23.

16. "If I Were Young Again . . . I'd Have an Aim," *Review and Herald* 138 (February 16, 1961): 14; Gary Land, *Historical Dictionary of Seventh-day Adventists* (Lanham, MD: Scarecrow, 2005), 344.

17. Ellen G. White, *The Great Controversy between Christ and Satan: The Conflict of the Ages in the Christian Dispensation* (Mountain View, CA: Pacific Press, 1911), 508, 678.

18. According to her obituary, Amelia and George both attended Battle Creek College. Obituary for Amelia Anna Nason Price, *Review and Herald* 131 (December 1954): 21. Numbers's account has only George attending, citing "Battle Creek College Records, 1876–94" (Adventist Heritage Center, Andrews University), 369, 383; Numbers, *Creationists*, 463n7.

19. George McCready Price, "Some Early Experiences with Evolutionary Geology," *Bulletin of Deluge Geology and Related Sciences* 1 (November 1941): 79, in *Early Creationist Journals*, ed. Ronald L. Numbers, vol. 9, *Creationism in Twentieth-Century America* (New York: Garland, 1995), 79.

20. Obituary for Amelia Anna Nason Price, *Review and Herald* 131 (December 1954): 21.

21. Numbers, *Creationists*, 91–92; Harold W. Clark, *Crusader for Creation: The Life and Writings of George McCready Price* (Mountain View, CA: Pacific Press, 1966), 14–16; Price, "Some Early Experiences with Evolutionary Geology," 151; George McCready Price, *Outlines of Modern Science and Modern Christianity* (Oakland, CA: Pacific Press, 1902).

22. Price, *Outlines*, 137.

23. George McCready Price, *Illogical Geology: The Weakest Point in the Evolution Theory* (Los Angeles: Modern Heretic, 1906), 30; Numbers, *Creationists*, 112–13.

24. Price, *Outlines*, 234, 252.

25. Price, 262.

26. "Evolution and Anarchy," *Advent Review and Sabbath Herald*, October 1, 1901.

27. Morgan, *Adventism and the American Republic*, 69.

28. Percy T. Magan, *Imperialism versus the Bible, the Constitution, and the Declaration of Independence; or, The Peril of the Republic of the United States* (Battle Creek, MI: National Co-operative Library Association, 1899), 81, emphasis in original.

29. Ronald D. Graybill, "The Abolitionist-Millerite Connection," in Numbers and Butler, *Disappointed*, 139–52.

30. Samuel G. London Jr., *Seventh-day Adventists and the Civil Rights Movement* (Jackson: University Press of Mississippi, 2009), 44–65.

31. Quoted in Numbers, *Creationists*, 102.

32. George McCready Price, *The Phantom of Organic Evolution* (New York: Fleming H. Revell, 1924), 106.

33. Quoted in Numbers, *Creationists*, 101.

34. Price, *Outlines*, 263.

35. Carlos A. Schwantes, "Labor Unions and Seventh-day Adventists: The Formative Years, 1877–1903," *Adventist Heritage* 4 (Winter 1977): 18–19.

36. Ellen G. White, "Our Duty to Leave Battle Creek," *General Conference Bulletin*, 35th Session, Oakland, CA, April 6, 1903, 87; See also K. C. Russell, "Seventh-day Adventists and Labor Unions," *Advent Review and Sabbath Herald*, January 26, 1905, 9.

37. New York State Census, First Election District, Block A, Fifth Assembly District, Borough of Manhattan, County of New York, State of New York, June 1, 1905, 9–10; 1904 Sanborn Atlas, v. 3, sheet 12a.

38. Gerald W. McFarland, *Inside Greenwich Village: A New York City Neighborhood, 1898–1918* (Amherst: University of Massachusetts Press, 2001), 58.

39. George McCready Price to Elder William Guthrie, from 95 Christopher Street, New York City, December 28, 1904, RG 11, box 20, President's Incoming Letters, 1905-D to 1905-P, Archives of the General Conference of Seventh-day Adventists, Silver Spring, MD.

40. George McCready Price to A.G. Daniells, from 422 W. 57th Street, New York, March 19, 1905, RG 11, box 20, President's Incoming Letters, 1905-D to 1905-P, Archives of the General Conference of Seventh-day Adventists, Silver Spring, MD.

41. *Advent Review and Sabbath Herald*, July 13, 1905.

42. "Civil War Threatened," *New York Times*, January 23, 1905, 1; "Revolution Party Here Hails News with Joy," *New York Times*, January 23, 1905, 2.

43. McFarland, *Inside Greenwich Village*, 122–23. Walling was not formally a member of the Socialist Party of America until 1910, but he helped write the party's 1904 platform and wrote widely for socialist publications during the decade. Mark Pittenger, *American Socialists and Evolutionary Thought, 1870–1920* (Madison: University of Wisconsin Press, 1993), 147.

44. Numbers, *Creationists*, 94–95, 98.

45. Eda A. Reid to Harold W. Clark, January 12, 1965, C2, box 2, folder 4, "Correspondence about George McCready Price's Biography," Price Papers.

46. *1908 Yearbook of the Seventh-day Adventist Denomination* (Washington, DC: Review and Herald, 1908), 139.

47. Ronald Numbers, "Reading the Book of Nature through American Lenses," in *Science and Christianity in Pulpit and Pew* (New York: Oxford University Press, 2007), 59–60.

48. Clark, *Crusader for Creation*, 32; Numbers, *Creationists*, 107.

49. On Price and Baconianism see Numbers, *Creationists*, 107–8; Malcolm Bull and Keith Lockhart, *Seeking a Sanctuary: Seventh-day Adventism and the American Dream*, 2nd ed. (Bloomington: Indiana University Press, 2007), 31.

50. George Marsden, *Fundamentalism and American Culture: The Shaping of Twentieth-Century Evangelicalism* (New York: Oxford University Press, 1982), 55–62, 111–16, 120–21; Jon H. Roberts, *Darwinism and the Divine in America: Protestant Intellectuals and Organic Evolution, 1859–1900* (Madison: University of Wisconsin Press, 1988), 41–42; Dwight Bozeman, *Protestants in an Age of Science: The Baconian Ideal and Antebellum American Religious Thought* (Chapel Hill: University of North Carolina Press, 1977); Reuben Torrey, "The Certainty and Importance of the Bodily Resurrection of Jesus Christ from the Dead," in *The Fundamentals: A Testimony to the Truth* (Chicago: Testimony, 1915), 5:83.

51. Charles Darwin, *On the Origin of Species* (1859; Cambridge, MA: Harvard University Press, 1964), ii. On Darwin and Baconianism see also Peter Novick, *That Noble Dream: The "Objectivity Question" and the American Historical Profession* (Cambridge: Cambridge University Press, 1988), 34–36.

52. George McCready Price, *God's Two Books: Or Plain Facts about Evolution, Geology, and the Bible* (Washington, DC: Review and Herald, 1911; 2nd ed., 1918).

53. Price, 59.

54. Price, 31, emphasis added.

55. Price, 32–33.

56. Philip S. Foner, *The AFL in the Progressive Era, 1910–1915*, vol. 5, *History of the Labor Movement in the United States* (New York: International, 1980), 7–31.

57. *Signs of the Times*, July 4, 1911.

58. *Signs of the Times*, May 23, 1911.

59. "A Condition and Not a Theory," *Signs of the Times*, September 26, 1911, 602–3.

60. See, for instance, the following, all from the *Advent Review and Herald*: "Christianity's Solution of the Problem of Capital and Labor," August 14, 1913, 7–9; "A World-Wide

Industrial Conflict," September 28, 1916, 1; "The Socialist Platform," June 11, 1908; "The Socialist Deluge," July 13, 1905; "Socialism versus Christianity," January 30, 1908; "Christianity versus Socialism," March 17, 1903; "Jesus and Socialism," May 28, 1908; "Recognizing Danger," November 5, 1908; "Man's versus God's Rule," May 1, 1913; "Christ and Socialism," September 23, 1909.

61. "A Disquieting Situation," *Advent Review and Herald*, March 7, 1912.

62. George McCready Price, *Back to the Bible: Or, The New Protestantism* (Washington, DC: Review and Herald, 1916), 170–72.

63. Price, 175–78.

64. Price, 168–69.

65. Price, 176–79.

66. George McCready Price and Robert B. Thurber, *Socialism in the Test-Tube* (Nashville: Southern, 1921), 29, 44, 113.

67. George McCready Price, *Poisoning Democracy: A Study of the Religious and Moral Aspects of Socialism* (New York: Fleming H. Revell, 1921), 14.

68. Price, *Poisoning Democracy*, 47, emphasis in original.

69. *Advent Review and Sabbath Herald*, April 3, 1919.

70. "Bolshevism in New York and Russian Schools," *Literary Digest*, July 5, 1919, 41.

71. Price, *Poisoning Democracy*, 70–72. On the early Bolshevik public school system see Larry E. Holmes, *The Kremlin and the Schoolhouse: Reforming Education in Soviet Russia, 1917–1931* (Bloomington: Indiana University Press, 1993), Lisa A. Kirschenbaum, *Small Comrades: Revolutionizing Childhood in Soviet Russia, 1917–1932* (New York: Routledge Falmer, 2001), and Yordanka Valkanova, "The Passion for Educating the 'New Man': Debates about Preschooling in Soviet Russia, 1917–1925," *History of Education Quarterly* 49 (May 2009): 211–21.

72. Rebecca Davis, "'Not Marriage at All, but Simple Harlotry': The Companionate Marriage Controversy," *Journal of American History* 94, no. 4 (March 2008): 1147.

73. *Advent Review and Sabbath Herald*, April 3, 1919.

74. Price, *Poisoning Democracy*, 51.

75. Pittenger, *American Socialists*, 139–40; Markku Ruotsila, *John Spargo and American Socialism* (New York: Palgrave Macmillan, 2006).

76. Price, *Poisoning Democracy*, 52.

77. Review of *Poisoning Democracy: A Study of the Religious and Moral Aspects of Socialism*, by George McCready Price, *Literary Review*, January 21, 1922, 372.

78. Malcolm Bissell, letter to the editor, *Literary Review*, March 25, 1922, 538; Malcolm Bissell to George McCready Price, April 6, 1922, C2, box 1, folder 2, Correspondence, 1906–1925, Price Papers.

79. Review of *Poisoning Democracy*, by George McCready Price, *Constitutional Review* 6 (January 1922): 64.

80. Advertisement for George McCready Price, *Poisoning Democracy*, citing *Sunday School Times*, January 21, 1922, in the *Canadian Watchman*, November 1922. On the *Sunday School Times* see Joel Carpenter, *Revive Us Again: The Reawakening of American Fundamentalism* (New York: Oxford University Press, 1997), 26.

81. Ronald L. Heinemann, "Joseph Eggleston," in *Encyclopedia of Virginia*, ed. Brendan Wolfe, http://www.encyclopediavirginia.org/Eggleston_Joseph_Dupuy_Jr_1867-1953; J.D. Eggleston to George McCready Price, November 29, 1921, C2, box 1, folder 2, Correspondence, 1906–1925, Price Papers.

82. Michael Kazin, *A Godly Hero: The Life of William Jennings Bryan* (New York: Alfred A. Knopf, 2006), 273–74.

83. Simon Baatz, *For the Thrill of It: Leopold, Loeb, and the Murder That Shocked Chicago* (New York: HarperCollins, 2008); Frank A. Pattie, "The Last Speech of William Jennings Bryan," *Tennessee Historical Quarterly* 6 (September 1947): 281; C. U. P. Smith, "Clever Beasts Who Invented Knowledge: Nietzsche's Evolutionary Biology of Knowledge," *Biology and Philosophy* 2 (January 1987): 65–91.

84. Quoted in William Jennings Bryan, *The Bible and Its Enemies: An Address Delivered at the Moody Bible Institute of Chicago* (Chicago: Bible Institute Colportage Association, 1921), 38. Bolshevism, according to Begbie, was "a religion founded in violence, and inspired by contempt for human freedom. It distrusts the human race; it hates the human soul." Harold Begbie, *The Glass of Fashion: Some Social Reflections* (New York: G. P. Putnam's Sons, 1921), ix, 9.

85. Bryan, *The Bible and Its Enemies*, 43.

86. *Advent Review and Sabbath Herald*, August 20, 1925.

87. William B. Riley, "'Choose You This Day': Between the Bible and Evolution," *Signs of the Times*, October 13, 1925.

88. W. E. Howell to George McCready Price, September 7, 1925, C2, box 1, folder 2, Correspondence, 1906–1925, Price Papers.

89. George McCready Price, *The Predicament of Evolution* (Nashville: Southern, 1925). The full text of the book is available at http://www.creationism.org/books/price/PredicmtEvol/.

90. "Books by George McCready Price," *Signs of the Times*, October 20, 1925, 15. For a recent use of illustrations from the book see Randall J. Stephens and Karl W. Giberson, *The Anointed: Evangelical Truths in a Secular Age* (Cambridge, MA: Harvard University Press, 2011), following 138.

91. Advertisement for *The Predicament of Evolution*, the *Present Truth*, December 1927, inside cover.

92. Price, *Predicament of Evolution*, 114–19.

93. "The Higher Criticism and Social Revolution," *New York Sun*, (Spring Literary Supplement), April 4, 1914, 8.

94. Beverly Gage, *The Day Wall Street Exploded: A Story of America in Its First Age of Terror* (New York: Oxford University Press, 2009).

95. Price, *Predicament of Evolution*, 119.

96. Price, 114–15.

97. Price, 116.

98. Price, 117–18, emphasis in original.

3. Blood Relationship, Bolshevism, and Whoopie Parties

1. Andrew Cunningham and Ole Peter Grell, *The Four Horsemen of the Apocalypse: Religion, War, Famine, and Death in Reformation Europe* (Cambridge: Cambridge University Press, 2000); R. L. Jones, "The Book of Revelation and War," *Advocate of Peace*, April 1874, 31.

2. Gerald B. Winrod, *The Red Horse* (Wichita, KS: Defender, 1932), 6–9, 12–14, box 25, FF 6, Rev. Dr. Gerald B. Winrod Papers, Wichita State University Libraries, Special Collections and University Archives, Wichita, KS (hereafter cited as Winrod Papers). The textbook cited by Winrod is D. E. Phillips, *Elementary Psychology: Suggestions for the Interpretation of Human Life* (Boston: Ginn, 1913).

3. "I Refuse to Claim a Blood-relationship with Such People—EVOLUTION is the Bunk!," *Defender*, July 1930, 8, in box 51, FF3, Winrod Papers.

4. George Barry O'Toole, *The Case against Evolution* (New York: Macmillan, 1925).

5. Emma Lou Thornbrough, *Indiana in the Civil War Era, 1850–1880* (Indianapolis: Indiana Historical Society, 1995); James Madison, *Hoosiers: A New History of Indiana* (Bloomington: Indiana University Press, and the Indiana Historical Society, 2014), 143–70.

6. Marie Acomb Riley, *The Dynamic of a Dream: The Life Story of Dr. William B. Riley* (Grand Rapids, MI: Wm. B. Eerdmans, 1938), 21. The author drew her information on Riley's birthplace from "A Sketch of My Life" done by Riley in 1931 for the Northwestern Bible School yearbook, wherein he identified his place of birth as "Green" County, Indiana. The correct spelling is Greene, named for Revolutionary War general Nathanael Greene. Rural Greene County (https://www.co.greene.in.us/) is located directly west of Monroe County, whose county seat is Bloomington, home of Indiana University.

7. Riley, *Dynamic of a Dream*, 197.

8. William Vance Trollinger, *God's Empire: William Bell Riley and Midwestern Fundamentalism* (Madison: University of Wisconsin Press, 1990), 11–13.

9. Trollinger, 14–18.

10. William Bell Riley, "Christ and Laboring Men," William Bell Riley Collection, Sermon and Pamphlet Collection, Haburn Hovda Archives, University of Northwestern–St. Paul (hereafter cited as Riley Collection).

11. William Bell Riley, *Christianity vs. Socialism* (Minneapolis, MN: L.W. Camp, c. 1911–12).

12. Riley, *Christianity vs. Socialism*, 10–13; "Dr. Riley on Socialism," *Duluth Herald*, February 23, 1912, reel 2: Scrapbooks, 1909–1923, Ephemera of William Bell Riley, Billy Graham Center, Wheaton College; "Challenge to Dr. Riley," *New Times*, n.d., reel 2: Scrapbooks, 1909–1923, Ephemera of William Bell Riley, Billy Graham Center, Wheaton College. For more on the *New Times* see John Haynes, "The New Times: A Frustrated Voice of Socialism, 1910–1919," *Minnesota History*, Spring 1991: 183–94.

13. Trollinger, *God's Empire*, 37–40.

14. William Bell Riley to Dr. Garten, June 14, 1916, Riley Collection; Trollinger, *God's Empire*, 89.

15. Trollinger, *God's Empire*, 93–95.

16. William Bell Riley, "The Theory of Evolution—Does It Tend to Anarchy?," in *Inspiration or Evolution*, 2nd ed. (Cleveland: Union Gospel, 1926), 51, 54, 61.

17. Riley, 55, 58, 62–63.

18. Riley, 61–62, 64. In the published version of the sermon, which appeared in *Inspiration or Evolution?* (1926), a "note" at the bottom of the last page directed readers who wanted to learn more to Price's *Poisoning Democracy*. Along with that note, the inclusion of the Bouck White "dynamite" story in Riley's sermon suggests he learned of it from reading Price, since it appeared first in Price's book. Among the books in Riley's personal library were five by Price. *Poisoning* was not among them, but they did include *The Predicament of Evolution* (1925), which also featured the White story. "Books from Dr. Riley's Library," *Riley Collection Index*, Riley Collection.

19. Riley, "Theory of Evolution," 64.

20. Ferenc M. Szasz, "William B. Riley and the Fight against Teaching of Evolution in Minnesota," *Minnesota History*, Spring 1969: 209, 212; "How and Why I Was Denied a Building on the University of Minnesota Campus for an Address on Evolution," March 7, 1926, Riley Collection.

21. Riley, "Evolution or Sovietizing the State through Its Schools," in *Inspiration or Evolution*, 94.

22. Riley, 95

23. Riley, 100.

24. Riley, 102–8.

25. Riley, 109–10.

26. "The Baptist Bible Union of America, First Conference in Kansas City, Mo," *Christian Fundamentals in School and Church*, April-May-June 1923, 6–7.

27. "Baptist Bible Union of America, First Conference."

28. T.T. Shields, "Why Some Individuals and Institutions Need to Be Blown Up with Dynamite," *Gospel Witness* 2 (January 20, 1924): 1–8; "Clergyman Calls Dr. Faunce Heretic," *New York Times*, December 4, 1923; Gerald L. Priest, "T.T. Shields, the Fundamentalist: Man of Controversy," *Detroit Baptist Seminary Journal* 10 (2005): 69–101.

29. "Burning the Image of Christ in Effigy," *Christian Fundamentals in School and Church*, April-May-June 1923, 12.

30. T.L. Blalock, *Experiences of a Baptist Faith Missionary for 56 Years in China* (Fort Worth, TX: Manney Printing, 1949).

31. T.L. Blalock, "The Cause of China's Confusion," *Christian Fundamentals in School and Church*, April-May-June 1923, 14.

32. James Reeve Pusey, *China and Charles Darwin* (Cambridge, MA: Council of East Asian Studies, Harvard University Press, 1983).

33. Blalock, "Cause of China's Confusion," 14–15.

34. Barry Hankins, *God's Rascal: J. Frank Norris and the Beginnings of Southern Fundamentalism* (Lexington: University Press of Kentucky, 1996), 9; Patsy Ledbetter, "Defense of the Faith: J. Frank Norris and Texas Fundamentalism, 1920–1929," in *Modern American Protestantism and Its World: Historical Articles on Protestantism in American Religious Life*, ed. Martin E. Marty, vol. 10, *Fundamentalism and Evangelicalism* (Munich: K.G. Saur, 1993), 173; Michael E. Schepis, *J. Frank Norris: The Fascinating, Controversial Life of a Forgotten Figure of the Twentieth Century* (Bloomington, IN: Westbow, 2012), 3, 10; E. Ray Tatum, *Conquest or Failure? Biography of J. Frank Norris* (Dallas: Baptist Historical Foundation, 1966).

35. David R. Stokes, *The Shooting Salvationist: J. Frank Norris and the Murder Trial That Captivated America* (Hanover, NH: Steerforth, 2011).

36. Nels Anderson, "The Shooting Parson of Texas," *New Republic*, September 1, 1926, 35–37.

37. Hankins, *God's Rascal*, 120.

38. Anderson, "Shooting Parson of Texas," 35.

39. Hankins, *God's Rascal*, 15, 18.

40. *Searchlight*, October 6, 1922.

41. "Address on Evolution before the Texas Legislature," *Searchlight*, February 23, 1923, 1–3.

42. Analyzing a similar performance at a mock trial of a Baylor professor whom Norris accused of teaching evolution, Michael Lienesch writes, "Its purpose was not legal but *political and psychological*, to encourage the audience to take more aggressive action against evolutionists." See Michael Lienesch, *In the Beginning: The Making of the Antievolution Movement* (Chapel Hill: University of North Carolina Press, 2007), 81, my emphasis.

43. "Address on Evolution before the Texas Legislature," 2.

44. Thomas E. Turner, "Neff, Pat Morris," Handbook of Texas Online (Denton: Texas State Historical Association, 2010), http://www.tshaonline.org/handbook/online/articles/fne05.

45. On the events in Herrin see Paul M. Angle, *Bloody Williamson: A Chapter in American Lawlessness* (New York: Alfred A. Knopf, 1952). On the UMWA in Illinois see Carl R. Weinberg, *Labor, Loyalty, and Rebellion: Southwestern Illinois Coal Miners and World War I* (Carbondale: Southern Illinois University Press, 2005).

46. Leo P. Ribuffo, *The Old Christian Right: The Protestant Far Right from the Great Depression to the Cold War* (Philadelphia: Temple University Press, 1983), 80–81.

47. Ribuffo, 88.

48. Gerald B. Winrod, *Christ Within* (Wichita, KS: Winrod Publication Center, 1925; 3rd ed. published by Christian Alliance, 1929).

49. Winrod, *Christ Within* (1925), 32–41.

50. Historians have paid increasing attention to the olfactory dimension of human experience. See Mark Smith, ed., *Smell and History: A Reader* (Morgantown: West Virginia University Press, 2019).

51. Winrod, *Christ Within* (1929), 24.

52. Winrod, *Christ Within* (1925), 54–69; *Christ Within* (1929), 91, on O'Toole.

53. Winrod, *Christ Within* (1929), 46–47; Gerald B. Winrod, "'Evolution' and the Great World Change," *Defender*, August 1926, 1.

54. Winrod, *Christ Within* (1925), 70–71. This passage is preserved in the third edition (1929), 111.

55. "Evolution Wrecks Youth," *Defender*, April 1926, 6.

56. "'Russia' in Universities," *Defender*, August 1927, 1.

57. "Russia's Mistake," *Defender*, October 1926, 5.

58. "Russia's Mistake."

59. Gerald B. Winrod, *Three Modern Evils: Modernism, Atheism, Bolshevism* (Wichita, KS: Defender, 1932).

60. Winrod, *Red Horse*, 6.

61. "Wild Joy in Moscow on Fall of Shanghai," *New York Times*, March 22, 1927, 1.

62. H. G. C. Hallock to R. Richards, March 22, 1927, Robert Dick Wilson Manuscript Collection, Presbyterian Church in America Historical Center, Covenant Theological Seminary, St. Louis, http://continuing.wordpress.com/tag/dr-h-g-c-hallock/.

63. "Bolshevism in China," *Defender*, September 1927, 3.

64. For a sample of Ham's preaching visit http://ia600403.us.archive.org/4/items/SER MONINDEX_SID0629/SID0629.mp3.

65. James T. Baker, "The Battle of Elizabeth City: Christ and Antichrist in North Carolina," *North Carolina Historical Review* 54 (October 1977): 393–94.

66. Baker, 394, 397.

67. Willard B. Gatewood Jr., *Preachers, Pedagogues and Politicians* (Chapel Hill: University of North Carolina Press, 1966), 43–48.

68. Gatewood, 42–43, 106–7; Lienesch, *In the Beginning*, 134–36; "Governor Vetoes Evolution and Board Cuts Out Books," *Raleigh News and Observer*, January 24, 1924, 1–2.

69. "No Hope in Welfare Work, Declares Evangelist Ham," *Raleigh News and Observer*, February 17, 1924, 1.

70. "Beheading 2,000 Teachers Would Have Saved Germany," *Raleigh News and Observer*, March 1, 1924, 1–2.

71. "Evangelist Ham Finds His Audience Fundamentalists," *Raleigh News and Observer*, February 29, 1924.

72. "Beheading 2,000 Teachers Would Have Saved Germany."

73. Keith Saunders, *The Independent Man: The Story of W. O. Saunders and His Delightfully Different Newspaper* (Edwards & Broughton, 1962); Michael Worthington, "W. O. Saunders: The Independent Man," http://highered411.com/Albemarle/Saunders.html.

74. Steven J. Zipperstein, *Pogrom: Kishinev and the Tilt of History* (New York: Liveright, 2018), 169–72.

75. Stephen Eric Bronner, *A Rumor about the Jews: Reflections on Antisemitism and the "Protocols of the Elders of Zion"* (New York: St. Martin's, 2000), 83–89.

76. Leo Ribuffo, "Henry Ford and *The International Jew*," *American Jewish History* 69 (June 1980): 437–77; Victoria Saker Woeste, *Henry Ford's War on Jews and the Legal Battle*

against Hate Speech (Stanford, CA: Stanford University Press, 2012); Joel Carpenter, *Revive Us Again: The Reawakening of American Fundamentalism* (New York: Oxford University Press, 1997), 102–3.

77. John Spargo, *The Jew and American Ideals* (New York: Harper & Row, 1921).

78. *The International Jew: The World's Foremost Problem*, vol. 1 (Dearborn, MI: Dearborn Publishing, 1920), 126.

79. *International Jew*, 1:191.

80. *International Jew*, 1:217–18.

81. *Jewish Influences in American Life*, vol. 3 (Dearborn, MI: Dearborn Publishing, n.d.), 65.

82. "Two Audiences of 5000 Hear Evangelist Sunday," *Daily Advance* (Elizabeth City, NC), October 20, 1924; "Ham Makes Direct Reply [to] Attack on Meeting Here," *Daily Advance*, November 12, 1924.

83. T. W. Callaway, "Father of Scopes Renounced Church; Was Staunch Socialist and Follower of Debs," *Chattanooga Daily Times*, July 10, 1925.

84. J. C. Schwarz, *Religious Leaders of America*, vol. 2 (New York: J. C. Schwarz, 1941), 182; B. J. W. Graham, *Baptist Biography*, vol. 1 (Atlanta: Index Printing, 1917), 63–66.

85. David Nelson Duke, *In the Trenches with Jesus and Marx: Harry F. Ward and the Struggle for Social Justice* (Tuscaloosa: University of Alabama Press, 2003); Doug Rossinow, "The Radicalization of the Social Gospel: Harry F. Ward and the Search for a New Social Order, 1898–1936," *Religion and American Culture: A Journal of Interpretation* 15 (Winter 2005): 63–106; Callaway, "Father of Scopes Renounced Church."

86. Quoted in Jeffrey Moran, *American Genesis: The Evolution Controversies from Scopes to Creation Science* (New York: Oxford University Press, 2012), 47.

87. Philip Hamburger, *Separation of Church and State* (Cambridge, MA: Harvard University Press, 2002), 406; T. W. Callaway, *Romanism vs. Americanism: The Roman Catholic System* (Atlanta: Index Printing, 1923).

88. "Evolution Question Looms as Important Question to Be Presented at Southern Baptists Convention Soon," *Kingsport (TN) Times*, May 2, 1926, 1.

89. "Bryan Memorial University to Graduate First Class," *Chattanooga News*, June 6, 1934; "Bryan Fundamentalist College Graduates Eight," *Fresno (CA) Bee*, June 14, 1934.

90. John Roach Straton, "Recent Books on Evolution," *American Fundamentalist*, September 27, 1925, 6.

91. Ronald Numbers, *The Creationists: From Scientific Creationism to Intelligent Design*, 2nd ed. (Cambridge, MA: Harvard University Press, 2006), 70–71.

92. Winrod, *Christ Within* (1925), 54–69; *Christ Within* (1929), 91, on O'Toole.

93. John D. Root, "The Final Apostasy of St. George Jackson Mivart," *Catholic Historical Review* 71 (January 1985): 1–25; Jacob W. Gruber, *A Conscience in Conflict: The Life of St. George Jackson Mivart* (New York: Columbia University Press, for Temple University Publications, 1960).

94. St. George Jackson Mivart, *On the Genesis of Species* (New York: Appleton, 1871), 16.

95. Patrick Allitt, *Catholic Converts: British and American Intellectuals Turn to Rome* (Ithaca, NY: Cornell University Press, 1997), 110.

96. J. A. Zahm, "Leo XIII and the Social Question," *North American Review* 161 (August 1895): 200–214.

97. David Mislin, "'According to His Own Judgment': The American Catholic Encounter with Organic Evolution, 1875–1896," *Religion and American Culture: A Journal of Interpretation* 22 (Summer 2012): 135, 148, 150–52; R. Scott Appleby, "Exposing Darwin's 'Hidden

Agenda': Roman Catholic Responses to Evolution, 1875–1925," in *Disseminating Darwinism: The Role of Place, Race, Religion, and Gender*, ed. Ronald L. Numbers and John Stenhouse (Cambridge: Cambridge University Press, 1999), 183–85.

98. R. Scott Appleby, "Between Americanism and Modernism: John Zahm and Theistic Evolution," *Church History* 56 (December 1987): 486–90; Appleby, "Exposing Darwin's 'Hidden Agenda,'" 185–94. See also Appleby, *Church and Age Unite! The Modernist Impulse in American Catholicism* (Notre Dame, IN: University of Notre Dame Press, 1992), 27–52; Pope Pius X, "Pascendi Dominici Gregis," September 8, 1907, Papal Encyclicals Online, http://www.papalencyclicals.net/Pius10/p10pasce.htm; C.J.T. Talar, "Pascendi dominici gregis: The Vatican Condemnation of Modernism," *U.S. Catholic Historian* 25 (Winter 2007): 1–12; William L. Portier, "Pascendi's Reception in the United States: The Case of Joseph McSorley," *U.S. Catholic Historian* 25 (Winter 2007): 13–30; Don O'Leary, *Roman Catholicism and Modern Science* (New York: Continuum, 2006), 116–18.

99. *Rerum Novarum Encyclical of Pope Leo XIII on Capital and Labor*, http://www.vatican.va/holy_father/leo_xiii/encyclicals/documents/hf_l-xiii_enc_15051891_rerum-novarum_en.html; Philip S. Foner, *History of the Labor Movement in the United States*, vol. 3, *The Policies and Practices of the American Federation of Labor, 1900–1909* (New York: International, 1964), 112–15.

100. David Goldstein and Martha Moore Avery, *Socialism: The Nation of Fatherless Children*, 2nd ed. (Boston: Thomas J. Flynn, 1911); Debra Campbell, "David Goldstein and the Rise of the Catholic Campaigners for Christ," *Catholic Historical Review* 72 (January 1986): 33–50; Owen Carrigan, "Martha Moore Avery: Crusader for Social Justice," *Catholic Historical Review* 54 (April 1968): 17–38; James M. O'Toole, *Militant and Triumphant: William Henry O'Connell and the Catholic Church in Boston, 1895–1944* (Notre Dame, IN: University of Notre Dame Press, 1992); Damien Murray, "'Go Forth as a Missionary to Fight It': Catholic Antisocialism and Irish American Nationalism in Post–World War I Boston," *Journal of American Ethnic History* 28 (Summer 2009): 43–65.

101. O'Toole, *Case against Evolution*, 126. As Ronald Numbers has noted, O'Toole's use of Price did not include an endorsement of flood geology, which is almost entirely absent from the book. Numbers, *Creationists*, 118.

102. O'Toole, *Case against Evolution*, 342.

103. O'Toole, 349–61; Giuseppe Tuccimei, "La teoria dell'evoluzione e le sue applicazioni," *Rivista Internazionale di Scienze Sociali e Discipline Ausiliarie* 11 (May 1896): 374–98.

104. Bertram C.A. Windle, "The Case against Evolution," *Commonweal*, June 10, 1925, 124–26; "The Case against Evolution," *Catholic Advance*, June 20, 1925, 8.

105. "Bryan, in His Fight with Evolution, Backed by a Professor of Biology, Scores Hard on American Museum," *Brooklyn Eagle*, May 17, 1925.

4. The Wolf Pack and the Upas Tree

1. Farrell Dobbs, *Teamster Rebellion* (New York: Pathfinder, 1972).

2. Donna T. Haverty-Stacke, *Trotskyists on Trial: Free Speech and Political Persecution since the Age of FDR* (New York: NYU Press, 2016).

3. William Bell Riley, "The Russian Boll-Weevil—Bolshevism," September 16, 1934, Riley Collection.

4. William Bell Riley, "Famishing Youth," c. 1935/36, Riley Collection.

5. At least we know that it appeared in Riley's personal library. See "Books from Dr. Riley's Library," *Riley Collection Index*, Riley Collection.

6. N.P. Foersch, translated from the Dutch by Mr. Heydinger, "Description of the Poison-Tree, in the Island of Java," *London Magazine*, December 1783, 511–17. Charles Darwin's grandfather Erasmus penned romantic, evolutionary verses devoted to this remarkable plant, whose gum was traditionally used by native peoples throughout South Asia and parts of Africa to make poison darts. See Erasmus Darwin, *The Botanic Garden: Part II, Loves of the Plants* (J. Johnson, St. Paul's Church-Yard, 1791).

7. Dan Gilbert, *Evolution: The Root of All Isms* (San Diego, CA: Danielle, 1935), 8, 24–25.

8. William Bell Riley, "'For Fear of the Jews,'" *Pilot*, July 1933, 299. The source of this quotation was not Hitler but Joseph Goebbels, a prominent Nazi Party leader who by 1933 was serving as Hitler's minister of propaganda. See Joseph Goebbels, "Warum sind wir Judengegner?" [Why are we enemies of the Jews?], in *Die verfluchten Hakenkreuzler. Etwas zum Nachdenken* (Munich: Franz Eher Nachfolger, 1930), 1–28.

9. William Bell Riley, "Why Recognize Russia and Rag Germany?," *Pilot*, January 1934, 110.

10. William Bell Riley, *Protocols and Communism* (Minneapolis, MN: L.W. Camp, 1934).

11. Matthew Avery Sutton, *American Apocalypse: A History of Modern Evangelicalism* (Cambridge, MA: Harvard University Press, 2014), 127–28; Michael Kazin, *A Godly Hero: The Life of William Jennings Bryan* (New York: Alfred A. Knopf, 2006), 273.

12. Riley, *Protocols and Communism*, 4.

13. Riley, 14, 15, 17.

14. Maria Mazzenga, ed., *American Religious Responses to Kristallnacht* (New York: Palgrave Macmillan, 2009), 1.

15. William Bell Riley, "The Nazi-Communist Battle," November 20, 1938, Riley Collection. Neither the source nor the significance of the number 541 is evident from Riley's sermon.

16. Matthew Bowman, "Persecution, Prophecy, and the Fundamentalist Reconstruction of Germany, 1933–1940," in Mazzenga, *American Religious Responses*, 183–204.

17. "Unmasking the 'Hidden Hand,'—A World Conspiracy," *Defender*, April 1933, 1; "Items," *Defender*, April 1933, 4.

18. Gerald Winrod, "Roosevelt, Hitler, and the Present Economic Collapse Considered in the Light of Prophecy," *Defender*, May 1933, 5.

19. "Items," *Defender*, November 1934, 4.

20. Matthew Avery Sutton, "Was FDR the Antichrist? The Birth of Fundamentalist Illiberalism in a Global Age," *Journal of American History* 98 (March 2012): 1052–74.

21. Gerald Winrod, *Adam Weishaupt: Human Devil* (Wichita, KS: Defenders, 1936); Gerald Winrod, "The Jew, International Finance, and the House of Rothschild," *Defender* 9, June 1934, 21. See also Robert Buroker Kemp, "The Political Theology of Hate: Gerald B. Winrod and Politicized Evangelicalism in America, 1927–1957," unpublished paper delivered at Organization of American Historians Midwestern Regional Conference, Ames, Iowa, August 5, 2000, in Winrod Papers.

22. "Nazi Propagandists Claim President Roosevelt Jewish," *Nevada State Journal*, June 8, 1934, 2.

23. "Roosevelt's Jewish Ancestry," *Revealer*, October 15, 1936, 1.

24. Jeffrey Herf, *The Jewish Enemy: Nazi Propaganda during World War II and the Holocaust* (Cambridge, MA: Harvard University Press, 2006), 106.

25. As historian Matthew Avery Sutton has noted, Gilbert has been "almost totally ignored by historians." See Sutton, "Was FDR the Antichrist?," 1070.

26. "Gilbert Family Selected for Lifetime Achievement," *Oakdale (CA) Leader*, January 13, 2016, http://www.oakdaleleader.com/archives/15748/; http://files.usgwarchives.net/ca/stanislaus/cemeteries/oakdalecitizens-dh.txt; Robert Gilbert, telephone conversation with

author, November 3, 2016; Monica Olmos, Admissions and Records, University of Nevada, Reno, to Anushka Mansukhani, email correspondence, November 14, 2016.

27. Dan Gilbert, "The Fundamentals Convention in Boston," *Sunday School Times*, July 4, 1942, 539, 548; "Radio Crusade Sues Evangelist's Estate," *Los Angeles Times*, June 8, 1963, B7; "Whither Are We Drifting? Extension of Remarks of Hon. Philip A. Bennett," 77th Congress, 2nd Sess., *Congressional Record* 88 (January 12, 1942): A97–98; Dan Gilbert, *Who Are the Enemies of Gospel Liberty in America?* (Washington, DC: Christian Press Bureau, 1942?); "Dan Gilbert," enclosure in Arnold Forster to ADL Regional Offices, February 4, 1949, box 26, Dan Gilbert folder, Jewish Community Relations Council Papers, Minnesota Historical Society, St. Paul.

28. Schmalhausen and Calverton clearly and publicly distinguished their politics from the liberalism of the *New Republic*. See Samuel D. Schmalhausen, V. F. Calverton, and Walter Lane (all editors of the *Modern Quarterly*), "Questions for Liberals," *New Republic*, November 9, 1927, 314.

29. Adam Shapiro, *Trying Biology: The Scopes Trial, Textbooks, and the Antievolution Movement in American Schools* (Chicago: University of Chicago Press, 2013), 67–75.

30. Upton Sinclair, *The Goslings: A Study of the American Schools* (Pasadena, CA: Upton Sinclair, 1924), 72–77; Teachers' Defense Fund, *The Trial of the Three Suspended Teachers of the De Witt Clinton High* School (New York: Teachers' Defense Fund, 1918), 40–42.

31. Leonard Wilcox, "Sex Boys in a Balloon: V. F. Calverton and the Abortive Sexual Revolution," *Journal of American Studies* 23 (April 1989): 11; "Deaths," *New York Times*, April 7, 1964, 35.

32. Dan Gilbert, *Crucifying Christ in Our Colleges*, 2nd ed. (San Diego, CA: Danielle, 1935), 44, 58, 64, 131.

33. Gilbert, 9, 132.

34. Donald Taylor, "The Initiation," *Antioch Review* 15 (Autumn 1955): 356–57.

35. Gould Wickey, "Trends Which Call for United Action," *Christian Education* 20 (December 1936): 122.

36. Gilbert, *Evolution*, 2, 4.

37. Gilbert, 28–29, 44–45.

38. William F. Dunne, *Why Hearst Lies about Communism: Three Open Letters to William Randolph Hearst* (New York: Workers Library, 1935), 2. The text of the pamphlet is available at http://www.redstarpublishers.org/#CPUSA. Dunne was a founding member of the Communist Labor Party and later edited the *Daily Worker*.

39. Despite a lack of documentation in Lenin's writings, it continues to appear on right-wing websites. See, for instance, http://www.sweetliberty.org/issues/wars/witness2history/3.html.

40. Gilbert, *Evolution*, 41.

41. Gilbert, 49, 51, 53, 54.

42. Gilbert, 55, 57–58, 61–62.

43. See Franklin H. Giddings, "The Social Lynching of Gorky and Andreiva," *Independent* 60 (April 26, 1906): 976–78.

44. Giddings, 65, 67.

45. Giddings, 70–73.

46. S. D. Schmalhausen, "Family Life: A Study in Pathology," in *The New Generation: The Intimate Problems of Modern Parents and Children*, ed. V. F. Calverton and S. D. Schmalhausen (New York: Macauley, 1930).

47. Dan Gilbert, "The Menace of 'Educational Totalitarianism,'" *Pilot*, December 1938, 73; Dan Gilbert, "The Modern Assault on the Home," *Pilot*, January 1939, 112–13.

48. Report on talk by Dan Gilbert, August 12, 1942, box 46, Rev. W. B. Riley, 1922–1942, Jewish Community Relations Council Papers, Minnesota Historical Society, St. Paul.

49. Margaret Lamberts Bendroth, *Fundamentalism and Gender, 1875 to the Present* (New Haven, CT: Yale University Press, 1993), 63.

50. Bendroth, 63; Joel Carpenter, *Revive Us Again: The Reawakening of American Fundamentalism* (New York: Oxford University Press, 1997), 282n60; Timothy Larsen, *Christabel Pankhurst: Fundamentalism and Feminism in Coalition* (Woodbridge, UK: Boydell, 2002), 126; Robert Wenger, *Social Thought in American Fundamentalism, 1918–1933* (Eugene, OR: Wipf and Stock, 1974), 278. As of March 1940, Knauss was fifty-five years of age, giving a birth year of c. 1885. See manuscript "Sixteenth Census of the United States: 1940," Davenport City, Iowa, 10A.

51. "Young People's Gospel Team Is Planning Meetings," *Davenport (IA) Democrat and Leader*, December 27, 1923, 11.

52. "Dr. Fields Talks in Fundamental Gathering Here," *Davenport Democrat and Leader*, April 27, 1928, 30; "Miss Knauss Has Most Successful Tour thru East," *Davenport Democrat and Leader*, February 17, 1928, 8; "At Hall's Grove Church," *Greene (IA) Recorder*, June 12, 1929, 8.

53. Walter T. Howard, ed., *Anthracite Reds: A Documentary History of Communists in Northeastern Pennsylvania during the Great Depression*, vol. 2 (Lincoln, NE: iUniverse, 2004), 65–66. The news report reprinted here misidentifies Knauss's state of origin as Louisiana (probably due to the similarity of IA and LA); Steve Nelson, James Barrett, and Rob Ruck, *Steve Nelson: American Radical* (Pittsburgh: University of Pittsburgh Press, 1981), 94–124.

54. Elizabeth Knauss, *The Menace of Bolshevism in America and throughout the World* (Chicago: Diligent, 1931).

55. Knauss, 3–4.

56. Knauss, 9.

57. Knauss, 9, 11.

58. Knauss, 11.

59. Ben Eklof, email to the author, October 10, 2014.

60. "The Question of Russian Recognition," 73rd Congress, 2nd Sess., *Congressional Record* 77 (April 12, 1933): S1538–39; "Senator Denounces Soviet Attacks on Religion and Home," *Catholic Advance* (Wichita, KS), March 14, 1931, 7; "Martyrs in Russia Estimated at 6,000," *New York Times*, March 18, 1930, 9.

61. "Preacher Tackles Fifth-Av. 'Heathen,'" *New York Times*, May 3, 1922, 19; "Former Moody Student Establishes Open Air Pulpit on Fifth Avenue, New York," *Moody Monthly*, July 1922, 1105; Edgar E. Strother, *A Bolshevized China—the World's Greatest Peril* (Shanghai: n.p., 1927).

62. Knauss, *Menace of Bolshevism*, 14.

63. Knauss, 14–23.

64. Knauss, 2.

65. Matthew Avery Sutton, *Aimee Semple McPherson and the Resurrection of Christian America* (Cambridge, MA: Harvard University Press, 2007), 4; Edith L. Blumhofer, *Aimee Semple McPherson: Everybody's Sister* (Grand Rapids, MI: Eerdmans, 1993); Daniel Mark Epstein, *Sister Aimee: The Life of Aimee Semple McPherson* (New York: Harcourt Brace, 1993); "Sister Aimee," American Experience (PBS), 2007, http://www.pbs.org/wgbh/amex/sister/.

66. Sutton, *Aimee Semple McPherson*, 44–46.

67. Sutton, 90–151.

68. Sutton, quoted on 207.

69. Sutton, quoted on 52.

70. Sutton, 248–54.

71. Sutton, 216; "Aimee Gets Overwhelming Decision as She Debates with Leader of Atheists," *Tampa Bay Times*, February 19, 1934.

72. "Reds Hammering at Our Gates," *Foursquare Crusader*, September 16, 1936, 1. Since the creature has only five, and not eight, tentacles, it is technically a quintipus.

73. Aimee Semple McPherson, *America, Awake!* (Los Angeles: Foursquare, c. 1934).

74. Sutton, *Aimee Semple McPherson*, 227–28.

75. Sutton, 7–8, 10, 27–32, 1–3.

76. Angela D. Dillard, *Faith in the City: Preaching Radical Social Change in Detroit* (Ann Arbor: University of Michigan Press, 2007), 132.

77. "Statement of Circulation of the Fundamentalist—Detroit, Michigan," 1944, box 15, F714, *Fundamentalist*, 1944, Norris Papers.

78. Ila Fleming to the *Fundamentalist*, May 28, 1936, box 15, folder 713, *Fundamentalist*, 1936, Norris Papers.

79. J. Frank Norris, *The Gospel of Dynamite* (n.p., 1935?); Bible Hub, http://biblehub.com/greek/1411.htm.

80. J. Frank Norris, "World-Wide Sweep of Russia Bolshevism, and Its Relation to the Second Coming of Christ," in *Gospel of Dynamite*.

81. Richard Wightman Fox, *Reinhold Niebuhr: A Biography* (New York: Pantheon Books, 1985), 88–110; Reinhold Niebuhr, *Leaves from the Notebook of a Tamed Cynic* (Chicago: Willett, Clark, & Colby, 1929).

82. Alan Brinkley, *Voices of Protest: Huey Long, Father Coughlin and the Great Depression* (1982; New York: Vintage Books, 1983), 199–203.

83. Quoted in Matthew Pehl, "'Apostles of Fascism,' 'Communist Clergy,' and the UAW: Political Ideology and Working-Class Religion in Detroit, 1919–1945," *Journal of American History* 99 (September 2012): 440; Zygmund Dobrzynski, "Raving Minister Uses Pulpit as a Mask to Promote Dictatorship," *United Automobile Worker*, June 11, 1938, 3.

84. J. Frank Norris, "Communistic Conspiracy of John L. Lewis to Destroy Present Economic System and Become the Joseph Stalin of U.S.A.," *Fundamentalist*, February 5, 1937.

85. Dan Gilbert, "The Rise of Beastism in America," *Moody Monthly*, September 1938, 14.

86. James M. Vincent, *The MBI Story: The Vision and Worldwide Impact of the Moody Bible Institute* (Chicago: Moody, 1911).

5. Beast Ancestry, Dangerous Triplets, and Damnable Heresies

1. "Texas Legislature Recognizes the Public Service Rendered by Dr. Norris," *Fundamentalist*, May 6, 1949. On Israel and anti-Semitism see J. Frank Norris to Harold Manson, American Zionist Emergency Council, December 19, 1947, box 22, folder 1009, Jews, 1947, Norris Papers; J. Frank Norris to Ben J. Goldman, Anti-Defamation League, February 19, 1948, box 22, folder 1010, Jews, 1948, Norris Papers; "A Message to Israel," signed by American Christian leaders including Norris, box 22, folder 1010, Jews, 1948, Norris Papers.

2. Barry Hankins, *God's Rascal: J. Frank Norris and the Beginnings of Southern Fundamentalism* (Lexington: University Press of Kentucky, 1996), 123.

3. Andrew Himes, *The Sword of the Lord: The Roots of Fundamentalism in an American Family* (Seattle: Chiara, 2011), 5.

4. Himes, 155–56.

5. Himes, 156.

6. Himes, 161–63, 204. See also John R. Rice, "Leaving All for Jesus," *Sword of the Lord* (newspaper, hereafter cited as *SotL*), April 11, 1952.

7. Himes, *Sword of the Lord*, 205; J. Frank Norris to William B. Riley, March 17, 1932, box 36, folder 1638, W. B. Riley, 1932, Norris Papers.

8. Judges 20; "Gideon," JewishEncyclopedia.com, http://www.jewishencyclopedia.com/articles/6664-gideon.

9. See, for instance, *SotL*, February 15, 1935.

10. John R. Rice, "Binghamton Revival Glorious," *SotL*, February 14, 1936; John R. Rice, "Norris Pentecostal as Rice in 1932," *SotL*, February 14, 1936; "Peace among Fundamentalists; How to Have It; Shall One Man with His Paper, Radio, Bible School and Paid Helpers Rule Fundamentalist Movement?," *SotL*, February 7, 1936. These issues include the new masthead language as well.

11. John R. Rice, "The Dance! Child of the Brothel, Sister of Gambling and Drunkenness, Mother of Lust—A ROAD TO HELL!," published sermon delivered June 9, 1935, box 14, folder 4, Papers of John R. Rice, Southwestern Baptist Theological Seminary, Fort Worth, Texas (hereafter cited as Rice Papers).

12. John R. Rice, "The Prevalence of Sex Sin," *SotL*, October 2, 1942; John R. Rice, "Divorce, Adultery, and Remarriage," *SotL*, October 2, 1936.

13. Rice, "Prevalence of Sex Sin"; Rice, "Divorce, Adultery, and Remarriage."

14. John R. Rice, *Bobbed Hair, Bossy Wives, and Women Preachers* (Murfreesboro, TN: Sword of the Lord, 1941), 13–14; John R. Rice, "Rebellious Wives and Slacker Husbands," *SotL*, July 22, 1938.

15. John R. Rice, "Editor's Family Moves to Wheaton, Ill.," *SotL*, April 19, 1940; Himes, *Sword of the Lord*, 218–19.

16. John R. Rice, "War: Human Wickedness and the Certain End of Civilization," *SotL*, June 2, 1940.

17. William Bell Riley, *Hitlerism; or, The Philosophy of Evolution in Action* (Minneapolis, MN: Irene Woods, 1941).

18. Riley, 2.

19. Howard Moore argues that J. Frank Norris and John R. Rice were fundamentally dissimilar: Norris was a "demagogue" who was primarily looking to boost his own fortunes, while Rice was a "theologically driven man." While there is some truth to both characterizations, this assessment misses the point that in terms of their politics, Rice and Norris were remarkably similar. As both men noted more than once, theology and politics were inextricably intertwined. The willingness of both Norris and Rice to collaborate, in the service of anticommunism, with Presbyterian separatist Carl McIntire in the 1940s and early 1950s is a good case in point. See Howard Moore, "The Emergence of Moderate Fundamentalism: John R. Rice and 'The Sword of the Lord'" (PhD diss., George Washington University, 1990), 110–15.

20. Matthew Avery Sutton, *American Apocalypse: A History of Modern Evangelicalism* (Cambridge, MA: Harvard University Press, 2014), 156–57; Darryl G. Hart, *The Lost Soul of American Protestantism* (Lanham, MD: Rowman & Littlefield, 2004), 88–104; Markku Ruotsila, *Fighting Fundamentalist: Carl McIntire and the Politicization of American Fundamentalism* (New York: Oxford University Press, 2016); John Fea, "Carl McIntire: From Fundamentalist Presbyterian to Presbyterian Fundamentalist," *American Presbyterian* 72 (Winter 1994): 253–68; and Heather Hendershot, "God's Angriest Man: Carl McIntire, Cold War Fundamentalism, and Right-Wing Broadcasting," *American Quarterly* 59 (June 2007): 373–96.

21. Ruotsila, *Fighting Fundamentalist*, 8, 291.

22. John R. Rice, "The McIntire Articles," *SotL*, April 8, 1949.

23. Carl McIntire, "Private Enterprise in the Scriptures," *SotL*, September 1945. See also Carl McIntire, "God's Proverbs Teach Free Private Enterprise; the Bible Answers Communists and Socialists," *SotL*, October 19, 1945.

24. See, for instance, "Dr. Bob Jones Says—the Business of the Church Is to Bear Fruit, That Is, Win Souls," *SotL* [September or December] 17, 1948; Evangelist James V. Lamb, "Gambling's Rotten Family Tree," *SotL*, May 6, 1949; William C. Irvine, "Modernism" (section titled "Its Fruitage"), *SotL*, September 5, 1952; W.C. Moore, "No Fruit—Cut It Down!," *SotL*, November 28, 1952.

25. John R. Rice, "More about Modernism," *SotL*, June 20, 1947. Darwin himself was well aware of the non-heritability of shortened foreskins: "With respect to Jews, I have been assured by three medical men of the Jewish faith that circumcision, which has been practiced for so many ages, has produced no inherited effect." Charles Darwin, *The Variation of Animals and Plants under Domestication*, vol. 1 (London: John Murray, 1905), 558. I am indebted to Ronald Numbers for pointing me to this reference.

26. John R. Rice, "Who for President?," *SotL*, September 1948.

27. Raymond Edman, "Karl Marx or Jesus Christ?," *SotL*, July 15, 1949.

28. John R. Rice, "Northwestern Schools and Minneapolis Celebrate Dr. W.B. Riley's 86th Birthday," *SotL*, April 11, 1947.

29. William Bell Riley, "Atheism, the Enemy of Civilization," *SotL*, July 25, 1952.

30. Billy Graham, *Just as I Am: The Autobiography of Billy Graham* (New York: HarperCollins, 1997), 22. Graham's truthful acknowledgment that Ham was known as a Jew-hater in the 1930s has unfortunately failed to be reflected in the most recent scholarly account of Graham's life. In his biography of Graham, historian Grant Wacker describes Ham as "a classic Southern barnstormer, *later* associated with anti-Semitism and right-wing causes but in 1934 known mostly for excoriating immorality, upbraiding lax clergy, and calling lost souls to Christ." Whatever Wacker's intention, the inaccurate dating of Ham's anti-Semitism to a period subsequent to the time he converted Billy Graham has the effect of obscuring the possible roots of Graham's own anti-Semitic views. See Grant Wacker, *America's Pastor: Billy Graham and the Shaping of a Nation* (Cambridge, MA: Harvard University Press, 2014), 6 (emphasis added).

31. Wacker, *America's Pastor*, 194–98.

32. Graham, *Just as I Am*, 22–27.

33. Graham, 61–63.

34. Molly Worthen, *Apostles of Reason: The Crisis of Authority in American Evangelicalism* (New York: Oxford University Press, 2014), 56.

35. See, for example, "Detailed Daily Program of Sword Conference on Evangelism, November 21–28, Chicago Gospel Tabernacle," *SotL*, November 5, 1948. In his autobiography, Graham refers to John R. Rice twice. First, in the context of a 1955 evangelistic trip the two men took to Scotland, Graham calls Rice "my old friend." Second, in a later passage about the break that the old guard fundamentalists made with Graham in 1956, Graham groups Rice with Bob Jones and Carl McIntire. From these brief mentions alone, one would never know just how close the two men had been during their time in Wheaton. Graham, *Just as I Am*, 253, 302.

36. "The Editor Says," *SotL*, September 24, 1948.

37. Billy Graham, "The American Crisis and the Needed Revival," *SotL*, January 7, 1949. Graham preached this address on November 22, 1948, at the Sword of the Lord Conference on Evangelism in Chicago.

38. Billy Graham, "Prepare to Meet Thy God," *SotL*, May 26, 1950; Donald E. Hoke, "Knowledge on Fire," *Christian Life*, July 1949, 9–12.

39. Telegram from J. Frank Norris to First Baptist Church, Jacksonville, Florida, February 27, 1950, box 17, folder 778, Billy Graham, 1950, Norris Papers; receipt of gift from

J. Frank Norris to Billy Graham Evangelistic Association, January 25, 1951, box 17, folder 779, Billy Graham, 1951, Norris Papers; J. Frank Norris to Dr. Billy Graham, March 1, 1950, box 17, folder 778, Billy Graham, 1950, Norris Papers.

40. Billy Graham to J. Frank Norris, July 7, 1950, box 17, F778, Billy Graham, 1950, Norris Papers.

41. Telegram from J. Frank Norris to First Baptist Church, Fort Worth, Texas, May 7, 1947, box 29, folder 1354, Louie D. Newton, 1947, Norris Papers; "Innocent Abroad," *Time*, August 26, 1926; "Communist Party Uses Oxnam-Newton Propaganda," *Christian Beacon*, January 15, 1948; "Louie D. Newton (1892–1986)," New Georgia Encyclopedia (2007), https://www.georgiaencyclopedia.org/articles/arts-culture/louie-d-newton-1892-1986.

42. J. Frank Norris to Carl McIntire, July 10, 1952, box 28, folder 1321, Carl McIntyre [*sic*], 1952, Norris Papers.

43. James H. Wigton, *Lee Roberson: Always about His Father's Business* (Maitland, FL: Xulon, 2010), 292, 355–56; Moore, "Emergence of Moderate Fundamentalism," 252; "Sword of the Lord Annual Conference on Revival and Soul Winning at Incomparable Lake Louise, Toccoa, Georgia, July 7–11," *SotL*, April 18, 1952; "The Editor's Travels," *SotL*, March 21, 1952.

44. John R. Rice, *Dangerous Triplets: 1. Russian Communism, 2. New-Deal Socialism, 3. Bible-Denying Modernism* (Murfreesboro, TN: Sword of the Lord, 1960).

45. Nikolai Krementsov, "Darwinism, Marxism, and Genetics in the Soviet Union," in *Biology and Ideology: From Descartes to Dawkins*, ed. Denis R. Alexander and Ronald L. Numbers (Chicago: University of Chicago Press, 2010), 242.

46. As one example of Rice's selective quotation style, the sermon quotes from Marx and Engels, as follows: "The bourgeois claptrap about the family and education, about the hallowed co-relation of parent and child, becomes all the more disgusting . . . by the action of Modern Industry." The original adds, "all family ties are torn asunder, and their children transformed into simple articles of commerce and instruments of labor." This clarifies the sense in which Marx and Engels are disgusted. Similarly, on the point about women, Rice quotes Marx and Engels as writing that "the communists have no need to introduce community of women; it has existed almost from time immemorial." He then paraphrases what they say next to mean that Marx and Engels would "legalize" the community of women and call it communism. What they actually write is that "it is self-evident that the abolition of the present system of production [capitalism] must bring with it the abolition of the community of women springing from that system, i.e., of prostitution both public and private." Karl Marx and Frederick Engels, *The Communist Manifesto* (New York: Pathfinder, 2008), 53–54.

47. Walter Hearn, "An Interview with Bernard Ramm and Alta Ramm," *Journal of the American Scientific Affiliation* 31 (September 1979): 179–86.

48. On the development of Henry's perspective see George Marsden, *Reforming Fundamentalism: Fuller Seminary and the New Evangelicalism* (Grand Rapids, MI: W. B. Eerdmans, 1987), 75–82; Worthen, *Apostles of Reason*, 18–19, 24–27.

49. Bernard Ramm, *The Christian View of Science and Scripture* (Grand Rapids, MI: Wm. B. Eerdman's, 1954), 28.

50. Ramm, 34, 69, 76, emphasis in the original.

51. Jeffrey Moran, *The Scopes Trial: A Brief History with Documents* (New York: Bedford / St. Martin's, 2002), 146–49.

52. Ramm, *Christian View*, 159–60.

53. Ramm, 175–76, 180, 222, 227–28.

54. Ramm, 272, 292–93.

55. Ramm, 262, 301.

56. Quoted in Moore, "Emergence of Moderate Fundamentalism," 235.

57. Moore, 248–56.

58. See Jeffrey Woods, *Black Struggle, Red Scare: Segregation and Anticommunism in the South, 1948–1968* (Baton Rouge: LSU Press, 2004), and George Lewis, *The White South and the Red Menace: Segregationist Anticommunism and Massive Resistance, 1945–1965* (Gainesville: University Press of Florida, 2004).

59. Jane Dailey, "Sex, Segregation, and the Sacred after *Brown*," *Journal of American History* 91 (June 2004): 119–44.

60. Ralph McGill, "One Southerner's Viewpoint," *Atlanta Constitution*, May 13, 1947, and Ranald McDonald to J. Frank Norris, May 22, 1947, both in box 29, folder 1354, Louie D. Newton, 1947, Norris Papers; Barbara Barksdale Clowse, *Ralph McGill: A Biography* (Macon, GA: Mercer University Press, 1998).

61. Dailey, "Sex, Segregation, and the Sacred," 125. Dailey writes that "the argument that God was against sexual integration was articulated across a broad spectrum of education and respectability, by senators and Ku Klux Klansmen, by housewives, sorority sisters, and Rotarians, and, not least of all, by mainstream Protestant clergymen."

62. Glenn Feldman, "Southern Disillusionment with the Democratic Party: Cultural Conformity and the 'Great Melding' of Racial and Economic Conservatism in Alabama during World War II," *Journal of American Studies* 43 (August 2009): 227–28; Glenn Feldman, *From Demagogue to Dixiecrat: Horace Wilkinson and the Politics of Race* (Lanham, MD: University Press of America, 1995).

63. In later decades the term was also spun into a branch of the sciences: baraminology. Ronald Numbers, *The Creationists: From Scientific Creationism to Intelligent Design*, 2nd ed. (Cambridge, MA: Harvard University Press, 2006), 150, 391.

64. "Noted Jurist Discusses This Matter of Segregation," *Alabama Baptist*, February 5, 1948, 8–9.

65. T. Laine Scales, *All That Fits a Woman: Training Southern Baptist Women for Charity and Mission, 1907–1926* (Macon, GA: Mercer University Press, 2000), 223–25.

66. Daniel Murph, "Daniel, Marion Price, Sr.," Handbook of Texas, https://tshaonline.org/handbook/online/articles/fda94; Dwonna Goldstone, *Integrating the 40 Acres: The 50-Year Struggle for Racial Equality at the University of Texas* (Athens: University of Georgia Press, 2006), 14–35.

67. Benjamin Marquez, *Democratizing Texas Politics: Race, Identity, and Mexican American Empowerment, 1945–2002* (Austin: University of Texas Press, 2014), 47.

68. Carey Daniel, *God the Original Segregationist* (Dallas: Carey Daniel, 1955); "Segregationist Tirade against 'Fools' and 'Traitors,'" *Life*, February 7, 1964; Edward H. Miller, *Nut Country: Right-Wing Dallas and the Birth of the Southern Strategy* (Chicago: University of Chicago Press, 2015), 71–74.

69. For scholarly speculation on the historical and/or mythological roots of Nimrod see K. van der Toorn and P. W. van der Horst, "Nimrod before and after the Bible," *Harvard Theological Review* 83 (January 1990): 1–29; Yigal Levin, "Nimrod the Mighty, King of Kish, King of Sumer and Akkad," *Vetus Testamentum* 52 (July 2002): 350–66.

70. Glenda Gilmore, *Defying Dixie: The Radical Roots of Civil Rights, 1919–1950* (New York: W. W. Norton, 2009).

71. Daniel, *God the Original Segregationist*, 2.

72. "Words & Works," *Time*, November 5, 1956, 70; "Church Offered as Dallas School," *Progress-Index* (Petersburg, VA), October 25, 1956, 1.

6. Flood, Fruit, and Satan

1. Ronald L. Numbers, *The Creationists: From Scientific Creationism to Intelligent Design* (Cambridge, MA: Harvard University Press, 2006), 234.

2. Numbers, 265.

3. John C. Whitcomb Jr., "The History and Impact of the Book, The Genesis Flood," manuscript sent by Whitcomb to author, also available online as "The History and Impact of the Book, The Genesis Flood, Part I," https://answersingenesis.org/ministry-news/ministry/the-history-and-impact-of-the-book-the-genesis-flood-part-1/.

4. Jared S. Burkholder, "Vigilance: The Fundamentalism of Herman Hoyt and John Whitcomb," in *Becoming Grace: Seventy-Five Years on the Landscape of Christian Higher Education in America*, ed. Jared S. Burkholder and M. M. Norris (Winona Lake, IN: BMH Books, 2015), 75.

5. Paul J. Scharf, "A Biographical Tribute to Dr. John C. Whitcomb, Jr.," in *Coming to Grips with Genesis: Biblical Authority and the Age of the Earth*, ed. Terry Mortensen and Thane H. Ury (Green Forest, AR: Master Books, 2008), 437–52; John C. Whitcomb, "Priorities in Presenting the Faith (The Conversion of an Evolutionist)," unpublished paper, c. 2007, http://www.whitcombministries.org/uploads/1/3/8/9/13891775/priorities_paper_the_conversion_of_an_evolutionist.pdf.

6. Numbers, *Creationists*, 217; Henry M. Morris, *History of Modern Creationism*, 2nd ed. (Santee, CA: Institute for Creation Research, 1993), 103.

7. "Rice Topics," *Thresher*, October 12, 1923.

8. Morris, *History of Modern Creationism*, 103–4; Numbers, *Creationists*, 217–18.

9. Numbers, *Creationists*, 218–19; Morris, *History of Modern Creationism*, 99–101.

10. Morris, *History of Modern Creationism*, 88.

11. John Morris, interview with the author, Dallas, May 22, 2013.

12. Henry Morris, *That You Might Believe* (Chicago: Good Books, 1946), iii, 1–2.

13. Morris, 2.

14. Henry Morris, *The Bible and Modern Science* (Chicago: Moody Institute, 1956).

15. Morris, *That You Might Believe*, 25–57; Morris, *Bible and Modern Science*.

16. Morris, *Bible and Modern Science*, 31.

17. "Mrs. Marrs Says Gilbert Father of Her Unborn Child," *Redlands (CA) Daily Facts*, August 25, 1962; "Jury Finds Marrs Innocent of Murder," *San Bernardino (CA) County Sun*, October 12, 1962.

18. Morris, *History of Modern Creationism*, 64n2; Numbers, *Creationists*, 220–21.

19. Numbers, *Creationists*, 195–98.

20. Morris, *History of Modern Creationism*, 164–65.

21. John C. Whitcomb, "The Genesis Flood: An Investigation of Its Geographical Extent, Geologic Effects, and Chronological Setting" (PhD diss., Grace Theological Seminary, 1957), iv–v.

22. Whitcomb, "Genesis Flood," 175–76, 197, 200, 267–68.

23. Numbers, *Creationists*, 213–17, 222.

24. Henry Morris to Harold Clark, November 19, 1964, C2, box 2, folder 4, Correspondence about George McCready Price's Biography, Price Papers.

25. Numbers, *Creationists*, 215–16, 223–24; John Morris interview. In contrast to Morris and Whitcomb's sketchy treatment of Price in *Genesis Flood*, Morris not only cited Price in the chapter on evolution in *That You Might Believe*, but cited books that were explicitly aimed at an Adventist audience, such as George McCready Price, *Genesis Vindicated* (Takoma Park, MD: Review and Herald, 1941).

26. Bethany Moreton, *To Serve God and Wal-Mart: The Making of Christian Free Enterprise* (Cambridge, MA: Harvard University Press, 2009), 163–67.

27. "Formation of Evolution Protest Movement," in Morris, *History of Modern Creationism*, appendix D, 412.

28. Numbers, *Creationists*, 174; Paul D. Haynie, "James David Bales (1915–1995)," CALS Encyclopedia of Arkansas, http://www.encyclopediaofarkansas.net/encyclopedia/entry-detail.aspx?entryID=4724.

29. James Bales to George McCready Price, January 8, 1954, and August 27, 1959, box 2, folder 2, AHC, Price Papers.

30. John C. Whitcomb and Henry M. Morris, *The Genesis Flood: The Biblical Record and Its Scientific Implications* (Philadelphia: Presbyterian and Reformed, 1961), 440.

31. Whitcomb and Morris, 440–41.

32. Whitcomb and Morris, 443–47.

33. John Morris interview; Henry M. Morris, *History of Modern Creationism*, 184.

34. Attachment in Henry Morris to George McCready Price, January 9, 1963, box 2, folder 2, AHC, Price Papers; Henry M. Morris, *Biblical Catastrophism and Geology* (Philadelphia: Presbyterian and Reformed, 1963), 12–13, emphasis added.

35. Morris, *History of Modern Creationism*, 178; Henry M. Morris, *The Twilight of Evolution* (Grand Rapids, MI: Baker Book House, 1963), 27. Ronald Numbers gives a slightly different account, based on Morris's correspondence, suggesting that there were no questions because the audience was, in Morris's words, "too stunned to speak." Numbers, *Creationists*, 235–36.

36. "Nationalist U-2 Downed by Reds over East China" and "Cubans in U.S. Ask Anti Castro Help," *New York Times*, September 1, 1962; "Reporter Fires Questions at Cuba; Castro Will Not Start Attack on U.S.," *El Paso Herald-Post*, September 10, 1962.

37. Henry Morris to George McCready Price, January 9, 1963, box 2, folder 2, AHC, Price Papers.

38. Morris, *Twilight of Evolution*, 13–24.

39. Morris, 77, 83, 94.

40. Numbers, *Creationists*, 240–41.

41. Morris, *History of Modern Creationism*, 195–212.

42. Morris, 209–10, 220–21.

43. Morris, 211.

44. "Don't Want Reds to Speak," *Holland (MI) Evening Sentinel*, January 7, 1961; "Prof Claims MSU Spreads Socialism," *News-Palladium* (Benton Harbor, MI), December 6, 1961; "Have Americans Lost Regard for the Stars and Stripes?," *News-Palladium*, June 15, 1962; "DAR Chapter Receives Honors at Conference," *Evening Independent* (Massillon, OH), March 18, 1967; "GOP Right Wing Studies Strategy," *Hillsdale (MI) Daily News*, July 29, 1963; "Dr. John N. Moore," *Hillsdale Daily News*, May 27, 1964.

45. John N. Moore, "Neo-Darwinism and Society," *Creation Research Society Quarterly* 2 (January 1966): 13–20.

46. John N. Moore, *Questions and Answers on Creation/Evolution* (Grand Rapids, MI: Baker Book House, 1976), 87; John N. Moore, "Should Evolution Be Taught?," *Creation Research Society Quarterly* (September 1970): 112. Oddly, the diagram included in the latter 1970 piece does not correspond to written material in the article itself. It does mesh nicely, however, with his "Neo-Darwinism and Society," containing as it does entries for Marx, Keynes, Beard, London, and Shaw, among others.

47. Zola Levitt and Thomas S. McCall, *The Coming Russian Invasion of Israel* (Chicago: Moody, 1987); Randall Balmer, "Zola Levitt," *Encyclopedia of Evangelicalism* (Louisville, KY: Westminster John Knox, 2002), 336–37; Yaakov Shalom Ariel, *Evangelizing the Chosen People: Missions to the Jews in America, 1880–2000* (Chapel Hill: University of North Carolina Press, 2000), 285.

48. Zola Levitt, *Creation: A Scientist's Choice* (Wheaton, IL: Victor Books, 1976).

49. Levitt, 8–11.

50. Numbers, *Creationists*, 247.

51. Numbers, 259–60.

52. Bolton Davidheiser, *Evolution and Christian Faith* (Grand Rapids, MI: Baker Book House, 1969), 203–5.

53. Julie J. Ingersoll, *Building God's Kingdom: Inside the World of Christian Recon-structionism* (New York: Oxford University Press, 2015), 119–39. Ingersoll does trace links between Rushdoony, Reconstructionism, and young-earth creationism but does not treat Rushdoony's writings in *CRSQ*.

54. Molly Worthen, "The Chalcedon Problem: Rousas John Rushdoony and the Origins of Christian Reconstructionism," *Church History* 77 (June 2008): 402. Worthen and others report only that Rushdoony received his BA in 1938, but since he turned eighteen four years earlier, in 1934, it is reasonable to infer he started that fall.

55. Robert E. Lerner, *Ernst Kantorowicz: A Life* (Princeton, NJ: Princeton University Press, 2017).

56. Michael J. McVicar, *Christian Reconstructionism: R.J. Rushdoony and American Religious Conservatism* (Chapel Hill: University of North Carolina Press, 2015), 19–25, 28–31, 34–40.

57. Worthen, "Chalcedon Problem," 409.

58. McVicar, *Christian Reconstructionism*, 46–78.

59. Worthen, "Chalcedon Problem," 407–9.

60. Numbers, *Creationists*, 224–25, 232; Morris, *History of Modern Creationism*, 173–75; John C. Whitcomb, "Priorities in Presenting the Faith (The Conversion of an Evolutionist)," in *Dispensationalism Tomorrow and Beyond: A Theological Collection in Honor of Charles C. Ryrie*, ed. Christopher Cone (Fort Worth, TX: Tyndale Seminary, 2008), 33–45.

61. For an insightful analysis of Freud's use of Lamarck see Eliza Slavet, "Freud's 'La-marckism,' and the Politics of Racial Science," *Journal of the History of Biology* 41 (2008): 37–80. Rushdoony's *CRSQ* article was based on R.J. Rushdoony, *Freud* (Nutley, NJ: Presbyterian and Reformed, 1965).

62. R.J. Rushdoony, "The Premises of Evolutionary Thought," *CRSQ* 2 (July 1965): 15–18.

63. R.J. Rushdoony, *The Necessity for Creationism* (Grand Rapids, MI: Puritan Reformed Theological Seminary, 1967?); Numbers, *Creationists*, 264.

64. George R. Howe, review of *The Necessity of Creation* (filmstrip, 1966), *CRSQ* 9 (June 1972): 80–82; Elizabeth E. Seittelman, "Teaching with the Filmstrip," *Classical World* 55 (January 1962): 112–14, 116.

65. Ingersoll, *Building God's Kingdom*, 79–118; McVicar, *Christian Reconstructionism*, 163–75.

66. "Fact-File: David A. Noebel, Tulsa, Oklahoma," file 48, David Noebel, box 3, Billy James Hargis Papers, University of Arkansas Libraries, Fayetteville (hereafter cited as Hargis Papers); "Lakeview Church Conference Scene," *Oshkosh (WI) Daily Northwestern*, September 25, 1963, 34; "Young Bereans Begin Broadcast Preparations," *Oshkosh Daily Northwestern*, September 9, 1954; "Young Women Are Accompanists for Capital Glee Club," *Oshkosh Daily Northwestern*, January 30, 1954.

67. Sarah Posner, "McCarthy, Born Again and Retooled for Our Time," *Religion Dispatches*, December 13, 2010, http://religiondispatches.org/mccarthy-born-again-and-re tooled-for-our-time/; Cornelis de Waal, "Arthur C. Garnett," in *Dictionary of Modern American Philosophers*, ed. John R. Shook (Bristol, UK: Thoemmes Continuum, 2005), 884–85. Among Noebel's favorite professors was historian of philosophy Julius R. Weinberg. David Noebel, email message to the author, May 3, 2013. On Weinberg see Keith E. Yandell, "Weinberg, Julius Randolph (1908–71)," in Shook, *Dictionary of Modern American Philosophers*, 2541–43.

68. "Faith Baptist Men Will Meet Tuesday," *Wisconsin State Journal* (Madison), February 7, 1961, 4; Rick Perlstein, *Before the Storm: Barry Goldwater and the Unmaking of*

the American Consensus (New York: Nation Books, 2001), 101–3; Seth Rosenfeld, *Subversives: The FBI's War on Student Radicals, and Reagan's Rise to Power* (New York: Farrar, Straus and Giroux, 2012), 94–97. For the original film and a filmic rebuttal to it from the northern California branch of the ACLU see John De Looper, "Operation Abolition and Operation Correction," Reel Mudd, Princeton University, October 19, 2010, https://blogs.prince ton.edu/reelmudd/2010/10/operation-abolition-and-operation-correction/.

69. "Rev. David Noebel Joins GOP Contest for Congress Seat," *Wisconsin State Journal*, May 12, 1962, 7; "Noebel on Commies," *Capital Times* (Madison, WI), August 21, 1962, 14.

70. "Kindschi Clinches Second District Bid," *Wisconsin State Journal*, September 12, 1962, 1.

71. Laura Jane Gifford, "'Girded with a Moral and Spiritual Revival': The Christian Anti-Communism Crusade and Conservative Politics," in *The Right Side of the Sixties: Reexamining Conservatism's Decade of Transformation* (New York: Palgrave Macmillan, 2012), 161–79; Hubert Villeneuve, *Teaching Anticommunism: Fred Schwarz and American Postwar Conservatism* (Montreal: McGill–Queen's University Press, 2020).

72. Fred Schwarz, "Philosophy of Communism, 1963," https://vimeo.com/18222133.

73. Gifford, "'Girded with a Moral and Spiritual Revival,'" 164–65.

74. US Congress, House, Committee on Un-American Activities, 85th Cong., 1st Sess., May 29, 1957, *International Communism (The Communist Mind), Staff Consultation with Frederick Charles Schwarz* (Washington, DC: Government Printing Office, 1957), 3, 8–9.

75. John Gurda, "One Union's Demise: The End of Local 1111 Should Prompt Serious Questions about the Economy," *Milwaukee Journal-Sentinel*, July 30, 2010, http://www.jsonline.com/news/opinion/99660119.html; Ronald L. Filippelli and Mark McColloch, *Cold War in the Working Class: The Rise and Decline of the United Electrical Workers* (Albany: SUNY Press, 1995), 156–57; Robert W. Ozanne, *The Labor Movement in Wisconsin: A History* (Madison: State Historical Society of Wisconsin, 1984), 86; Kim Phillips-Fein, *Invisible Hands: The Businessmen's Crusade against the New Deal* (New York: W. W. Norton, 2009), 60–67; John Gurda, *The Bradley Legacy: Lynde and Harry Bradley, Their Company, and Their Foundation* (Milwaukee: Lynde and Harry Bradley Foundation, 1992), 114–18; Jane Mayer, *Dark Money: The Hidden History of the Billionaires behind the Rise of the Radical Right* (New York: Doubleday, 2016), 112–19.

76. "Will You Be Free to Celebrate Christmas in the Future?," full-page ad containing the text of Fred Schwarz's testimony, *Milwaukee Sentinel*, December 22, 1957, 4; Kevin M. Kruse, *One Nation under God: How Corporate America Invented Christian America* (New York: Basic Books, 2015), 150; Fred C. Schwarz, "Operation Testimony," *CACC Newsletter*, May 1958.

77. Kruse, *One Nation under God*, 151–58; Steven J. Ross, *Hollywood Left and Right: How Movie Stars Shaped American Politics* (New York: Oxford University Press, 2011), 163.

78. "Billy James Hargis, 79, Pastor and Anticommunist Crusader," *New York Times*, November 29, 2004; Billy James Hargis, *Why I Fight for a Christian America* (Nashville, TN: Thomas Nelson, 1974), 12–13.

79. Hargis, *Why I Fight*, 26–31; Fernando Penabaz, *"Crusading Preacher from the West": The Story of Billy James Hargis* (Tulsa, OK: Christian Crusade, 1965), 67.

80. Penabaz, *"Crusading Preacher,"* 12–36, 147–49.

81. David Settje, *Lutherans and the Longest War: Adrift on a Sea of Doubt about the Cold and Vietnam Wars, 1964–1975* (Lanham, MD: Lexington Books, 2007), 76–83; "Cleric Tells of Communist Torture," *New York Times*, May 7, 1966; Richard Wurmbrand, "In a Certain City," *Christian Crusade*, August–September 1967, 11; "By Billy James Hargis," *Christian Crusade*, August–September 1967, 14–28; "Wurmbrand in Washington," *Christian Crusade*, December 1967, 15.

82. Hargis, *Why I Fight*, 24–25.

83. Jobe Martin and Don Baker, "Noebel Cause: A Brief History of Summit Ministries," in *A Summit Reader: Essays and Lectures in Honor of David Noebel's 70th Birthday*, ed. Michael Bauman and Francis Beckwith (Manitou Springs, CO: Summit, 2007), 5–17; "Director Reports on First Sessions of Anti-Communist Youth University," *Christian Crusade*, June–July 1963, 18, folder 2, "Christian Crusade," box 48, Winrod Papers; "Why You Should Act Now about the Summit," *Christian Crusade*, May 1966, 24, folder 4, box 54, Hargis Papers.

84. Martin and Baker, "Noebel Cause," 12. For an earlier, non-fundamentalist reading of the same passage see John Frederick Vichert, "The Contribution of Church History to Ministerial Efficiency," *Biblical World* 48 (November 1916): 280–81. For a recent, more conservative one see http://issachartraining.org/about/.

85. Edward Hunter, *Brain-Washing in Red China: The Calculated Destruction of Men's Minds* (New York: Vanguard, 1951); David Seed, *Brainwashing: The Fictions of Mind Control: A Study of Novels and Films* (Kent, OH: Kent State University Press, 2004).

86. "Echoes from the Summit," *Christian Crusade*, November 1963, 2, folder 2, box 48, Christian Crusade, Winrod Papers; D.J. Mulloy, *The World of the John Birch Society: Conspiracy, Conservatism, and the Cold War* (Nashville, TN: Vanderbilt University Press, 2014), 117–23.

87. Quoted in Matthew Avery Sutton, *American Apocalypse: A History of Modern Evangelicalism* (Cambridge, MA: Harvard University Press, 2014), 188.

88. Kenneth E. Lawson, *W. O. H. Garman: Patriarch of Christian Fundamentalism within the United States Military Chaplaincy* (Taylors, SC: Associated Gospel Churches, 2000), 1–17; W. O. H. Garman, *What Is Wrong with the Federal Council?* (New York: American Council of Christian Churches, 1950), 3.

89. James Bales, *Communism, Its Faith and Fallacies: An Exposition and Criticism* (Grand Rapids, MI: Baker Book House, 1962), 36, 55–67; David Noebel, "Echoes from the Summit," *Christian Crusade*, January 1967, 21.

90. W. Cleon Skousen, *The Naked Communist: Exposing Communism and Restoring Freedom* (Salt Lake City: Ensign, 1958), 36.

91. David Noebel, email message to the author, May 28, 2013.

92. Glen H. Utter and James L. True, *Conservative Christians and Political Participation: A Reference Handbook* (Santa Barbara, CA: ABC-CLIO, 2004), 240.

93. Debra Ann Pawlak, *Farmington and Farmington Hills* (Charleston, SC: Arcadia, 2013), 90–91, 118, 121; Bureau of the Census, *Fifteenth Census of the United States: 1930 Population Schedule*, Farmington Township, Oakland County, MI, Sheet 8B.

94. Jane Lampman, "The End of the World," *Christian Science Monitor*, February 18, 2004, http://www.csmonitor.com/2004/0218/p11s01-lire.html.

95. There are conflicting accounts of the nature of Tim LaHaye's World War II service. According to journalist Robert Dreyfuss, LaHaye was a machine gunner on a bomber in Europe. In contrast, Nancy Shepherdson claims that LaHaye was "stateside." Robert Dreyfuss, "Reverend Doomsday," *Rolling Stone*, February 19, 2004, 46–50; Nancy Shepherdson, "Writing for Godot," *Los Angeles Times*, April 25, 2004.

96. Utter and True, *Conservative Christians and Political Participation*, 240.

97. Quoted in Heather Hendershot, *What's Fair on the Air? Cold War Right-Wing Broadcasting and the Public Interest* (Chicago: University of Chicago Press, 2011), 1–2.

98. "Are Communists behind the Obscenity Explosion?," *San Bernardino (CA) County Sun*, September 7, 1969, 18; "Sparling Gets Anti-Obscenity Group Post," *Chula Vista (CA) Star-News*, August 30, 1964; "E. Richard Barnes, 79, Dies; Former California Legislator," *New York Times*, August 24, 1985; Daniel K. Williams, *God's Own Party: The Making of*

the Christian Right (New York: Oxford University Press, 2010), 82; Whitney Strub, *Perversion for Profit: The Politics of Pornography and the Rise of the New Right* (New York: Columbia University Press, 2011), 123–25.

99. Williams, *God's Own Party*, 85.

100. Ted Vollmer, "Christian Schools Expanding Rapidly," *Los Angeles Times*, May 17, 1981, A2.

101. Tim LaHaye, *Spirit-Controlled Temperament* (Wheaton, IL: Tyndale House, 1966).

102. Martin Bobgan and Deidre Bobgan, *Four Temperaments: Astrology and Personality Testing* (Santa Barbara, CA: EastGate, 1992), 50–66.

103. "Don't Miss the FAMILY LIFE Seminar!," *Lincoln (NE) Evening Journal*, August 3, 1975, 35; "Religious Best Sellers," *New York Times*, March 12, 1978, 92.

104. "Evangelical Group Invites Dr. Tim LaHaye to Greeley," *Greeley (CO) Daily Tribune*, November 19, 1971, 10.

105. Morris, *History of Modern Creationism*, 258–59, 414.

106. Morris, 259–61.

7. Trees, Knees, and Nurseries

1. Visit to the Museum of Creation and Earth History, Santee, CA, November 18, 2012. The museum is now operated by Tom Cantor of the Life and Light Foundation. See http://www.lifeandlightfoundation.org/. On the Santee tree of evil see Elizabeth Anderson, "If God Is Dead, Is Everything Permitted?," in *Philosophers without Gods: Meditations on Atheism and the Secular Life*, ed. Louise M. Antony (New York: Oxford University Press, 2007), 215–30, and Robert T. Pennock, *Tower of Babel: The Evidence against the New Creationism* (Cambridge, MA: MIT Press, 1999), 315–16.

2. "Gutierrez-Laughton," *San Bernardino (CA) County Sun*, October 15, 1978. They were married at San Bernardino First Baptist.

3. "Pastor of the Week," *San Bernardino County Sun*, March 15, 1979; Joel Carpenter, *Revive Us Again: The Reawakening of American Fundamentalism* (New York: Oxford University Press, 1997).

4. Henry M. Morris, *History of Modern Creationism*, 2nd ed. (Santee, CA: Institute for Creation Research, 1993), 260–61; "The Relationship of ICR to other organizations," *Acts & Facts*, September 1978, 3.

5. Morris, *History of Modern Creationism*, 211, 275.

6. Morris, 265–73.

7. Morris, 278; "Two City Teachers Win Fellowships," *Racine (WI) Journal Times Sunday Bulletin*, March 21, 1965; Ken Ham, "Interview with Dr. Richard Bliss, Ed.D. Curriculum Director Institute for Creation Research, U.S.A.," *Creation 5* (June 1982): 14–17, http://creation.com/interview-with-dr-richard-bliss-usa.

8. Morris, *History of Modern Creationism*, 273.

9. Morris, 273.

10. "The 'Acts & Facts' Survey," *Acts & Facts*, June 1984, 2.

11. Donald B. DeYoung, "Defects in the Jupiter Effect," *Acts & Facts*, June 1979, i–iv; John Gribbin and Stephen Plagemann, *The Jupiter Effect* (New York: Walker, 1974).

12. "The 'Acts & Facts' Survey," *Acts & Facts*, June 1984, 2.

13. John N. Moore, "The Impact of Evolution on the Social Sciences," *Acts & Facts*, October 1977, i–iv; Moore, "The Impact of Evolution on the Humanities and Science," *Acts & Facts*, November 1977, i–iv.

14. Henry Morris, "The Religion of Evolutionary Humanism in Public Schools," *Acts & Facts*, September 1977, 9.

15. "Third Miami Creation Conference Stresses Social Impact of Evolution," *Acts & Facts*, March 1979, 1–3.

16. "Creation Issue Excites Florida," *Acts & Facts*, May 1980, 4.

17. "Proposed California Science Framework," *Acts & Facts*, December 1988, ii.

18. Henry M. Morris, *The Troubled Waters of Evolution* (San Diego, CA: Creation-Life, 1974), 5; "Announcing New Video by Henry M. Morris, PhD., *Evolution in Turmoil*," *Acts & Facts*, March 1984 (indicates that *Turmoil* is a "sequel" to *Troubled Waters*).

19. Morris, *Troubled Waters of Evolution*, 6.

20. Morris, 40–41.

21. Morris, 59–60.

22. "Director's Column," *Acts & Facts*, February 1984, 3. Morris shared with readers that "the following dirge for humanism was written almost 50 years ago when I was a freshman student in college."

23. John Milton, *Paradise Lost*, book 12, lines 33–36.

24. Morris, *Troubled Waters of Evolution*, 72–75.

25. Morris, 168, 184–85.

26. Jerry Adler and John Carey, "Enigmas of Evolution," *Newsweek*, March 29, 1982, 44–49. See also Michelle Green, "Stephen Jay Gould," *People Weekly*, June 2, 1986, 109–14.

27. Henry M. Morris, *Evolution in Turmoil* (San Diego, CA: Creation-Life, 1982).

28. Stephen Jay Gould, *The Structure of Evolutionary Theory* (Cambridge, MA: Harvard University Press, 2002), 1018.

29. Niles Eldredge and Stephen Jay Gould, "Punctuated Equilibria: An Alternative to Phyletic Gradualism," in *Models in Paleobiology*, ed. Thomas J. M. Schopf (San Francisco: Freeman, Cooper, 1972), 84.

30. Eldredge and Gould, 83.

31. Stephen Jay Gould and Niles Eldredge, "Punctuated Equilibria: The Tempo and Mode of Evolution Reconsidered," *Paleobiology* 3 (Spring 1977): 145.

32. Gould and Eldredge, 145–46.

33. Eldredge came from a Baptist and Republican family background. David Prindle, *Stephen Jay Gould and the Politics of Evolution* (Amherst, NY: Prometheus Books, 2009), 21.

34. Ann T. Keene, "Gould, Stephen Jay," *American National Biography Online*, http://www.anb.org/articles/13/13-02671.html; Louis P. Masur, "Stephen Jay Gould's Vision of History," *Massachusetts Review* 30 (Autumn 1989): 480.

35. Myrna Perez, "Evolutionary Activism: Stephen Jay Gould, the New Left and Sociobiology," *Endeavor* 37 (2013): 2–3.

36. Phil Gasper, "A Scientist of the People: The Radical Politics of the Biologist Stephen Jay Gould," *Socialist Worker*, June 7, 2002.

37. "Nerve Gas Next Door," *Science for the People*, August 1970, 9; "Bobby Seale at Cold Spring Harbor," *Science for the People*, August 1970, 3; "Birth Control for Amerika," *Science for the People*, December 1970, 28–32; "Help for Science Education in Cuba and Vietnam," *Science for the People*, May 1971, 28; Donna Haraway, "The Transformation of the Left in Science: Radical Associations in Britain in the 30's and the U.S.A. in the 60's," *Soundings: An Interdisciplinary Journal* 58 (Winter 1975): 441–62; John Walsh, "Science for the People: Comes the Evolution," *Science*, New Series, 191 (March 12, 1976): 1033–35; Kelly Moore, "Organizing Identity: The Creation of Science for the People," in *Social Structure and Organizations Revisited*, ed. Marc Ventresca and Michael Lounsbury (Amsterdam: Elsevier, 1996).

38. Morris, *Evolution in Turmoil*, 111, 148, 150.

39. E. O. Wilson, *Sociobiology: The New Synthesis* (Cambridge, MA: Harvard University Press, 1975).

40. "Sociobiology: A Tool for Social Oppression," *Science for the People*, March 1976, 7–9; Neil Jumonville, "The Cultural Politics of the Sociobiology Debate," *Journal of the History of Biology* 35 (Autumn 2002): 569–93.

41. Julia L. Mickenberg, *Learning from the Left: Children's Literature, the Cold War, and Radical Politics in the United States* (New York: Oxford University Press, 2006), 188.

42. Richard Levins, "Living the 11th Thesis," in *Biology under the Influence: Dialectical Essays on Ecology, Agriculture, and Health*, ed. Richard Lewontin and Richard Levins (New York: Monthly Review, 2007), 365.

43. Levins, 366–67.

44. Richard Levins and Richard Lewontin, *The Dialectical Biologist* (Cambridge, MA: Harvard University Press, 1985); Lewontin and Levins, *Biology under the Influence*.

45. Gould, *Structure of Evolutionary Theory*, 985, 1018.

46. "Stephen Jay Gould, 60, Is Dead; Enlivened Evolutionary Theory," *New York Times*, May 21, 2002, A1. In his chapter on Gould in *The New Celebrity Scientists: Out of the Lab and into the Limelight* (Lanham, MD: Rowman & Littlefield, 2015), 87–110, Declan Fahy foregrounds Leonard Gould's politics and Stephen Jay Gould's political activism, but his brief discussion of PE oddly omits any mention of Marxism.

47. Gould, *Structure of Evolutionary Theory*, 985.

48. Morris, *Evolution in Turmoil*, 91–94.

49. Margaret A. Fay, "Did Marx Offer to Dedicate *Capital* to Darwin? A Reassessment of the Evidence," *Journal of the History of Ideas* 39 (January–March 1978): 133–46; Stephen Jay Gould, *Ever since Darwin: Reflections in Natural History* (New York: W. W. Norton, 1977), 26. Gould corrected his mistake in subsequent editions of the book. See also Stephen Jay Gould, "A Darwinian Gentleman at Marx's Funeral," *Natural History* 108 (September 1999): 64.

50. Morris, *Evolution in Turmoil*, 99–100.

51. Quoted in Morris, 95.

52. Stephen Jay Gould, *The Mismeasure of Man* (1981; New York: W. W. Norton, 1996), 261; Edward J. Larson, *Summer for the Gods: The Scopes Trial and America's Continuing Debate over Science and Religion* (Cambridge, MA: Harvard University Press, 1997), 113–14.

53. Morris, *Evolution in Turmoil*, 98.

54. Morris, 104.

55. Tim LaHaye, *The Battle for the Public Schools: Humanism's Threat to Our Children* (Old Tappan, NJ: Fleming H. Revell, 1983), 207; Tim LaHaye, *The Battle for the Mind: A Subtle Warfare* (Old Tappan, NJ: Fleming H. Revell, 1980).

56. On the close collaboration between Schaeffer and Whitehead see Michael J. McVicar, *Christian Reconstruction: R. J. Rushdoony and American Religious Conservatism* (Chapel Hill: University of North Carolina Press, 2015), 173–75, and Barry Hankins, *Francis Schaeffer and the Shaping of Evangelical America* (Grand Rapids, MI: Eerdmans, 2008), 193–96.

57. Hankins, *Francis Schaeffer*, 51–73.

58. Francis A. Schaeffer, *How Should We Then Live? The Rise and Decline of Western Thought and Culture* (Old Tappan, NJ: Fleming H. Revell, 1976); *How Should We Then Live? The Rise and Decline of Western Thought and Culture*, directed by John Gonser, produced by Franky Schaeffer V, 1977.

59. McVicar, *Christian Reconstruction*, 163–65.

60. John W. Whitehead and John Conlan, "The Establishment of the Religion of Secular Humanism and Its First Amendment Implications," *Texas Tech Law Review* 10 (1978): 1–65. Born in Oak Park, Illinois, in 1930, Conlan was trained at Harvard Law School, served in

the Arizona Senate from 1965 to 1972, then as a US Representative from the Fourth District of Arizona from 1973 to 1977. "Conlan, John Bertrand, (1930–)," *Biographical Directory of the United States Congress, 1774–Present,* http://bioguide.congress.gov/scripts/biodisplay.pl?index=C000682; Sara Diamond, *Roads to Dominion: Right-Wing Movements and Political Power in the United States* (New York: Guilford, 1995), 173.

61. Whitehead and Conlan, "Establishment of the Religion of Secular Humanism," 13.

62. Whitehead and Conlan, 20.

63. Whitehead and Conlan, 24, 27, 29–31, 37–46, 50, 51, 53.

64. Whitehead and Conlan, 55, 59, 64. On Borisov see S.P. DeBoer, E.J. Driessen, and H.L. Verhaar, eds., *Biographical Dictionary of Dissidents in the Soviet Union, 1956–1975* (The Hague, Netherlands: Martinus Nijhoff, 1982), 63.

65. Francis A. Schaeffer, *A Christian Manifesto* (Wheaton, IL: Crossway Books, 1981), 17.

66. Schaeffer, 10, 25–26, 53.

67. Schaeffer, *How Should We Then Live?*, 124–28.

68. Schaeffer, *Christian Manifesto*, 13.

69. Schaeffer, 25.

70. Darren Dochuk, *From Bible Belt to Sunbelt: Plain-Folk Religion, Grassroots Politics, and the Rise of Evangelical Conservatism* (New York: W.W. Norton, 2011), 384–85; LaHaye, *Battle for the Public Schools,* 199–200; "Californians for Biblical Morality State Goals," *Los Angeles Times,* March 20, 1981, G1.

71. LaHaye, *Battle for the Mind,* 5, 241.

72. As historian Adam Laats has noted of the violent controversy over public school textbooks in Kanawha County, West Virginia, in the mid-1970s, the growing preference for "secular humanism" and declining invocation of "communism" did not mean "a simple one-for-one replacement of bogeymen." See Adam Laats, *The Other School Reformers: Conservative Activism in American Education* (Cambridge, MA: Harvard University Press, 2015), 205. Michael Lienesch writes of LaHaye and others, "As these authors see it, humanism and communism are virtually synonymous." Michael Lienesch, *Redeeming America: Piety and Politics in the New Christian Right* (Chapel Hill: University of North Carolina Press, 1993), 162.

73. LaHaye, *Battle for the Mind,* front and back covers.

74. LaHaye, 227, 233, 235, 239–40.

75. LaHaye, 80.

76. LaHaye, 59–83.

77. LaHaye, 64–65.

78. Tim LaHaye, *Battle for the Public Schools,* 99–172.

79. LaHaye, 83.

80. Kenneth E. Woodward and Eloise Salholz, "The Right's New Bogyman," *Newsweek,* July 6, 1981, 48.

81. Woodward and Salholz, 46–47.

82. Quoted in Christopher Toumey, *God's Own Scientists: Creationists in a Secular World* (New Brunswick, NJ: Rutgers University Press, 1994), 83. Citing Woodward and Salholz, "The Right's New Bogyman," 48, Toumey writes that Revell publishing had "sold" 375,000 copies. This may be true, but the *Newsweek* article says of *Battle for the Mind* only that 350,000 copies "are currently in print."

83. Jerry Falwell, *Strength for the Journey: An Autobiography* (New York: Simon & Schuster, 1987), 139.

84. Ronald L. Numbers, *The Creationists: From Scientific Creationism to Intelligent Design* (Cambridge, MA: Harvard University Press, 2006), 237.

85. Dirk Smillie, *Falwell Inc.: Inside a Religious, Political, Educational, and Business Empire* (New York: St. Martin's, 2008), 57, 61.

86. Daniel K. Williams, "Jerry Falwell's Sunbelt Politics: The Regional Origins of the Moral Majority," *Journal of Policy History* 22 (April 2010): 131.

87. Williams, 131; Smillie, *Falwell, Inc.*, 78; Numbers, *Creationists*, 237; Falwell, *Strength for the Journey*, 274.

88. "Jerry Falwell Reports Blessings," *SotL*, November 12, 1975; "Added to Sword Cooperating Board," *SotL*, December 17, 1971; "Lynchburg Baptist College," advertisements, in *SotL*, June 11, June 18, June 25, July 2, July 9, August 13, September 10, October 8, November 5, November 12, November 19, November 26, December 10, all in 1971; "Jerry Falwell Announces Pastors' Conference," advertisement in *SotL*, May 28, 1971.

89. Andrew Himes, *The Sword of the Lord: The Roots of Fundamentalism in an American Family* (Seattle: Chiara, 2011), 281–82. Falwell's appearance in *The Grim Reaper* is at 9:15; see https://www.youtube.com/watch?v=xzxtN_mraUk.

90. Jerry Falwell, "The Divine Mandate for All Christians," *SotL*, December 27, 1974 (delivered at National Sword of the Lord Conference, Indianapolis, August 13, 1974). Plus the following, all in *SotL*: "A Day of Many Solomons," September 26, 1975; "Seven Things Corrupting America," December 31, 1976; "Winning the Race That Is Set before Us," February 13, 1976; "'The Establishment,'" October 6, 1978; "I Love America," October 22, 1976; "Let's Reach the World Together," September 2, 1977; "'And the Hand of the Lord Was with Them,'" June 2, 1978; "Living by Faith," October 6, 1978; "Earnestly Contending for the Faith," October 20, 1978; "The Role of the Churches in the Last Quarter of the Twentieth Century," August 24, 1979; "Let's Promote, Upgrade, Increase FUNDAMENTALISM," January 12, 1979.

91. Falwell, "Role of the Churches in the Last Quarter of the Twentieth Century," *SotL*, August 24, 1979.

92. Jerry Falwell, "Abortion-on-Demand: Is It Murder?," *SotL*, March 31, 1978.

93. All the following in *SotL*: Bob Ware, "Evolution, a Hoax," February 27, 1970; "About . . . the New Scofield Reference Bible," July 2, 1971; "Now—Teach Science without Evolution," November 26, 1971; John R. Rice, "Our Triple Enemies," July 21, 1972; John R. Rice, "Help Save America by Safeguarding the Schools of America," August 15, 1975; John R. Rice, "The Evolution Fantasy," October 29, 1976; R.I. Humberd, "Are Evolutionists Intelligent?," October 7, 1977; Robert E. Kofahl, "EVOLUTION—Fact or Fraud?," March 10, 1978; Arthur I. Brown, "Must Young People Believe in Evolution?," August 10, 1979.

94. Himes, *Sword of the Lord*, 5–6.

95. Quoted in Smillie, *Falwell, Inc.*, 85; Elmer Towns, interview with Lowell Walters, Cline E. Hall, and Abigail Ruth Sattler, July 15, 2010, http://digitalcommons.liberty.edu/ohp_towns_e/2/.

96. John R. Rice, "School Integration and American Freedom," *SotL*, February 6, 1970; "Editor's Moody Invite Cancelled," *SotL*, January 15, 1971, 1.

97. William L. Banks to John R. Rice, March 15, 1970, and John R. Rice to William F. Banks, March 19, 1970, both in folder 34, Segregation at Bob Jones University, box 2, Rice Papers; Randall J. Stephens, "'It Has to Come from the Hearts of the People': Evangelicals, Fundamentalists, Race, and the 1964 Civil Rights Act," *Journal of American Studies* 50 (August 2016): 559–85.

98. Williams, "Jerry Falwell's Sunbelt Politics," 131–32; Falwell, *Strength for the Journey*, 279–99.

99. Towns interview.

100. Falwell, *Strength for the Journey.*

101. "Cast Your Vote for Creation or Evolution," advertisement, *Moral Majority Report* 2 (April 20, 1981): 11; Henry M. Morris, *The Remarkable Birth of Planet Earth* (San Diego, CA: Creation-Life, 1972).

102. Hal Lindsay, *The Late, Great Planet Earth* (Grand Rapids, MI: Zondervan, 1970); Paul Boyer, *When Time Shall Be No More: Prophecy Belief in Modern American Culture* (Cambridge, MA: Harvard University Press, 1992), 126–27.

103. Morris, *Remarkable Birth of Planet Earth*, 4.

104. Morris, 75, back cover.

105. Lane P. Lester, "Evolution AND Creation Best Science but, America's Youth Brainwashed," *Moral Majority Report* 1 (June 30, 1980): 16; Toumey, *God's Own Scientists*, 55; Morris, *History of Modern Creationism*, 278; list of ICR staff, *Acts & Facts*, August 1980, vi–vii.

106. "Scientific Creation vs. Evolution Is Not Religion vs. Science Debate," *Moral Majority Report* 2 (October 19, 1981), 11; "'. . . God Created the Heaven and the Earth,'" *Moral Majority Report* 3 (April 26, 1982): 7.

107. James L. Hall, *History of Life* (Lynchburg, VA: Simbiosys, c. 1975). Thanks to Liberty University archivist Abigail Sattler and to the Liberty University Biology Department for sharing this material with me. *Liberty Baptist College Bulletin* 9 (January 1979).

108. Hall, *History of Life*, 1–6.

109. Jerry Falwell, "Why the Moral Majority?," *Moral Majority Capitol Report* (August 1979): 1, General Materials of the Moral Majority, Publications, Capitol Report Newsletter, MOR 1:1–4, box 1, Liberty University Archives.

110. Jerry Falwell, "Abortion-on-Demand: Is It Murder?," *OTGH*, February 26, 1978, Falwell Ministries, Old Time Gospel Hour Transcripts, Liberty University Archives.

111. Jerry Falwell, "Communism," *OTGH* #399, Falwell Ministries, Old Time Gospel Hour Transcripts, Liberty University Archives.

112. Jerry Falwell, "Sexual Promiscuity in America," *OTGH* #717, July 13, 1986, Falwell Ministries (FM 3:4), Old Time Gospel Hour Transcripts, Liberty University Archives; Jerry Falwell, "Strengthening Families in the Nation," address to Christian Life Commission's Strengthening Families seminar in Atlanta, enclosure in Harry Hollis to Falwell, April 6, 1982, FAL 4–1, series 1, folder 1, Speeches 1982, Liberty University Archives.

113. Jerry Falwell, *Spiritual Heartburn* (Lynchburg, VA: Old Time Gospel Hour Press, 1981), FAL 4–2, series 4, folder 2, Falwell Sermons, 1981, Liberty University Archives.

114. Jerry Falwell, *Listen, America!* (New York: Doubleday, 1980).

115. Falwell, 5, 235–37; "Evangelicals Press Political Activities," *New York Times*, September 29, 1980, D13.

116. Falwell, *Listen, America!*, 56.

117. Falwell, 71.

118. Daniel K. Williams, *God's Own Party: The Making of the Christian Right* (New York: Oxford University Press, 2010), 188; Thomas W. Evans, *The Education of Ronald Reagan: The General Electric Years and the Untold Story of His Conversion to Conservatism* (New York: Columbia University Press, 2006); Kevin M. Kruse, *One Nation under God: How Corporate America Invented Christian America* (New York: Basic Books, 2015), 148.

119. Reagan Bush Committee, "Address by the Honorable Ronald Reagan, the Roundtable National Affairs Briefing, Dallas, Texas," August 22, 1980, http://digitalcollections.library.cmu.edu/awweb/awarchive?type=file&item=684006; "Reagan Tries to Cement His Ties with TV Evangelicals," *Los Angeles Times*, August 23, 1980, A1; "Reagan Denies

Gaffe in Statement," *Galveston Daily News*, September 19, 1980, 10; Stephen Hayward, *The Age of Reagan: The Fall of the Old Liberal Order, 1964–1980* (New York: Three Rivers, 2001), 681.

120. "Reagan's Visits with Evangelicals," *Los Angeles Times*, August 29, 1980, F10; "Carter Aide: Reagan Gaffes Are 'What We've Looked For,'" *Christian Science Monitor*, September 4, 1980; "Governor Reagan Backs Two-Model Approach," *Acts & Facts*, October 1980, 3.

121. Brenda D. Hofman, "The Squeal Rule: Statutory Resolution and Constitutional Implications—Burdening the Minor's Right of Privacy," *Duke Law Journal* 1984 (December 1984): 1325–57.

122. Ronald Reagan, "Evil Empire" speech, Orlando, Florida, March 8, 1983, http://millercenter.org/president/speeches/speech-3409. For video of the speech see "Evil Empire Speech" by Ronald Reagan—Address to the National Association of Evangelicals, https://www.youtube.com/watch?v=FcSm-KAEFFA&t=163s.

123. Reagan, "Evil Empire" speech. For the original text of Lenin's speech see V.I. Lenin, "The Tasks of the Youth Leagues," in *Collected Works*, vol. 31, *April–December 1920* (Moscow: Progress, 1977), 291.

124. Paul Kengor, *God and Ronald Reagan: A Spiritual Life* (New York: HarperCollins, 2004), 249–54.

125. Reagan, "Evil Empire" speech.

126. As of February 1981, the executive board comprised Falwell; Greg Dixon; Tim LaHaye; Curtis Hutson (an independent Baptist who succeeded Rice as editor of *Sword of the Lord*); Charles Stanley, pastor of the First Baptist Church of Atlanta, who would be elected president of the SBC in 1984; and Kennedy.

127. "Rev. D. James Kennedy, Broadcaster, Dies at 76," *New York Times*, September 6, 2007; "Empire Builder D. James Kennedy Dies at 76," *Christianity Today*, September 6, 2007; "Politically Powerful TV Evangelist D. James Kennedy," *Washington Post*, September 6, 2007.

128. Ronald L. Numbers, *Darwinism Comes to America* (Cambridge, MA: Harvard University Press, 1998), 61–66.

129. Luke E. Harlow, "The Long Life of Proslavery Religion," in *The World the Civil War Made*, ed. Gregory P. Downs and Kate Masur (Chapel Hill: University of North Carolina Press, 2015), 138; James Oscar Farmer Jr., *The Metaphysical Confederacy: James Henley Thornwell and the Synthesis of Southern Values* (Macon, GA: Mercer University Press, 1986).

130. D. James Kennedy, *Evangelism Explosion*, 3rd ed. (Wheaton, IL: Tyndale House, 1983).

131. John Morris, interview with the author, Dallas, May 22, 2013.

132. Michael Foust, "D. James Kennedy Dead at 76," Baptist Press, September 5, 2007, http://www.bpnews.net/26374/d-james-kennedy-dead-at-76; E. Russell Chandler, *The Kennedy Explosion* (Elgin, IL: David C. Cook, 1971), 93.

133. D. James Kennedy, *Why I Believe* (Waco, TX: Word Books, 1980), 53.

134. Kennedy, 84.

135. D. James Kennedy, *Reconstruction: Biblical Guidelines for a Nation in Peril* (Fort Lauderdale, FL: Coral Ridge Ministries, 1982).

136. Kennedy, i, emphasis in original.

137. Kennedy, 6.

138. Kennedy, 13–16.

139. Kennedy, 31–32.

140. C. Gregg Singer, *A Theological Interpretation of American History*, 3rd ed. (Greenville, SC: A Press, 1994), 316; Steven Alan Samson, "A Theological Interpretation of American History Study Guide" (1985), Faculty Publications and Presentations, Paper 198, http://digitalcommons.liberty.edu/cgi/viewcontent.cgi?article=1204&context=gov_fac_pubs.

141. Henry M. Morris, *The Long War against God* (Grand Rapids, MI: Baker Book House, 1989), 9–10.

142. Morris, 59.

143. Morris, 82–92. Morris had called attention to this article shortly after it appeared in 1980. See Henry Morris, "Evolution and Socialism," *Acts & Facts*, February 1981, 2; Cliff Conner, *Evolution vs. Creationism: In Defense of Scientific Thinking* (New York: Pathfinder, 1981). Conner later published *A People's History of Science* (New York: Nation Books, 2005).

144. Morris, *Long War*, 80.

145. "1990 Summer Institutes—*Register Now!*," *Acts & Facts*, March 1990, 1.

146. Morris, *Long War*, 160, 168, 180.

147. Morris, 173–75; Martin Fichman, *An Elusive Victorian: The Evolution of Alfred Russel Wallace* (Chicago: University of Chicago Press, 2010). Among the few scholars to note Morris's interpretation of Wallace's "aha" moment in the jungle as satanically inspired are Randall J. Stephens and Karl W. Giberson, *The Anointed: Evangelical Truth in a Secular Age* (Cambridge, MA: Harvard University Press, 2011), 260.

148. Morris, *Long War*, 182–83; Richard Wurmbrand, *Marx and Satan* (Barlesville, OK: Living Sacrifice Book Co., 1986), 15. For the text of "The Player" and *Oulanem* see Robert Payne, *The Unknown Karl Marx* (New York: NYU Press, 1971), 59–60, 65–94.

149. Morris, *Long War*, 182; "The Long War against God," *Acts & Facts*, November 1989, 1.

150. "Ken Ham Joins ICR Staff," *Acts & Facts*, December 1986, 1.

151. Kenneth A. Ham, "Creation Evangelism: A Powerful Tool in Today's World," *Acts & Facts*, January 1987 ("Impact" feature), ii; Ken Ham, "Where Are All the Godly Offspring?," *Acts & Facts* (Back to Genesis feature), December 1989.

152. "It's Time to Tell the Truth about Dinosaurs," advertisement, *Acts & Facts*, May 1987, back cover.

153. "The ICR Museum of Creation and Earth History," *Acts & Facts*, November 1980, 4–5; "Museum Receives New Display," *Acts & Facts*, September 1981, 2; "Dedication Message," *Acts & Facts*, Impact feature, August 1986, iii; panel on history of Museum of Creation and Earth History, Creation and Earth History Museum, Santee, California, author's notes on visit, November 17, 2012.

154. "Video Tour of the ICR Museum Now Available," *Acts & Facts*, September 1994, 1–2; "ICR Museum Marks Fifth Anniversary," *Acts & Facts*, November 1997, 1; author's notes on visit to Creation and Earth History Museum, November 17, 2012.

8. The Nightcrawler, the Wedge, and the Bloodiest Religion

1. Dave Welch, "From Darwin to Marx to Kinsey to Obama," WorldNetDaily, January 1, 2011, http://www.wnd.com/2011/01/246053/.

2. See, for instance, https://iloveyoubutyouregoingtohell.org/tag/sensuous-curmudgeon/ and https://pandasthumb.org/archives/2012/05/the-sensuous-cu.html.

3. "WorldNetDaily: Darwin, Marx, Kinsey & Obama," January 1, 2011, https://sensuouscurmudgeon.wordpress.com/2011/01/01/worldnetdaily-darwin-marx-kinsey-obama/.

4. WorldNetDaily, "About us," http://go.wnd.com/aboutwnd/.

5. David Edwin Harrell Jr., *Pat Robertson: A Life and Legacy* (Grand Rapids, MI: William B. Eerdmans, 2010), 1–61.

6. Harrell, 82–97.

7. Pat Robertson, "A Presidential Bid Launched," Pat Robertson.com, accessed December 6, 2016, http://www.patrobertson.com/Speeches/PresidentialBidLaunched.asp.

8. Dave Welch, email correspondence with author, June 7, 2016; E. B. Mittleman, "The Gyppo System," *Journal of Political Economy* 31 (December 1923): 840–41; "Statement of Receipts and Expenditures at General Headquarters, Industrial Workers of the World," *Industrial Union Bulletin*, August 24, 1907, 3.

9. Welch email correspondence, June 7, 2016; "Gary Allen, 50, Dies in West; Spread Conservatives' View," *New York Times*, December 2, 1986.

10. Gary Allen, with Larry Abraham, *None Dare Call It Conspiracy* (Rossmoor, CA: Concord, 1971), 39–40, 76, 80, 88, 124–25, back cover. On Allen's conspiracy theorizing and Christian rightists in the American West see James Alfred Aho, *The Politics of Righteousness: Idaho Christian Patriotism* (Seattle: University of Washington Press, 1990), 255–59.

11. Pat Robertson, *The New World Order* (Dallas: Word Publishing, 1991); Robert Goldberg, *Enemies Within: The Culture of Conspiracy in Modern America* (New Haven, CT: Yale University Press, 2001), 91–93.

12. Welch email correspondence, June 7, 2016; Dave Welch, "The America Plan," http://uspastorcouncil.org/america-plan/; Dave Welch, "Mobilizing God's Generals," http://uspastorcouncil.org/wp-content/uploads/2015/07/Mobilizing_God_s_Generals_9.101.pdf.

13. Dave Welch, "Mobilizing God's Generals," 14.

14. Welch email correspondence, June 7, 2016.

15. Dave Welch, "Making a Monkey out of Christians," WorldNetDaily, February 17, 2009, http://www.wnd.com/2009/02/89090/.

16. Jobe Martin and Don Baker, "Noebel Cause: A Brief History of Summit Ministries," in *A Summit Reader: Essays and Lectures in Honor of David Noebel's 70th Birthday*, ed. Michael Bauman and Francis Beckwith (Manitou Springs, CO: Summit, 2007), 16.

17. David A. Noebel, *Understanding the Times: The Story of the Biblical Christian, Marxist/Leninist, and Secular Humanist Worldviews* (Manitou Springs, CO: Summit, 1991), 37.

18. David A. Noebel, *Understanding the Times: The Religious Worldviews of Our Day and the Search for Truth*, abridged ed. (Colorado Springs, CO: Association of Christian Schools International and Summit Ministries, 1995), 10; David A. Noebel, *Understanding the Times: The Collision of Today's Competing Worldviews*, rev. 2nd ed. (Manitou Springs, CO: Summit, 2006).

19. David A. Noebel, *Understanding the Times: The Story of the Biblical Christian, Marxist/Leninist and Secular Humanist Worldviews* (Manitou Springs, CO: Summit, 1991), 134, 220–23, 297–302, 382, 527, 741–61.

20. Noebel, *Understanding the Times: The Collision of Today's Competing Worldviews*, 191.

21. Milton Gaither, *Homeschool: An American History* (New York: Palgrave Macmillan, 2008), 108.

22. Gaither, 175–200.

23. "Understanding the Times 2006 Homeschool Edition, DVD/Book Curriculum," Cathy Duffy Reviews, last updated 2010, http://cathyduffyreviews.com/worldview/understanding-the-times.htm; Cathy Duffy, "How to Teach Worldview" (originally published in *Practical Homeschooling* #13, 1996), https://www.home-school.com/Articles/how-to-teach-worldview.php.

24. "ICR Meetings," *Acts & Facts*, June 1995; "Meeting Highlights: Manitou Springs, Colorado," *Acts & Facts*, October 1997, 5.

25. Henry M. Morris and John D. Morris, *The Modern Creation Trilogy*, vol. 3, *Society and Creation* (Green Forest, AR: Master Books, 1996), 107, 116.

26. "Creation Trilogy Published," *Acts & Facts*, November 1996, 1.

27. "Creation Seminar at Pensacola Christian College," *Acts & Facts*, September 1995, 4.

28. Michael Behe, *Darwin's Black Box: The Biochemical Challenge to Evolution* (New York: Free Press, 1996); William Dembski, *The Design Inference: Eliminating Chance through Small Probabilities* (Cambridge: Cambridge University Press, 1998).

29. Chris Mooney, *The Republican War on Science* (New York: Basic Books, 2005), 170–74.

30. Stephen Meyer, "Danger: Indoctrination. A Scopes Trial for the '90s," *Wall Street Journal*, December 6, 1993, A14.

31. Transcript of "Intelligent Design vs. Evolution," Think Tank with Ben Wattenberg, October 12, 2006, http://www.pbs.org/thinktank/transcript1244.html.

32. Kitzmiller v. Dover Area School District, 400 F. Supp. 2d 707 (W.D. Pa. 2005).

33. Philip E. Johnson, *Defeating Darwinism by Opening Minds* (Downers Grove, IL: InterVarsity, 1997), 103–5.

34. Johnson, 92.

35. Phillip E. Johnson, *Darwin on Trial* (Downers Grove, IL: InterVarsity, 1991), 150.

36. Johnson, *Defeating Darwinism*, 113.

37. Barbara Forrest and Paul R. Gross, *Creationism's Trojan Horse: The Wedge of Intelligent Design* (New York: Oxford University Press, 2004), 267; Michael J. McVicar, *Christian Reconstruction: R.J. Rushdoony and American Religious Conservatism* (Chapel Hill: University of North Carolina Press, 2015), 174–75, 220, 264n136.

38. Johnson, *Defeating Darwinism*, 106.

39. Forrest and Gross, *Creationism's Trojan Horse*, 271.

40. While Discovery Institute leaders first refused to acknowledge that the document represented official policy, the publication of *Creationism's Trojan Horse* by Barbara Forrest and Paul Gross in 2004 led Discovery to own up, in 2005, in a piece titled "The 'Wedge Document': So What?": http://www.discovery.org/scripts/viewDB/filesDB-download.php?id=349.

41. "'Wedge Document': So What?"

42. "'Wedge Document': So What?," emphasis added.

43. Jay W. Richards, *Money, Greed, and God: Why Capitalism Is the Solution and Not the Problem* (New York: HarperCollins, 2009), 1.

44. "Major Expansions at Ark Encounter as It Enters Year Four; Auditorium, Zoo Expansion, Playground Added," *Northern Kentucky Tribune*, June 27, 2019, https://www.nkytribune.com/2019/06/major-expansions-at-ark-encounter-as-it-enters-year-four-auditorium-zoo-expansion-playground-added/.

45. Susan L. Trollinger and William Vance Trollinger Jr., *Righting America at the Creation Museum* (Baltimore: Johns Hopkins University Press, 2016), 10–11, 13, 228, 233.

46. Dave Cullen, *Columbine* (New York: Twelve, 2009), 41, 184, 235, 260; Ralph W. Larkin, *Comprehending Columbine* (Philadelphia: Temple University Press, 2007), 1, 133.

47. Ken Ham, "The 'Missing Link' to School Violence," https://answersingenesis.org/sanctity-of-life/mass-shootings/the-missing-link-to-school-violence/. On AiG and school shootings see Trollinger and Trollinger, *Righting America*, 154–57.

48. *Congressional Record*, June 16, 1999, H4366; "House Undertakes Days-Long Battle on Youth Violence," *New York Times*, A26; "Culture Wars Erupt in Debate on Hill," *Washington Post*, June 18, 1999, A1; "U.S. Congressional Leader Castigated for Creation Comments," Answers in Genesis, April 25, 2002, https://answersingenesis.org/college/us-congressional-leader-castigated-for-creation-comments/. In 2012, in the aftermath of the Aurora, Colorado, movie theater shootings, which took place during a showing of *The Dark Knight*

Rises (the third installment of Christopher Nolan's *Batman* trilogy), Dawson published a similar letter, in which he lamented the breakdown of the family and the passing of classic TV shows that (in contrast to *Batman* films) "promoted the moral order," such as *Leave It to Beaver*, *Ozzie and Harriet*, and *The Beverly Hillbillies*. See Addison L. Dawson, "Our Society Has Degraded," *San Angelo (TX) Standard-Times*, July 26, 2012, http://www.gosanangelo. com/opinion/letter-our-society-has-degraded-ep-439198174-356171891.html.

49. Author's notes on visit to Creation Museum, Petersburg, KY, April 11, 2015; Trollinger and Trollinger, *Righting America*, 31–32, 52–53.

50. Trollinger and Trollinger, *Righting America*, 191. As Susan and William Trollinger write, the museum is best understood as "a Christian Right arsenal in the culture war."

51. Author's notes on talk by Tony Perkins, "Advancing a Culture of Life in the U.S.," Answers in Genesis Mega-Conference, July 22, 2013, Sevierville, TN.

52. Ray Comfort, dir., *Evolution v. God: Shaking the Foundations of Faith*, 2013; author's notes on Answers in Genesis Mega-Conference, July 22, 2013, Sevierville, TN.

53. "The Battle for the Truth at the Answers Mega Conference," *Answers in Genesis Blog*, July 23, 2013, https://answersingenesis.org/blogs/outreach-events/2013/07/23/the-battle-for-the-truth-at-the-answers-mega-conference/.

54. Trollinger and Trollinger, *Righting America*, 64–65; "Bill Nye Debates Ken Ham," February 4, 2014, https://www.youtube.com/watch?v=z6kgvhG3AkI.

55. Not that mainstream scientists or philosophers of science accept this distinction. See Robert T. Pennock, *Tower of Babel: The Evidence against the New Creationism* (Cambridge, MA: MIT Press, 1999), 147–50.

56. "Bill Nye Debates Ken Ham," quote begins at 56:10.

57. "Bill Nye Debates Ken Ham." Nye begins this line of argument at 1:28:03. The quotation starts at 1:28:26. In the aftermath of the debate, Nye published a book-length version of his argument for evolution. In the chapter on his debate with Ham, Nye did not mention Ham's culture war arguments. Nye did write proudly, however, that he did not "disparage anyone's religion" and that "I did not mention anything about the *Bible*." See Bill Nye, *Undeniable: Evolution and the Science of Creation* (New York: St. Martin's, 2014), 9.

58. Author's notes on visit to Creation Museum; Trollinger and Trollinger, *Righting America*, 49.

59. "Student Kills 7 Classmates, Principal at School in Finland," *Washington Post*, November 8, 2007, https://www.washingtonpost.com/wp-dyn/content/article/2007/11/07/AR20 07110700663.html.

60. "Student Kills 7 Classmates, Principal"; Bodie Hodge, "Finland School Shootings: The Sad Evolution Connection," Answers in Genesis, November 8, 2007, https://answersingene sis.org/sanctity-of-life/mass-shootings/finland-school-shootings-the-sad-evolution-connection/; "Pentti Linkola: Ideas," http://www.penttilinkola.com/pentti_linkola/ecofascism/.

61. Hodge, "Finland School Shootings."

62. Raymond Hall, "Darwin's Impact—the Bloodstained Legacy of Evolution," Answers in Genesis, March 1, 2005, https://answersingenesis.org/charles-darwin/racism/darwins-impact-the-bloodstained-legacy-of-evolution/. This piece was also published in AiG's *Creation* 27 (March 2005): 46–47.

63. Bodie Hodge, "The Results of Evolution: Could It Be the Bloodiest Religion Ever?," Answers in Genesis, July 13, 2009, https://answersingenesis.org/sanctity-of-life/the-results-of-evolution/; Bodie Hodge, "Evolutionary Humanism: The Bloodiest Religion Ever," in *A Pocket Guide to Atheism*, Answers in Genesis (Petersburg, KY: Answers in Genesis, 2014), 21.

64. Answers in Genesis, *A Pocket Guide to Atheism*. I purchased the book there in April 2015. It was still for sale in the museum bookstore when I last visited on March 1, 2020.

65. Jerry Bergman, *The Criterion: Religious Discrimination in America* (Richfield, MN: Onesimus, 1984); Jerry Bergman, *Slaughter of the Dissidents* (Southworth, WA: Leafcutter, 2008).

66. Jason Rosenhouse, *Among the Creationists: Dispatches from the Anti-Evolutionist Front Line* (New York: Oxford University Press, 2012), 199.

67. Jerry Bergman, "The Darwinian Foundation of Communism," *Creation* 15 (April 2001): 89, also available at https://answersingenesis.org/charles-darwin/racism/the-darwinian-foundation-of-communism/.

68. Jerry Bergman, "Darwin's Influence on Ruthless Laissez-Faire Capitalism," *Acts and Facts*, Impact feature article, March 2001, i.

69. Jerry Bergman, *The Darwin Effect: It's [sic] Influence on Nazism, Eugenics, Racism, Communism, Capitalism and Sexism* (Green Forest, AR: Master Books, 2014), 253.

70. Kent Hovind's reputation has tarnished fellow creationists to the degree that even Chad Hovind has made a point of publicly distancing himself from his infamous uncle. See Chad Hovind, "You Can Pick Your Friends, but Not Your Family," Beliefnet, http://www.beliefnet.com/columnists/godonomics/2011/09/you-can-pick-your-friends-but-not-your-family-chad-hovind-and-kent-hovind.html.

71. Chad Hovind, *Godonomics: How to Save Our Country—and Protect Your Wallet—through Biblical Principles of Finance* (Colorado Springs, CO: Multnomah Books, 2013), 6–38, 40–42.

72. Hovind, 169–70.

73. "Government Master" card, obtained at *Godonomics* booth, AiG Mega-Conference, Sevierville, TN, July 23, 2013.

Epilogue

1. Author's notes on visit to Ark Encounter, Williamstown, Kentucky, July 7, 2016.

2. Ken Ham, "The 'Donald Trump Phenomenon,'" Answers in Genesis, October 28, 2015, https://answersingenesis.org/ministry-news/ministry/donald-trump-phenomenon/.

3. "Donald Trump-Liberty University Convocation," Liberty University, January 18, 2016, YouTube video, https://www.youtube.com/watch?v=xSAyOlQuVX4.

4. Jon Ward, "Transcript: Donald Trump's Closed-Door Meeting with Evangelical Leaders," Yahoo News, June 22, 2016, https://www.yahoo.com/news/transcript-donald-trumps-closed-door-meeting-with-evangelical-leaders-195810824.html; Sarah McCammon, "Inside Trump's Closed-Door Meeting, Held to Reassure 'the Evangelicals,'" National Public Radio, http://www.npr.org/2016/06/21/483018976/inside-trumps-closed-door-meeting-held-to-reassure-the-evangelicals.

5. "A Born-Again Donald Trump? Believe It, Evangelical Leader Says," *New York Times*, June 25, 2016.

6. "The World According to Trump," *Time*, March 1, 2004, 48–57; "Donald Trump: Macho Man of 2016," CNN Politics, August 27, 2015, http://www.cnn.com/2015/08/27/politics/donald-trump-macho-elections-2016/index.html.

7. "How Norman Vincent Peale Taught Donald Trump to Worship Himself," Politico Magazine, October 6, 2015, https://www.politico.com/magazine/story/2015/10/donald-trump-2016-norman-vincent-peale-213220.

8. "Trump Says Very Curious Things about God, Church and the Bible," *Washington Post*, January 19, 2016, https://www.washingtonpost.com/news/morning-mix/wp/2016/01/19/trump-says-very-curious-things-about-god-church-and-the-bible/?utm_term=.610587c081d8.

9. For a study that makes a similar point about modern Christian conservatism and gender politics see Kristin Kobes Du Mez, *Jesus and John Wayne: How White Evangelicals Corrupted a Faith and Fractured a Nation* (New York: Liveright, 2020), 298.

10. John Eidsmoe, *God and Caesar: Biblical Faith and Political Action* (Eugene, OR: Wipf and Stock, 1997), 217.

11. Ryan Lizza, "Leap of Faith: The Making of a Republican Front-Runner," *New Yorker*, August 15, 2011; Katy Steinmetz, "Michele Bachmann's Reading List," *Time*, August 9, 2011; "For Bachmann, God and Justice Were Intertwined," *New York Times*, October 13, 2011; "Summit Forum with Michelle Bachmann," afikomag, March 7, 2016, http://afikomag.com/event/summit-forum-with-michele-bachmann/; Coral Ridge Ministries, *Socialism: A Clear and Present Danger* (Fort Lauderdale, FL: Coral Ridge Ministries, 2010).

12. Edward H. Miller, *Nut Country: Right-Wing Dallas and the Birth of the Southern Strategy* (Chicago: University of Chicago Press, 2015), 69–85; Samuel K. Tullock, "'He, Being Dead, Yet Speaketh': J. Frank Norris and the Texas Religious Right at Midcentury," in *The Texas Right: The Radical Roots of Lonestar Conservatism*, ed. David O'Donald Cullen and Kyle G. Wilkison (College Station: Texas A&M University Press, 2014), 63.

13. Robert Jeffress, *Countdown to the Apocalypse: Why ISIS and Ebola Are Only the Beginning* (New York: FaithWords / Hachette Book Group, 2015), 78, 100.

14. "ICR's Discovery Center: Reaching New Generations," Institute for Creation Research, April 26, 2017, https://www.icr.org/article/icrs-discovery-center-reaching-new.

15. "Robert Jeffress Wants a Mean 'Son of a Gun' for President, Says Trump Isn't a Racist," *Dallas Observer*, April 5, 2016, http://www.dallasobserver.com/news/robert-jeffress-wants-a-mean-son-of-a-gun-for-president-says-trump-isnt-a-racist-8184721; John Fea, *Believe Me: The Evangelical Road to Donald Trump* (Grand Rapids, MI: Eerdmans, 2018), 39, 120–21, 124–29, 160–62.

16. "Strong Faith Shapes Mike Pence's Politics," *Indianapolis Star*, July 14, 2016; "Mike Pence: A Conservative Proudly Out of Sync with His Times," *New York Times*, July 15, 2016.

17. Mike Pence of Indiana speaking on "Theory of the Origin of Man," 107th Cong., 2nd sess., *Congressional Record* 148 (July 11, 2002): H4527.

18. "5 Faith Facts about Ben Carson: Retired Neurosurgeon, Seventh-day Adventist," Religion News Service, February 1, 2016, http://religionnews.com/2016/02/01/ben-carson-religion-adventist-evangelical/; Ben Carson, with Candy Carson, *America the Beautiful: Rediscovering What Made This Nation Great* (Grand Rapids, MI: Zondervan, 2012).

19. "Celebration of Creation / Ben Carson, MD," IDquest, YouTube video, February 14, 2015, https://www.youtube.com/watch?v=Z6ChFtIDUbg&t=44m16s.

20. Lawrence M. Krauss, "Ben Carson's Scientific Ignorance," *New Yorker*, September 28, 2015, http://www.newyorker.com/news/news-desk/ben-carsons-scientific-ignorance.

21. "The Doctor Is Out: Ben Carson Doubles Down on Satan-Clinton Connection," *Vanity Fair*, July 20, 2016, http://www.vanityfair.com/news/2016/07/ben-carson-clinton-lucifer; "Who Is Saul Alinsky? Ben Carson Claims He Was Hillary Clinton's 'Role Model,'" NBC News, July 20, 2016, http://www.nbcnews.com/storyline/2016-conventions/who-saul-alinsky-ben-carson-claims-he-was-hillary-clinton-n613341; "Dr. Ben Carson Speech at Republican National Convention July 19, 2016 Cleveland, Ohio," Association V.A.A., YouTube video, July 20, 2016, https://www.youtube.com/watch?v=ZjbQ6w2KuxU; Hillary Rodham to Saul Alinsky, July 8, 1971, in "The Hillary Letters: Hillary Clinton, Saul Alinsky Correspondence Revealed," Washington Free Beacon, September 21, 2014, http://freebeacon.com/politics/the-hillary-letters/; Stanley Kurtz, "Why Hillary's Alinsky Letters Matter," *National Review*, September 22, 2014, http://www.nationalreview.com/corner/388560/why-hillarys-alinsky-letters-matter-stanley-kurtz; Saul Alinsky, *Rules for Radicals: A Pragmatic Primer for Realistic Radicals* (New York: Random House, 1971).

22. "Coral Ridge Presbyterian Church," Conservative Transparency, http://conservative transparency.org/recipient/coral-ridge-presbyterian-church/.

23. Katherine Stewart, "Betsy DeVos and God's Plan for the Schools," *New York Times*, December 13, 2016.

24. "DeVos' Code Words for Creationism Offshoot Raise Concerns about 'Junk Science,'" ProPublica, January 29, 2017, https://www.propublica.org/article/devos-education-nominees-code-words-for-creationism-offshoot-raise-concerns.

25. "Influential Conservative Group: Trump, DeVos Should Dismantle Education Department and Bring God into Classrooms," *Washington Post*, February 15, 2017, https://www.washingtonpost.com/local/education/influential-conservative-group-trump-devos-should-dismantle-education-department-and-bring-god-into-classrooms/2017/02/15/196bf872-f2df-11e6-8d72-263470bf0401_story.html; Council on National Policy Education Committee, "Education Reform Report," https://web.archive.org/web/20170210044252/http:/www.cfnp.org/file/ERR-CNP-Site.pdf. The report was originally posted on the CNP site and then removed. According to the *Washington Post* article, CNP staff confirmed its authenticity.

26. Dan Smithwick, *The PEERS Story* (Lexington, KY: Pillars, 2013), 2, 5.

27. The Nehemiah website features this endorsement of Summit: "Their flagship publication, Understanding the Times (UTT) is excellent. Many youth have been put 'on the right track' through this ministry": http://www.nehemiahinstitute.com/links.php; Smithwick, *PEERS Story*, 20.

28. Steve Deckard and Daniel Smithwick, "High School Students' Attitudes toward Creation and Evolution Compared to Their Worldview," Impact feature in *Acts & Facts*, May 2002, i–iv, http://www.icr.org/article/high-school-students-attitudes-toward-creation-evo/.

29. Dan Smithwick, "Where Are We Going?," Nordskog Publishing (blog), September 28, 2010, https://www.nordskogpublishing.com/where-are-we-going/.

30. Ken Ham and Britt Beemer, with Todd Hillard, *Already Gone: Why Your Kids Will Quit Church and What You Can Do to Stop It* (Green Forest, AR: Master Books, 2009).

31. "White Evangelicals Voted Overwhelmingly for Trump, Exit Polls Show," *Washington Post*, November 9, 2016, https://www.washingtonpost.com/news/acts-of-faith/wp/2016/11/09/exit-polls-show-white-evangelicals-voted-overwhelmingly-for-donald-trump/?utm_term=.949e8af6f770.

32. Ken Ham, "How to Make America Great Again," Answers in Genesis, January 19, 2017, https://answersingenesis.org/blogs/ken-ham/2017/01/19/how-to-make-america-great-again/.

33. Ken Ham, "The Real Fake News," Answers in Genesis, March 21, 2017, https://answersingenesis.org/blogs/ken-ham/2017/03/21/real-fake-news/.

34. Author's notes on visit to AiG Creation Museum, Petersburg, KY, April 1, 2017.

35. Ken Ham to AiG mailing list, May 23, 2017. Copy in author's possession. On Huxley's politics see Joanne Woiak, "Designing a Brave New World: Eugenics, Politics, and Fiction," *Public Historian* 29 (Summer 2007).

36. John Hevener, *Which Side Are You On? The Harlan County Coal Miners, 1931–1939* (Urbana: University of Illinois Press, 1978).

37. Wikipedia, s.v. "Cumberland, Kentucky," last modified July 24, 2020, 20:02 UTC, https://en.wikipedia.org/wiki/Cumberland,_Kentucky#.

38. "Unpaid Miners Blocked a Coal Train in Protest. Weeks Later, They're Still There," *New York Times*, August 19, 2019, https://www.nytimes.com/2019/08/19/us/kentucky-coal-miners.html.

39. "Still without Pay, Kentucky Coal Miners Continue Month-Long Railroad Blockade," WBUR, August 29, 2019, https://www.wbur.org/onpoint/2019/08/29/kentucky-coal-miners-railroad-blockade.

40. Ken Ham, "Teachers Union Endorses Killing Unborn Children," *Ken Ham's Blog*, July 25, 2019, https://answersingenesis.org/sanctity-of-life/teachers-union-endorses-killing-unborn-children/, emphasis added.

41. "The Historic Strikes and Protests by Teachers across the Country Aren't Over," *Washington Post*, April 10, 2019, https://www.washingtonpost.com/education/2019/04/10/historic-strikes-protests-by-teachers-around-country-arent-over/.

42. John R. Rice, "Rebellious Wives and Slacker Husbands," *Sword of the Lord*, July 22, 1938.

43. Mark Bray, *Antifa: The Anti-Fascist Handbook* (New York: Melville House, 2017); "One Author's Controversial View: 'In Defense Of Looting,'" National Public Radio, August 27, 2020, https://www.npr.org/sections/codeswitch/2020/08/27/906642178/one-authors-argument-in-defense-of-looting.

44. Charles Creitz, "Ingraham: Biden 'Propped Up' by 'Unholy Alliance of Billionaires and "Bolsheviks,"'" Fox News, September 16, 2020, https://www.foxnews.com/media/in graham-biden-propped-up-by-unholy-alliance-of-billionaires-and-bolsheviks-to-turn-amer ica-into-baltimore; Donald J. Trump, @realDonaldTrump, "The Real Polls are starting to look GREAT! We will be having an even bigger victory than that of 2016. The Radical Left Anarchists, Agitators, Looters, and just plain Lunatics, will not be happy, but they will behave!," Twitter, September 7, 2020, 10:23 p.m., https://twitter.com/realDonaldTrump/status/1303156993847328771; "Trump Rips Athletes Kneeling for Anthem, Calls Black Lives Matter a 'Marxist Group,'" Fox News, August 5, 2020, https://www.foxnews.com/media/trump-anthem-kneeling-black-lives-matter-marxist.

45. "Trump Says QAnon Followers Are People Who 'Love Our Country," *New York Times*, August 19, 2020, https://www.nytimes.com/2020/08/19/us/politics/trump-qanon-conspiracy-theories.html.

46. "The Prophecies of Q," *Atlantic*, June 2020, https://www.theatlantic.com/magazine/archive/2020/06/qanon-nothing-can-stop-what-is-coming/610567/.

47. Seth Cohen, "The Troubling Truth about the Obsession with George Soros," *Forbes*, September 12, 2020, https://www.forbes.com/sites/sethcohen/2020/09/12/the-troubling-truth-about-the-obsession-with-george-soros/#555ae17a4e2e.

48. Robert H. Knight, *The Coming Communist Wave* (Fort Lauderdale, FL: D. James Kennedy Ministries, 2020), 5, 23–27.

INDEX

Page numbers in italics refer to figures.

9 781501 759291